CHARCUTERÍA

The Soul of Spain

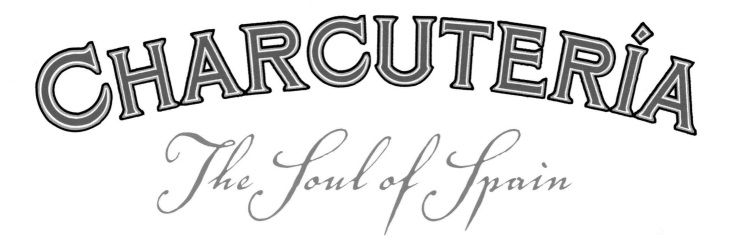

CHARCUTERÍA
The Soul of Spain

JEFFREY WEISS

FOREWORD BY JOSÉ ANDRÉS

PHOTOGRAPHS BY NATHAN RAWLINSON

ILLUSTRATIONS BY SERGIO MORA

SURREY BOOKS

AN AGATE IMPRINT

CHICAGO

Printed in China.

All photos by Nathan Rawlinson except:
Photo on page 170 by Jeff Koehler / copyright ICEX
Photo on page 322 by Juan Manuel Sanz / copyright ICEX
Photo on page 389 by Amador Toril / copyright ICEX
Photo on page 396 by Félix Lorrio / copyright ICEX
Photo on page 430 by Toya Legido / copyright ICEX

Weiss, Jeffrey
Charcuteria : the soul of Spain / Jeffrey Weiss.
    pages cm
Text in English; some chapter headings and parts of table of contents, in Spanish.
Includes index.
Summary: "A guide to Spanish charcuteria, with recipes"-- Provided by publisher.
ISBN-13: 978-1-57284-152-9 (hardcover)
ISBN-10: 1-57284-152-4 (hardcover)
. Cooking (Meat) 2. Cooking, Spanish. I. Title.
TX612.M4W45 2013
641.5946--dc23
          2013018755

14 15 16 17   10 9 8 7 6 5 4 3 2 1

Surrey Books is an imprint of Agate Publishing. Agate books are available in bulk at discount prices.
For more information, go to agatepublishing.com.

To the *maestros, matanceros, sabias,* and chefs of Spain
who welcomed me into their kitchens, families, and homes.

Though your delicious good works speak for themselves,
I hope that this book serves as a testament to your passion and traditions.

CONTENTS

FOREWORD

GROWING UP IN Asturias, in the north of Spain, my life was flavored with *chorizo*. My favorite memories as a small boy were coming home to the rich smell of garbanzo stew, big bowls of warm chickpeas and *chorizo*. There was *chorizo a la sidra,* cooked with the apple cider famous in the region. And of course, there was always *fabada,* the most classic dish of Asturias. This hearty stew, made with buttery beans called *fabes,* is studded with *chorizo, morcilla* (blood sausage), and thick bacon. Then, when my family moved to Barcelona, most holidays were scented with the smell of *butifarra,* the Catalan fresh pork sausage, as it was grilled in the town squares. As an adult now, in America, I'm never far from a whole leg of the prized *jamón Ibérico de bellota,* Spain's most celebrated ham from the *pata negra* pig. Meat has real meaning for someone like me.

To most of the world, Spain looks like a single country, but it is really a fascinating mix of people, languages, cultures, and food. Regional differences are many, and nowhere is this more obvious than when you look at our tradition of cured meats, *carnes y embutidos.* From each corner of the country to its heart, in Madrid, you'll find a huge variety of fresh sausage, cured sausage, hams, and so many other preserved cuts of meat. Many of them, however, rarely make it outside of the country. But this is beginning to change.

I could not be more proud to see how Spanish cuisine is honored and recognized throughout the world today. Spanish restaurants top nearly every list of the world's best, thanks to many of my friends, including Ferran and Albert Adrià, Juan Mari Arzak, and Joan and Jordi Roca. Spanish ingredients and dishes are becoming part of every day life. For nearly 25 years, I have been cooking the foods of my country and sharing the story of Spain here in my home of

America. In my restaurants in Washington, DC, Los Angeles, Las Vegas, and Miami, diners are fascinated by the smoky, spicy, and sweet flavors of *embutidos: chistorra, chorizo, lomo, sobrasada, butifarra, jamón Serrano* and of course *jamón Ibérico*. Whether I'm guiding TV viewers, chefs, and culinary students through La Alberca, near Salamanca, to see the pigs of my friend Santiago Martin (whose Fermín *jamón Ibérico* was the first to legally come to the US), or taking friends like Stanley Feder of Simply Sausage through the Catalan countryside to taste traditional *butifarra* he makes for some of my restaurants, I love to see the amazement newcomers experience when learning the stories behind these flavors and traditions.

But the great story of preserving and curing in Spain has yet to be truly told, until now. *Charcutería* brings to life—with real heart, history and technique—an astonishing look at the legacy of Spain's flavorful meats. More and more Americans today are enjoying superb cured meats. Some are imported, and many are made locally, following European traditions. Chefs across the country offer amazing displays of artisan-made or house-cured *charcuterie*. Even home cooks are learning the skills of the butcher and understanding how to use the whole animal. This book is special, and perfectly timed to introduce these stories of salt and smoke and Spain.

What you will find here in *Charcutería* is a love letter; it is filled with the passion and perseverance of a young man on a mission to discover the soul of Spain. Right out of a top university, with no formal training in Spanish cuisine but eager to learn and fascinated with Spain, Jeffrey Weiss came to cook with us at Jaleo, my tapas restaurant in downtown Washington, DC. We could tell right away that he had more than just a fascination with Spanish

cooking. It was a true calling for him. He was able to use his experience to win a spot in the Spanish government's amazing gastronomy training program, tasting and traveling across Spain. His time there—spent cooking, learning, eating, drinking—allowed him to open a window into the heart of Spanish culture; to become part of families and hear their stories; to watch masters at work; and to participate in the cycle of life on the farms and in the fields.

This book perfectly marries the necessary techniques of brining, salting, fermenting, and drying with the exceptional stories of Spain's particular animals, such as the *Ibérico* pig; special cuts of meat like *secreto;* and traditions of old-world dishes like tongue, tails, and salted bones. It connects the past to the present. And I know it will open up the door to new possibilities for what you can create at home.

—*José Andrés*

ON *CARMEN,* PORK BLOOD, AND *POLLA* JOKES

Understanding the Spanish gastronomy of today and how it continues to influence the American gastronomy of tomorrow

MADRID

MANOLO, A FAITHFUL friend and barkeep at our favorite dive bar for cooks, rasps in his cigarette smoke-laden Spanish baritone:

"Nunca has probado una morcilla como esta, Americano."

Translation: "You have never tasted a *morcilla* like this one, American."

His steady gaze dares me to disagree but, sadly, I realize he's right. That's all he wants to hear anyway.

Fiercely partisan culinary fightin' words like these are typical in bars all around Madrid, often spoken over *cañas* of beer and plates of cured meats. In this instance, Manolo is talking about a *ración* of *morcilla achorizada,* a blood sausage from Jaén, that he's just thrown in front of me like a gauntlet. This *morcilla,* unlike other blood sausages from around the world, is a mixture of *chorizo masa* mixed with pig's blood, cooked potatoes, rice, onions, and spices. The stuffed *morcilla* is then smoked and dry cured and the result is utterly delicious; it's *sabroso* in a way that makes me angry you can't find anything like it where I live, since it's difficult to find *morcilla achorizada* outside Spain and completely impossible to find it in the United States.

"Thanks for rubbing it in, *cabrón.*"

I manage a sarcastic smile and stab the last slice with a toothpick. It's one of the best *embutidos* I have ever tried—the intense smokiness of the *morcilla* is amazing—and my brain immediately starts rationalizing, with the calculating obsession of a heroin junkie getting helter-skelter for another hit, how I can sneak some of these wrinkled black delights back home to California without causing an international incident.

ANDALUCÍA

IN GRANADA THERE is a small, inconspicuous alleyway that houses one of the best *flamenco* clubs in Spain. You'd never know it, though, since the only advertisement for the club is some black graffiti lettering on a dimly lit, sunwashed wall. A scraggly, hastily painted arrow points deeper into the abyss. To make matters worse, the club is open only at night, which was where I found myself late one summer evening, staring into this great unknown.

The sign says: "Eshavira Club."

Standing there in the moonlight, confronted by the deafening stillness of this portal leading to God-knows-where, I realized that at times like these, there are two types of people in the world: Some look down that alley and, acknowledging their lack of the requisite testicular fortitude, quickly sprint away with their tail between their legs; and others, spurred on by a chemical courage borne of the local inebriant of choice, plod down that alley and onward, toward destiny.

With a few hesitant steps made easier by said inebriant, I joined the latter group.

The sun had long since rose again before I emerged from that passage to a bright new day in southern Spain. I stumbled into the light with shaky, hungover steps, but I had sufficient faculties to notice that something was very different: This Andalucían culture, with a veneer of Moorish influence found everywhere—from the food to the architecture to the people themselves—finally made so much beautiful sense.

What did I find in the depths of that alley, you might ask?

I found a confluence of cultures; a place lost in time, yet wholly comfortable in the present. I found a consortium for *flamenco* and the people who cling to the practice of an ever-evolving art; the sort of place where old and older aren't afraid to mingle with new, modern, and tragically hip. I found an ancient wooden door; a bouncer with only one name; a bar that serves beer or sangria *y nada mas;* and a universe centered on a dusty, worn stage manned by men and women who stomp, clap, and sing the spirit of Gitano pain and pride.

Down that alley, I found a small piece of the Andalucían soul.

EXTREMADURA

IT'S A GOOD day to die, little piggies.

Here, 45 minutes away from the nearest city, herds of *Ibérico* pigs roam from tree to tree, searching for acorns to eat. They also do their best to avoid the butchers in blue coveralls—poised with knives in hand—who stalk the herd to cull three pigs for our *matanza,* a wintertime ritual slaughter/alcohol-and-pork-fueled party that has deep roots in Spanish antiquity.

The *Ibérico* pigs are anything but pretty—they are closely related to wild boars—but they possess a unique characteristic: They store large quantities of their fat intramuscularly—what we in America would call marbling. This naturally high amount of marbling makes *Ibérico* meat highly coveted and expensive, in large part because the fat is monounsaturated, like olive oil, and flavored by the acorns my delicious little friends gorge themselves on during the *montanera* (the acorn-feeding months prior to slaughter).

At the Rocamador, a gorgeous converted monastery and four-star hotel situated in the

THE SPAIN I KNOW

THIS IS MY Spain, a place of transcendent memories centered on the diverse foods, rich culture, and welcoming people whom I have come to adopt as my own—if they'll have me.

These memories are the staccato sounds of the *flamenco bailaora's* footfalls, the multicolored sights of *pintxo* platters laid out on bars in San Sebastián, and the unmistakable smells of *charcutería*[1]—that smoky aroma of cured pork mixed with *pimentón*, which permeates much of Spain's cuisine, culture, history, and regional pride.

But while Spain stands porky cheek to jowl with other great cured meat–producing nations, such as Italy and France, the *charcuterie* traditions of Spain are perhaps the least understood of all three. That's because Spain has an almost infinite number of regional variances to its *charcuterie* delicacies and woefully little exportation of these products, least of all to the United States. In fact, only a handful of Spanish producers have overcome the strenuous regulations set forth by the US Department of Agriculture and, as a result, only a fraction of the products available in the Spanish market actually make it to American shores.

These restrictions—coupled with a general misunderstanding of Spanish gastronomy in the '70s and '80s that lumped it under the general heading of Hispanic cooking alongside the cuisines of Cuba, Puerto Rico, and much of Central and South America—mean that a niche product like traditional *charcutería* from Spain is just now coming into popularity.

This limited distribution also means that you probably haven't tasted the sheer eye-rolling deliciousness that is *morcilla achorizada, fuet,* or *sobrasada*—birthrights for any Spaniard, but for *extranjeros* like you or me, simply items for the top of our bucket lists.

Fortunately for us, however, Spanish cuisine is thriving. For one, there's been a globalization of world cuisines over the past decade or so that has planted the seeds of classic Spanish culinary traditions everywhere (thanks to people like Penelope Casas and chefs like José Andrés and Ferran Adrià). Likewise, culinary awareness has grown dramatically thanks to the foodie culture that permeates our collective consciousness via the Internet, food television, blogs, and other media. And last, truly artisan foods like *charcuterie,* fermentation, baking, and preservation have seen a resurgence in the American culinary lexicon as many of us demand to know the history, craft, and quality of our food.

WHAT WAS OLD IS NEW AGAIN

THAT LAST PIECE of the puzzle—a return to artisan cuisine (meaning food that is hand-crafted, small-scale, made with an eye for quality and detail, and as far as possible from a Domino's pizza, a Frito Lay chip, or any other corporation that dares to co-opt the term)—has been the catalyst for the rebirth of *charcuterie* traditions that are now so popular in the United States.

As Chef Thomas Keller perfectly phrased it in his foreword to Michael Ruhlman and Brian Polcyn's book *Charcuterie,* these foods have been all around us for years, but mostly in commercially manufactured forms, like supermarket bolognas, hot dogs, and other meats of dubious origin. The butchery and *charcuterie* traditions of our forefathers became misplaced in our fast-paced, commercially driven food culture, and these pseudofoods are living proof.

Today's chef- and butcher-driven *charcuterie* programs, by contrast, are much smaller in scale. The process often includes breaking down the animals in-house and attempts to vertically integrate the food's progress from farm to table, all in the name of producing the highest-quality product possible—as opposed to cutting corners for the sake of profit.

This streamlined farmer-to-consumer supply chain, which includes such luminaries as producers Armandino Batali, Allan Benton, and Paul Bertolli and chefs April Bloomfield, Jamie Bissonnette, Chris Cosentino, and Brian Polcyn, among so many others, has pushed American gastronomy forward by leading us back to our culinary past.

THAT BRINGS ME to why I wrote this book.

My journey of a thousand meat-curing miles began with a single obsession that we, the American public, have yet to fully be exposed to: the wide array of cured meats available in Spain. Sure, our collective *charcuterie* IQ has increased over the past ten or so years. We've become generally acquainted with popular cured meats like *mortadella* and *pepperoni.* Hell, we even know various forms of *prosciutto, kielbasa,* and *saucisson.* But *chorizos? Morcillas? Butifarras? No tenemos ni puta idea, amigos.*

Spanish-style *charcuterie* is underrepresented, misunderstood, and largely unheard of in this country, as any expat Spaniard who has searched the United States from coast to coast for a taste of home well knows. While the blame for this state of affairs largely rests with stifling restrictions that keep out far too many Spanish cured goodies, our historical misunderstanding of the traditions of Spanish cuisine is also a culprit. Case in point: No self-respecting Spanish *abuela* I know makes chicken with green olives, but for some reason countless American cookbooks from the '70s, '80s, and '90s erroneously call this dish "Spanish Chicken."

That said, American gastronomy may be a melting pot that now correctly represents our love affair with Spanish flavors and techniques, but the traditions of *charcutería* are still woefully absent from it.

That is why I embarked on a culinary odyssey on the topic—to introduce Americans to regional and national *charcutería* specialties that permeate the hearts and souls of the Spanish people. In doing so, I hope to provide a road map for producing and using these recipes in your own kitchen, just in case you don't feel like braving a transatlantic flight or US smuggling ordinances for your own slice of Extremadura, País Vasco, or Castilla–La Mancha.

During my hands-on education, I participated in the opportunity of a lifetime in 2009, when I was awarded one of only two Spanish Institute for Foreign Trade (*Instituto Español de Comercio Exterior,* known as ICEX) scholarships given that year to American cooks. The ICEX scholarship program is a culinary grant sponsored by the Spanish government, which allowed me to cook

Presa   Pluma   Secreto

★ CERDO IBÉRICO ★

LOS MEJORES CORTES

in kitchens all over the Spanish countryside and provided me with the chance to make *charcutería* elbow-to-elbow with the *maestros* (or, more often than not, the *maestras)* of this craft.

In this book, I'll share much of my journey, the times I learned and practiced the art of *charcutería* and *la cocina Española.* You'll meet the cast of characters who helped me understand why the cuisine and culture of Spain are so unique, and why they deserve a place on the world stage.

First, I'll discuss the history of *chorizos, jamones,* and other forms of *charcutería* as they evolved through the ritual pig slaughters known as *matanzas,* finishing up with the modern age of industrialized *charcuterie* and the restrictions placed on it in the United States.

From there, I will introduce you to something you've likely never seen before: Spanish pork butchery, which differs significantly from methods used in the United States. Specifically, I'll cover the *cerdo Ibérico*—the famous black *Ibérico* pigs—and discuss the *matanza* ritual, as well as Spanish-specific butchery cuts for pigs, including the *secreto, pluma, presa, aguja,* and others. I hope you have the opportunity to seek out these cuts on your own, but this information will also serve you well later in the book, when I discuss the different parts required for different types of *charcutería.*

Next, I'll continue with the basics of *charcutería,* including the steps involved in making fresh, semicured, dry-cured, and whole-muscle *charcuterie.* I'll also cover equipment and ingredient options that will help you get the job done, including the best ways for weighing, measuring, buying, and practicing as you start curing your own meat.

Then I'll get to *el alma* of the book: the recipes and techniques that I learned from my time cooking, traveling, and learning with the chefs, *sabias,* and *matanceros* of Spain.

The recipes are divided into chapters according to the techniques involved for making specific types of *charcuterie.* For many of these preparations, I'll also include my favorite ways to use the *charcuterie* when preparing delicious traditional dishes. These cherished recipes were given to me by the talented Spaniards I've come to call *mi familia.*

➡ I start with the most basic of preparations: *salmuera* (brine) and *salazón* (salt cure), two of the oldest and simplest preservation techniques that yield some of the best-known *charcutería* recipes.

➡ Next comes *adobo,* a technique that involves using a marinade as the primary means of preservation.

➡ Then, you'll learn about the *escabeche* technique, a method of hot pickling for proteins, fish, or vegetables that allows them to be stored for a period of time to ripen and mature.

➡ *Conservas y Confits* introduces recipes for preserving meats, vegetables, and seafood in the style of Spain's renowned canned foods. Surprisingly, in Spain canned foods are considered a luxury and can actually cost more than identical foods in their fresh form.

➡ The largest section of the book covers *embutidos,* the various sausages and other stuffed meats found throughout Spain. There, you'll learn about fresh, cooked, semicured, and dry-cured *charcuterie,* as well as recipes for using each style of sausage in traditional and modern preparations.

➡ Next, I'll share with you my personal favorite corner of the world of *charcutería: pâtés* and *terrines.* This subject has become very popular in the past several years, as *terrine* boards of different sizes and shapes have begun appearing on restaurant menus across the United States.

➡ Last, but certainly not least, you'll learn about the traditional sauces and garnishes typically served with *charcutería,* followed by a short chapter of traditional desserts and *licores* that incorporate *charcuterie* recipes and techniques.

CHAPTER

1

WHO'S YOUR PAPÍ CHULO?

Rediscovering our porcine forefathers
of the Iberian Peninsula

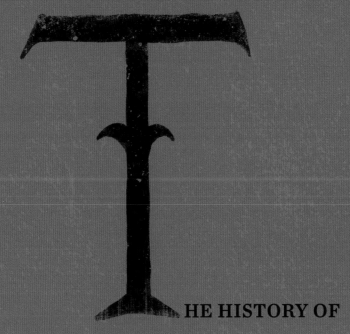

HE HISTORY OF *charcuterie* in Spain, like much of the historical record of Europe's cured meats legacy, is difficult to trace. The details have been lost to time, but it's safe to say that the evolution of *charcutería* is closely linked to the life and death of the local pig population, Celtic-Iberian tribes, the march of the Holy Roman Empire, and the oral traditions passed down as early man (and woman) discovered that their snouted, floppy-eared, porcine companions were in fact a delicious means of survival.

EL TOCINO LA OLLA, EL HOMBRE LA PLAZA, Y LA MUJER LA CASA.[2]

THE PORK BELLY IN THE POT, THE MAN IN THE PLAZA, AND THE WOMAN IN THE HOUSE.

—An old, and decidedly chauvinistic, Spanish refrain

PIGS, WARS, CELTS, AND COWS

A NUMBER OF cultures occupied Spain in early times, and together they helped lay the foundations for many regionally unique food cultures.

Phoenicians crossed the Strait of Gibraltar and founded colonies in the southern Iberian cities Cadiz and Tartessus in the ninth century BC. Carthaginians took up residence in the Balearic Islands before setting out to conquer most of Iberia in the third century BC, with Carthago Nuevo (located in modern-day Murcia) serving as their base of power and capital.[3]

Around the fifth century BC, northern Spain was occupied by the ancestors of today's Basque people and tribes of early Celtic colonists. The latter is credited with establishing a strong pork-centric culture, and possibly the initial regular practices of the *matanza*,[4] as they marched across the Pyrenees and eventually settled north of the Rio Duero and Rio Ebro.[5] These early Celtic settlers viewed the pig as more than a simple food source, however; animals like pigs, bulls, and lambs were worthy of sacrificial rituals, *verracos* (statuettes) for worship, and representation on ancient coins from the era.[6]

After the conquest of Spain by the Roman Empire in 19 BC, the country was renamed Hispania. The cosmopolitan Romans brought ideas, goods, and technology to the Iberian Peninsula via the Roman roads. Concepts became a shared commodity between lands through the spread of Christian teachings, architectural fundamentals for building marvels like aqueducts, new farming methods, and—most important to the culinary world—new *charcuterie* and meat-preserving techniques, many of which likely arose from Rome's conquest of Gaul around 51 BC.

Those meat-curing secrets begat *jamones* and *embutidos* coveted throughout the Empire, as towns like Pompeiopolis (known today as Pamplona) structured whole economies around exportation of their *jamones*.[7] These culinary specialties were solely the province of Rome's social elite, since only the wealthy upper class had the financial means to procure such luxuries from faraway Hispania.[8]

THE CHURCH TAKES CHARGE

AS ROMAN RULE declined in the third century AD, Visigoths from the north gradually extended their control over Hispania, while the porcine-centric culture of the peninsula's emerging gastronomy continued to flourish. Specifically, acorn-fed pigs roamed even more freely in their native oak forests once they became protected by new Gothic laws codified in the *Liber Iudiciorum*. In this Visigothic book of laws, Iberian pigs received specified protections for their feeding habitat in oak-tree laden *dehesas,* and protections were also put into place for the gradual transformation of freshly harvested Iberian hams into succulent *jamones.*[9]

At the same time, the Catholic Church took a leading role in maintaining formal education, preserving historical records, and managing government in the region. As a result, Hispania was one of the few centers of culinary development that survived both the Visigothic reign and the 800-year Moorish occupation in later years. For example, monasteries like the order of San Fructuoso maintained large herds of pigs for making *embutidos*. The cured meats kept the resident monks and traveling pilgrims fed, and the monks were also able to study the pigs and products made from them. In fact, the order maintained archives about medicinal uses for pork. Their work makes up a good portion of the archival information about *charcutería* that survives today.[10]

800 YEARS OF MOORISH RULE

BEGINNING IN AD 711 and continuing for the next eight centuries until the completion of the *Reconquista* in 1492, most of the Iberian Peninsula was ruled by Moorish and Berber invaders from Africa—all except the northern states, which defiantly resisted occupation and remained the last vestiges of Christian freedom in Spain. During these centuries of occupation, Spain was renamed yet again (Al-Andalus) and religious tolerance was granted, up to a certain point, by the conquering Moors. Specifically, the populace's cultural and religious practices were allowed to continue under the conditions of stiff taxation, which eventually became a rallying point for revolution for many native Spaniards.

Fomenting anger over both the occupation and the new taxes shifted the pig from a symbol of wealth or family survival to one of Spanish pride and defiance. Since eating pork or even

The *matanza* became the ultimate act of defiance, as Spanish families moved their pig-slaughtering practices away from the hidden corrals or inner courtyards of their homes to front yards and village squares.

handling pigs by Muslims was forbidden by strict religious laws, the only people who could farm and eat pigs in Spain during this period were Christians opposed to the occupation.

Thus, the *matanza* became the ultimate act of defiance, as Spanish families moved their pig-slaughtering practices away from the hidden corrals or inner courtyards of their homes to front yards and village squares. These actions loudly proclaimed their political and culinary leanings, while also delivering a great big, passive-aggressive, porky middle finger to the ruling Moors.[11]

NEW WORLD, NEW INGREDIENTS, NEW *CHARCUTERÍA*

IN 1492, CHRISTOPHER Columbus "sailed the ocean blue" to the New World with the blessing and sponsorship of Spain's King Fernando and Queen Isabel. It may be debatable whether Columbus was in fact the first European to discover these shores, but we can thank him for bringing eight special passengers along for the ride on his second voyage. Part of his precious cargo included eight *Ibérico* pigs, the forerunners of our swine population, which were released to multiply and proliferate on the island of Hispaniola, where the nations of Haiti and the Dominican Republic exist today (Hernando de Soto later brought more pigs to continental North America in 1539).[12]

Columbus's voyages, and those of subsequent explorers, not only planted some very important culinary seeds here in the Americas; they also forever changed European gastronomy. Specifically, the culinary heritage of the Old World became enriched with new ingredients like corn, peppers, potatoes, chocolate, and other discoveries; notably, some very important red-pepper-tinged changes occurred in the *charcutería* of the period.[13]

Peppers from the Americas eventually made their way into recipes for *chorizos, lomos, morcillas,* and other forms of Spanish *charcuterie*—sometimes in a smoked, dried, and powdered form *(pimentón)* and other times as a mashed paste (from ñora and choricero peppers). As a result, all manner of sausages became tinted a vibrant red, and Spanish *embutidos* were imbued with a heretofore unknown smoky, spicy quality. Quite simply, it was one of the greatest developments in Spanish culinary history.

As described by noted author and Spanish culinary expert Teresa Barrenechea, this truly was a golden age for Spain, from both a culinary and cultural perspective:

> Spain served as a center for Europe and the gateway to the newly conquered lands of the Americas. The grandson of the Catholic Kings, Carlos I, ruled the sprawling Hapsburg Empire and became Holy Roman Emperor Charles V. With the seat of the Hapsburgs now in Spain, food traditions traveled back and forth all over Europe,

affecting the eating habits of the entire continent. Then, in the mid-eighteenth century, the Hapsburgs gave way to the Bourbons, who introduced French styles to the Spanish court and upper classes.[14]

COMFORT IN DARK TIMES

THE RICH, MULTICULTURAL gastronomic landscape of Spain—a collective harmony of so many countries and cultures dating back over a thousand years—came into serious jeopardy in modern times. During the Spanish Civil War (1936–1939) and dictator Francisco Franco's reign (1939–1975) thereafter, the autonomy of Spain's provinces faced oppression and near annihilation. Everything from local linguistic dialects to bullfights to *flamenco* came under scrutiny by a regime hell-bent on uniting the proudly regionalist Spanish people under a single banner: one language and one rule.[15]

And the dark times continued thereafter. An economic depression ensued until the 1950s, leading to a strong black market for everyday luxuries like coffee, sugar, and tobacco. The cuisine of the era, which of course was focused on wasting absolutely nothing, showcased necessary innovations Spanish households had to employ to make the most of what they had. Stale bread provided a means for thickening soups. Families found ways to use every odd and end from slaughtered animals, and many people turned to preserved foods like *charcutería* to get the protein and fat they needed, since fresh meat was far too expensive.

Coincidentally, as suppression and economic woes swept across Spain during and after the Franco era, *charcutería* production soared. In fact, its popularity reached its highest levels to date at this point in time, as businesses with fiercely guarded family recipes for curing meat expanded.[16] Franco's suppressive regime opposed overt regionalist displays of language, culture, and the arts, so the people of Catalonia, País Vasco, Galicia, and other regions turned to their native *charcutería* recipes and traditions—a source of comfort to their souls—as a means of expression, freedom, and economic sanctuary from oppression.

MODERN SPAIN ON THE WORLD STAGE

THESE DAYS, CATALONIA—just one of seventeen distinct *comunidades autónomas* (autonomous communities) in Spain—alone recognizes seventeen different versions of *chorizo.* And while the Catalan viewpoint on cuisine is certainly valid (they gave the world Ferran Adrià, after all), add to this number the various local, regional, and national *charcutería* specialties of the rest of the country, and you'll begin to understand the dizzying scope and depth of Spain's cured meat lineage.

Generally considered the national sausage of Spain, more than 65,000 tons of *chorizo* are made by Spanish producers every year, which amounts to about 40 percent of Spain's entire sausage production.[17] In fact, *chorizo* is such an innate part of the Spanish soul that every February, a festival is held in its honor in the small town of Vila de Cruces in Galicia. *Charcutiers*

from all over Spain bring their products, and festival goers devour and debate over whose is the best in the land.

But for all of the hype about *chorizo* within Spain's borders, very little information about this delicacy has still yet to make it out of the country. Sure, you might see *chorizo* at your local Whole Foods or spot José Andrés, one of the most vocal proponents of all things Spanish, waxing poetic about it on TV. But only in the last few years have Spanish producers gained access to the American market, and even then only a small fraction of pork-laden goodies have trickled through.

In October 2002, for example, Palacios, a producer of *chorizo* based in La Rioja, became the first to be allowed by the United States Department of Agriculture (USDA) to import *chorizo* into the United States.[18] And while the Palacios product is certainly a decent *chorizo,* it is hardly representative of the multitude of other *charcutería* products enjoyed every day by the average Spaniard. Fortunately, in recent years, companies like La Tienda, Fermín, 5J, Wagshal's, and a few others have been allowed to either import some Spanish *charcutería* products or put their traditional recipes to work here in the United States, creating authentic reproductions.

Likewise, *jamón Ibérico* is only a recent revelation here in the United States. For years, Embutidos Fermín fought to educate the USDA about its products and the process involved—but it wasn't until 2007 that the first legal *jamones* were ceremonially sliced at José Andrés's Washington, DC, restaurant, Jaleo. In a *New York Times* article by Amanda Hesser, exporter Jesús García

commented on the issue, "The problem is that American authorities do not recognize the European Union's standards for production. They want companies to follow their own standards. And some companies do not want to change."[19]

Fast forward to today. Americans still have very few options when it comes to trying the lesser-known varieties of *charcutería* available to the Spanish people. Hope exists, however, that this may be soon rectified, as evidenced by the sweeping acquittal of many Italian cured-meat imports in April 2013.[20]

For now, anyway, we can travel to Spain and consume to our heart's content. We can buy what precious little is available in our country. We can make it ourselves. Or we can make a futile attempt at stuffing contraband pork into our suitcases and pray, with the wide-eyed, guilt-laden face of a Colombian drug mule, not to get busted by the Department of Homeland Security.

Just know that on this point, dear reader, I can offer a bit of personal advice: Getting caught is an epic fail of disastrous proportions, even if it's not your fault.

Case in point: After a trip to Madrid and the surrounding countryside, my Spanish "family" thought that they'd surprise me with a little package of *morcilla* secreted away in my suitcase. It was a gesture borne of more heart than brains, as ultimately it truly *was* a great surprise—especially when I found myself tagged for an agricultural check at a particularly thorough US Customs checkpoint.

I simply didn't understand. I'd filled out my Customs card and done everything right. Yet there I was, unloading my dirty unmentionables on a counter for God, curious passersby, and the TSA to look over and admire. And that's when I caught a waft of something familiarly porky, and my heart sank: There, in the gloved hands of an agent, was a gift-wrapped package of undeniably dubious origin wrapped in orange-tinted, grease-stained butcher paper and covered with a plastic baggie.

And all I could stammer was: "Awwww, *shit.*"

Fortunately, after some quick explaining to the Customs agents who took pity on me (and thankfully didn't stick me with either the $50,000 fine or the 10-year jail sentence), I was let off with a warning. Of course, that warning entailed having my name entered into a national database as a "person of concern" with notes of the encounter attached. That's why to this very day, whenever I go through US Customs, I get the same question: "Sir, do you have any meats in your possession? Perhaps some *'ham-own'?*"

Trust me, folks...until the USDA pulls that ginormous stick (or maybe it's a big ham bone?) out of its ass, it's going to be much easier to make *charcutería* yourself or buy it from a legal vendor. And I'm here to help!

CHAPTER

2

THE SECRETO OF THE SECRETO

Surviving bad hangovers, great pork, and crazy butchers with hatchets

 IT UP ANY

self-respecting *restaurante* serving the elusive, uniquely Spanish cut of pork called *secreto Ibérico*—be it a temple of *alta cocina* in the heartland of the Costa Brava or a roadside bar in the outskirts of Badajoz—and you will notice something peculiar; something that runs counter to the government-advised method of pork cookery we Americans have come to know.

DEL CERDO HASTA LOS ANDARES.
OF THE PIG, EVERYTHING CAN BE EATEN.

The pork will arrive to your table properly *poco hecho*—that's medium rare in American culinary parlance. Yes, medium-rare pork...but relax, my fellow Americans. You have nothing to fear but fear itself.

Ibérico pork is raised with love, slaughtered with respect, marbled like the finest Kobe beef, and loaded with the sort of flavor you simply cannot find anywhere else. And that is why Spain's four-legged national porcine treasure is simply not meant for the kind of indistinguishable, chalk-dry, cooked-to-160°F ending that the USDA has historically recommended as the only means of "safe" pork consumption since the last century.

But this chapter is not merely a diatribe against the way in which we have been told to incinerate our pork in this country since the 1950s (thankfully, we have progressively rebelled against The Man in recent years and recommendations have relaxed...a little[21]). Rather, I am not-so-subtly hinting at a different way of doing things. My goal for this chapter is to present to you the perspective and practices of porcine husbandry, butchery, and cookery of the Iberian people to compare and contrast with our own.

What follows in this chapter, therefore, will be an introduction of sorts into the life, the death, and the afterlife of the famed *Ibérico* pig. In the Spanish countryside, these meadow-dwelling, acorn-grazing herds live out their lives before being dispatched with honor for their one-of-a-kind musculature and delectable, acorn-imbued fat.

I'll start by discussing our four-legged friends' final days of life as part of a typical *matanza*, including defining the ritual and explaining how it's performed. I'll then take you through the laborious task of breaking down *Ibéricos* following the Spanish method, including identifying the various commercial cuts of pork you're likely to find in Spanish markets. Last, I'll contrast the economy and methodology of Spanish pork butchery with our own American system of breaking pigs down.

PORCOPHOBES AND PORCOPHILES

ANTHROPOLOGIST MARVIN HARRIS theorized that the human race can be divided into porcophobes and porcophiles; his theory differentiates between cultures that consider the pig forbidden in all its forms and those that consider all things swine-related to be righteous, delicious, and true.[22]

If we take this theory at face value, the Spanish decidedly fall into the latter group: They have consistently ranked as one of the top five pork-consuming nations in the world for the past 10 years,[23] and they are the world's largest producers and consumers of ham with production at around 40 million hams a year (roughly 1 ham per Spaniard!). Most Spanish hams never leave Spain, however, since Spain exports only about 10 percent of its annual production due to high demand within its own borders.[24]

And the Spaniards display that fervent love and respect for their native Babe most at the inception point: the age-old *matanzas del cerdo* ("pig slaughters"), which are held every winter across the entire country. While this ritual originated as a burden of necessity—a means for entire regions of people to survive harsh winter conditions while living on the cusp of poverty—these days, most *matanzas* are parties that serve a festive, social purpose. Modern *matanzas* are familial gatherings where attendees are plied with enough free-flowing alcohol and soul-satisfying food to keep them blissfully elbows-deep in piggy blood, guts, and gore.

In years past, *matanzas* took place around holidays honoring local saints, thereby allowing entire extended families to get together in their households for the occasion. Families would employ the services of some *matanceros* (slaughterers, typically men), a *matancera* (a person in charge of meat processing, typically a woman), and a group of *sabias* (the word literally means "wise women," as they are the ones with the knowledge of various *embutido* recipes and techniques). The team would work in concert—with the family doing the grunt work—to ensure the quality standards for the slaughter and to make enough fresh and cured meats to sustain all for the year to come.

In cities like La Alberca in Salamanca, home of Embutidos Fermín, the entire populace waits yearlong with giddy excitement for its *Ibérico*-based celebration during the festival weekend of San Anton. Throughout the year, the town pet—an *Ibérico* pig, of course—wanders the streets with abandon and is fed and fattened up by everyone in town. Then, in January, a great *matanza* is held during the festival weekend and the pig is raffled off to its lucky new owner. (In years past, the pig always went to a needy family in the area to ensure the family's survival. Today, the raffle benefits local charities.)

Nowadays, however, all pigs meant for commercial distribution and many pigs destined for individual families are brought to a local slaughterhouse and killed in a controlled environment, inspected by an on-site veterinarian for any diseases, and only then released for further processing.[25] These are modern changes for modern times.

> The Spanish produce around 40 million hams a year (roughly 1 ham per Spaniard!).

MORE THAN ONE WAY TO SKIN A PIG

ASIDE FROM THE relatively new, factory-based slaughter system (begrudgingly set in place years ago to help prevent the spread of trichinosis, which is now nearly nonexistent), the *matanza* thankfully varies little from time-honored methods.

For example, some *matanzas* last for two days, and some for three. Some families make *embutidos* on the first day, and some on the last. The type of *embutidos* made will vary depending on what region you hail from, but generally speaking, the cuts of pork and schedules of production at most *matanzas* are all similar to those you'll learn about in this chapter.

My experiences with the *matanza* ritual, which mostly took place in Extremadura, a rugged region in the western part of Spain, taught me that it is for neither the faint-hearted nor the weak-willed. *Matanzas* are intense, draining, and days-long affairs requiring early and constant fortification from food as well as drink. Both come in copious amounts, and the latter is preferably a strong alcoholic variety served before, during, and especially after a long day's work.

Following is an account of a typical first day of a two-day *matanza* on a working farm in the rural countryside of Spain—the sort of place that harvests and processes all of their own animals on-site. I participated in many such *matanzas* over the course of a few seasons, with shifts starting in the early morning before the sun rises and continuing on until the early evening. The schedule often varied a bit depending on the temperature of the farmhouse and surrounding countryside, the time of day, the degree of porcine stubbornness we faced, and the state of hangover the butchers and the rest of us were in. But this is more or less how it went.

BEFORE THAT, HOWEVER, let me give you the first, last, and only disclaimer of this book: What follows is not for those who prefer to think of meat as a nameless, faceless, sterile, Styrofoam-packaged product available at your local supermarket.

The *matanza* ritual—at its core—is dedicated to following techniques handed down through the centuries to minimize animal suffering and maximize usable yields—in addition to being unacceptably cruel, upsetting the animal ruins the meat.

Nonetheless, slaughtering pigs is a messy business that cannot be accomplished without a sharp implement, a capacity to witness death, and a good amount of blood...so read on only if you accept these truths to be self-evident.

DARKNESS FALLS

OUR SLEEPY GROUP of *matanza* participants begins the day with a bumpy, motion sickness–inducing, off-road excursion to an ancient farmhouse in the Extremeñan countryside, a place that time has clearly excused from the burdens of change. The sun is slowly rising above the surrounding ridgeline, tracing outlines of a large valley we'd entered under the cover of darkness. The new morning light betrays the shadows of acorn meadows, orange groves, and herds of roaming, mammoth, black

pig-monsters on a desperate and noisy search for some breakfast.

Inside the farmhouse, we enter a simple, gigantic room with portents of things to come literally hanging over our heads: Row upon row of *morcillas, chorizos, lomos,* and other *charcutería* hang from ceiling racks, drying in the morning breeze.[26] The room includes a small kitchen area with ancient water faucets and a large dining area. In the dining space, giant wood tables are covered in plastic tarps, ready for the messy work to come. An ancient, hand-crank sausage stuffer and grinder dominates the room's eastern wall. Next to that, open doors lead out to the backyard slaughter area, providing a magnificent view of the valley and the sunrise. On the opposite side of the building—protected from the morning sun—lies a salting room, where *jamones* and other large slabs of meat from previous *matanzas* quietly sit and cure.

It's the smells that get you, though. The essence of *pimentón* emanates from the very walls, the floors, and the rafters; there's the faint, sweet funk of drying *chorizo*; the comforting waft and "pop!" of an oak-wood fire in the small fireplace that serves as a body-warming station and hot line for cooking; and the hunger-inducing smells of a delicious, hearty, high-calorie breakfast meant to sustain us for the day's grueling work.

Strong coffee percolates in a coffeepot older than most of us, set directly in the fire. On the table, platters contain slices of *chorizo* and *lomo* from a prior *matanza* (inspiration for our work to come, perhaps?); *cachuela,* a specialty local *pâté* made from pork blood and other offal, smeared on grilled slices of bread; and *migas,* a peasant's dish of crumbled bread, *chorizo,* peppers, and garlic fried in lard and topped with a fried farm egg. (Of course, the *migas* varies for those of us needing to sop up a hangover—two

From left: El Maestro, Elvis, and Little Joey

eggs, in that case. One portly chef visiting from Barcelona avails himself of four eggs...Spanish cooks really love their fried eggs.)

Against the clatter of fork-on-plate, we can hear *them* through the open doors leading to the slaughtering area. They are our reason for coming, the reason for all of this. Three little *cerdos Ibéricos*—and by little, I mean upward of 350 pounds each—were rounded up earlier on this, the morning of their final day on this cosmic plane. They look happy—or at least as happy as pigs can look—as they lounge in their spacious, warm Death Row shelter.

We meet our *matanceros,* the taskmasters of our day's work. The three large men in blue coveralls who will handle the slaughter and butchery duties have the serious faces and dark eyes typical of men who dispense death professionally. I mentally give them names, so I can remember them: *El Maestro,* the oldest one, does the talking;

Elvis, so named for his sideburns and comparable good looks to The King in his prime, is the middle-sized one; and Little Joey, the shortest of the group, looks like Joe Pesci from the *Goodfellas* years and even has the same staccato speech.

We are then introduced to three deceptively diminutive women in floral, plastic-lined aprons—our *sabias*—who will handle pretty much everything else, including setting up our fortifying morning shots of Chinchón, a potent anise-flavored liquor from the town of the same name.

The shots *(chupitos)* are poured and the *matanceros* offer a toast to the local saints. A second round honors our soon-to-be-dead piggy friends, and then a third honors the start of the *matanza.* Three shots in, out come the sharp knives and hatchets—wood-handled, medieval-looking, blood-stained axes for splitting bones—and suddenly I'm glad I didn't share Elvis's nickname with the group.

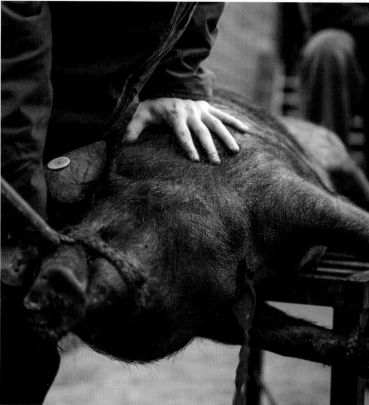

DEAD PIG WALKING

IT'S TIME.

Piggie Number 1 is encouraged (read: shoved, pushed, and cajoled with food) out of the relative comfort of his shed onto a dirt-filled runway with three sequential gates. As he unknowingly walks past the first gate, a *matancero* closes it with a "CLANG." The process continues until Piggie is poised in the slaughter area next to an ominous-looking stainless steel table. At this point, much too late to make a difference, Piggie realizes something is up and the *matanceros* make their move.

Piggie is quickly corralled to a corner and pushed to the ground. He is hog-tied and then carried—relatively quickly, given his size—to the table by all three huffing, puffing, swearing, and now cigarette-smoking *matanceros*.

The process is not a silent one, but once Piggie is on the table, he grows a little quieter, perhaps accepting his fate as *El Maestro* speaks a few words to him (is there such thing as a Pig Whisperer?). As Piggie's neck dangles over a large bowl *(lebrillo)* destined to catch blood for making *morcilla,* Little Joey uses a very sharp knife to quickly sever its carotid and jugular. Piggie immediately bleeds out, and brain death occurs within seconds.

At this point, I want to clarify something for those of you squirming in your seat or wetting these pages with tears of mourning.

I witnessed and participated in the slaughter of dozens of Piggies during these *matanzas*, and each time this part—the act of taking a life—was

of the utmost importance to everyone involved. The most serious step in the ritual, it is absolutely intense and almost always occurs in relative silence, save the protesting sounds of Piggie and occasional swear words from the *matanceros.*

El Maestro explained it all to me over a *caña* later that day. Not only does the quality of the slaughter rely on the animal's emotional state in this moment and in the preceding moments of its impending sacrifice, but a quick death is a matter of respect. Anything less would be disrespectful to the ancient custom, to its modern practice, to the sacrifice of these majestic animals, and to the family that raised the animals over the past years. In fact, even the verb that *El Maestro* used to describe the action—*sacrificar*—denotes the accepted responsibility and gravity of his work. Using that term acknowledges that the *matanza* is as much a celebration of life's basic elements of living, dying, and life going on as it is about acquiring meat, fat, and calories for sustenance.

In other words, these men in blue, these dealers of death, specifically work in the service of ending life quickly and honorably. *El Maestro* said it best: "That is the most important thing that we do." It's something that, on this day, they demonstrate twice more without error.

> Not only does the quality of the slaughter rely on the animal's emotional state in this moment and in the preceding moments of its impending sacrifice, but a quick death is a matter of respect.

Death's Warm Embrace

DEAD PIGS ARE warm pigs. This is a little un-settling to me.

One of the *sabias* is stirring the blood with her hand—to keep it from coagulating—while the men use a gigantic, propane-powered blowtorch to remove coarse bristles from the skin of our recently departed future meal. The smell of burning hair permeates the air.

After laying the now-hairless Piggie backside down on one of the steel tables, *El Maestro* slits the belly open and passes to the *sabias* the intestines (*intestinos*), the stomach (*tripa*), and the kidneys (*riñones*). Next, he carefully inspects the liver (*hígado*) for any signs of potential disease.

But it's the steam rising off of the carcass that catches my eye. For most *extranjeros* working under a clear winter sky in temperatures hovering around 30°F, the sight of steam—of the very essence of warmth—flowing from the body of an animal that was alive just five minutes before is one of the most arresting moments of this experience. This is, very possibly, the first freshly killed *any-thing* that most participants may have ever seen, a stark reminder of the disconnect perpetuated by our daily routines in modernity. Seeing evidence of life drain away is a stark contrast to the emotionally safe world we live in, where "meat" is a faceless commodity stashed neatly behind psychological and physical barriers.

El Maestro slices into the *hígado*, grunts affirmation of the animal's health, and passes the *hígado* inside to be taken to the local vet. He then turns to me and motions with a glistening, blood-stained hand. It's time for me to learn how to turn hundreds of pounds of solid Piggie into the traditional primal and sub-primal pork cuts of Spain.

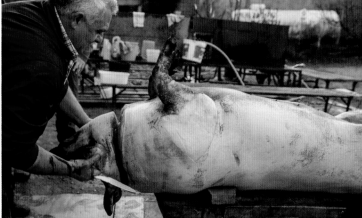

From left: *El Maestro* inspecting a pig's liver; harvesting the head; starting the breakdown process.

SPANISH BUTCHERY 101

CLASS IS IN session.

First, we twist off the head *(cabeza), Exorcist*-style, spinning it round and round till it detaches. Decapitation complete, we pass the *cabeza* to Elvis, who harvests the tongue *(lengua),* ears *(orejas),* cheeks *(carrilladas),* brains *(sesos),* snout *(morro),* and the bones *(huesos)* for stock. At the same time, we remove the tail *(rabo),* the tenderloin *(solomillo),* and the jowls from the head *(papada).* The *rabo* will be used for braises, the *solomillo* for grilling, and the *papada* will either be covered in an *adobo* and tied up in the farmhouse's large chimney for smoking or chopped up to flavor many sausages to come.

Working from the belly toward the back, the skin *(piel)* is slowly peeled back from the carcass until it lays open like a book cover. What's left is the entire torso of the animal *sans* skin: a warmish, thickish mountain of fat obscuring meat and bone.

Grabbing his hatchet, *El Maestro* starts hacking away at the top of the rib cage inside of the cavity, working his way downward with short chopping motions that elicit a satisfying "thwack." He cuts through the chest cavity, striking on either side of the spine *(espinazo),* until he reaches the bottom of the spinal

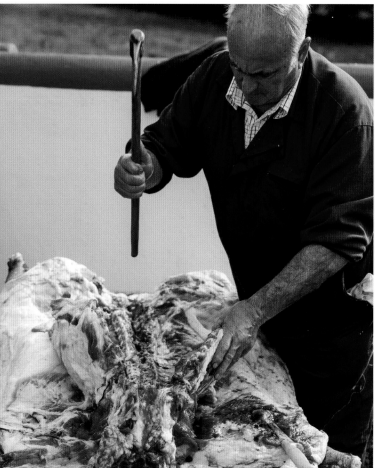
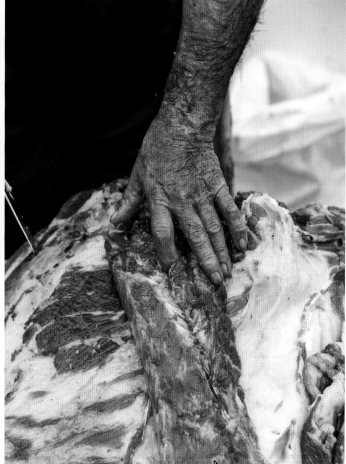

Harvesting the spine; harvesting the *pluma*.

column and eventually frees the spine from the rib cage. This allows us to open the cavity completely.

The majority of his work now complete for Piggie Number 1, *El Maestro* moves on. He heads over to the other animals to repeat the process while I join Elvis and Little Joey in finishing the breakdown. This begins with removing the ribs *(costillas);* we trace the underside of the rib bones where they meet the loin, exposing the entire loin–belly section. The King and I then pass whole rib sections to Joey for further breakdown by the *sabias.*

Next, we break down the loins *(lomo)* and surrounding muscles. The *lomo* itself is freed from the midsection by a slice along a natural membrane. The amount of quivering fat that remains on the *lomo* makes it impossible to handle and to eventually be cured for *lomo embuchado,* so we scrape the fat off and later melt it down for dual uses: cooking medium and awesome boot polish.

The *lomo* is then further divided by slicing along a membrane on the underside of the head of the muscle, removing a triangular, feathery piece of meat called the *pluma* (the word literally means "feather"). Elvis remarks, with the

gravitas of someone who has consumed many *plumas,* that it is the best cut on the pig.

We then remove the entire ham *(jamón)* followed by the shoulder *(paletilla).* Both will keep their black hooves *(patas negras)* throughout the curing process. Now exposed are most of the back fat *(tocino),* the belly *(panceta),* and the famous quasi-skirt steak known as the *secreto,* a piece of meat sandwiched in layers of fat between the shoulder and the belly. The *secreto,* a heavily marbled piece of meat, is something everyone gathers around to salivate over. Conspiratorially, Little Joey whispers that the *secreto,* and not the *pluma,* is the best cut from the pig. Elvis enthusiastically disputes the claim by questioning Joey's sexual desire for certain farm animals.

A few important muscles from the shoulder and neck region still need to be broken down. The *presa,* another favorite cut of the *matanceros,* is located where the top of the loin meets the neck, and the more heavily marbled *aguja* is in the collar area. Together, these pieces make up the *cabecero,* a major group of muscles located at the head of the loin.

Now that the major and minor muscles are removed from the carcass, we clean any lean meat *(magro)* from the remaining mound of *tocino.* The monstrous, off-white, gelatinous mass of fat will be used for making sausages or *tocino salado,* the Spanish analog to Italy's *lardo.* The table is cleared. The entire pig is broken down and ready for further processing inside.

THE *SABIAS'* GOOD WORKS

MEANWHILE, THE *SABIAS* have been busy inside the farmhouse with the bounty of our toil. Three *artesas*—giant wooden, bathtub-like troughs—are assigned numbers and set up in the middle of the room. The meat from each pig is then segregated into its respective trough. Doing so is both monument to our hard work and also insurance against cross-contamination, as the three livers must be inspected by the local veterinarian and cleared of any signs of disease.[27]

Once *El Maestro* checks the *hígados* and gives the all clear to begin working, the *sabias* quickly get to work directing participants on their various duties. While the intestines—future home of our *embutidos*—soak in a bath of citrus peels and vinegar, a group of participants begin sorting and weighing a pile of lean meat trimmed from the carcass for sausages. All three animals' trim will eventually be combined and ground with meat from the neck, belly, shoulders, and jowls, along with a good amount of fat. In the end, two entire *artesas* will be filled with ground goodness for later stuffing.

Elvis and I get to work removing the aitchbones—a part of the pig's pelvis—from the *jamones* and cleaning up the *paletillas,* which will both be salted and put away for a yearslong transformation. Little Joey sets to work

preparing *secretos, plumas, presas,* and other cuts for our lunch; *El Maestro,* working on an ancient tree stump, takes his axe to the *espinazo, codillos,* and *costillas,* while the remaining *sabias* get to work preparing the *riñones, hígados, tripa,* and other offal.

The *matanza* is in full swing now. Sounds of laughter mix with the percussive "thwack!" of *El Maestro's* hatchet, knives being honed on steels, and the slow gurgle of simmering lard. Scents of braising tripe, caramelizing onions, and an ever-present waft of *pimentón* fill the air. Beer is passed around, and the group sings songs honoring the *cerdo Ibérico,* regional Extremeñan pride, and all things Spanish.

Wine is poured and a bowl appears filled with deliciously sticky, spicy *callos Extremeños* simmered with *chorizo* and *morcilla.* Everyone dives in with spoons or hunks of bread; somewhere outside, a grill is fired up for lunch, sending in smells of burning wood and grilling meat.

More beer and food appears: ribs marinated in lemon, bay leaves, and cinnamon, and then *confited* in *Ibérico* lard; perfectly *poco hecho*-grilled *secreto, pluma,* and *presa,* adorned with only coarse salt, are laid out on platters side by side for comparison and consumption with a basket of grilled bread. The bread was key, as the fat of the meat mixed with salty, juicy drippings was the work of God's personal *saucier,* just begging to be sopped up.

The group erupts into more song as the ancient stuffer is set into place; sausage making will be the last Herculean push of the day.

While half the group cleans the intestines, prepares the machines, or sneaks out for a cigarette, the other half begins dumping garlic, salt, *pimentón,* and other spices into an *artesa* to make *chorizo.* Freshly ground Medici spices (clove, allspice, and white pepper), salt, and black pepper are added to the other *artesa* for making *salchichón.* Then the hand mixing of the sausage begins, using a rhythmic motion akin to a cat pawing a blanket.

Numerous changes of sore-armed volunteers later, it's time to make the *prueba,* the test *picadillo* of ground sausage meat. A pan is prepared, the meat is seared, and the *sabias* each take a taste right out of the pan using superhuman, heat-resistant fingers. The ladies close their eyes as they retreat into memories

The ladies close their eyes as they retreat into memories of their grandmother's, great-grandmother's, and great-great-grandmother's sage advice on flavor, texture, meat-to-fat ratio, and other data stored—unquestionably—in their very DNA.

of their grandmother's, great-grandmother's, and great-great-grandmother's sage advice on flavor, texture, meat-to-fat ratio, and other data stored—unquestionably—in their very DNA. One says add more salt, the other more pepper. A third suggests maybe some oregano? They hold an informal caucus while the rest of us try the *prueba*—to us, it's delicious—and finally

everyone agrees on a plan. More seasonings are added, more mixing is done, and it's time to begin stuffing.

Or, more accurately, it's time to begin dancing, since the act of sausage making with these women is more akin to a *tango* or *paso doble*. Little Joey sets the tempo by slowly cranking down on the stuffer's plunger and a *sabía* slowly forms the sausage. Another coils the filled intestine until she reaches the end, at which point she ties it off in a blur of fingers. I join a third *sabía* in cutting the sausage into lengths large enough to form hoops, while another ties off the hoops in a series of knots before quickly using a needle stuck into a cork to prick the sausages.

Form, pass, tie, cut, prick-prick-prick. It is the rhythm of an ancient dance. Sometimes we tag out—cigarette, beer, and bathroom breaks are hazards of a job rich with great food and a limitless supply of liquor—but by the end of the day, fresh new rows of *embutidos* join their brethren in the rafters, gently fermenting and drying in the warm breeze of the Extremeñan countryside.

As dusk falls, we are treated to more shots of *Chinchón,* plates of *perrunillas* (delicious, warm cookies made from *Ibérico* fat, cinnamon, and lemon), and a fresh pot of strong coffee. *El Maestro* raises a toast to our hard work and the *cerditos* who gave their lives for our physical, social, and spiritual nourishment.

A TALE OF TWO PIGGIES

THROUGHOUT HISTORY, COUNTLESS cultures have found new or varied ways to butcher and cook pigs for reasons of tradition, economy, or luxury. One man's garbage became another man's treasure; one man's pork belly became another man's *panceta*. Meanwhile, here in America, we developed a butchery lexicon uniquely our own—a porcine perspective resulting from a cultural mishmash of societal and gastronomic viewpoints and our arms-wide-open culinary worldviews in melting-pot cities like New York and San Francisco.

But *secreto? Pluma? Presa?* Most Americans won't recognize these as some of the best cuts of pork in the world, even though in Spain they command high prices, lengthy discourse, and even near-fisticuffs over which is the most *delicioso.*

AMERICA AND EATING HIGH ON THE HOG

AS I COMPARED the style of butchery that we have developed in America with the style in Spain (which is similar in technique to much of Europe), a major difference became evident: The majority of butchery in America—meaning the meat that has historically lined the refrigerated shelves of supermarkets—has almost always been cut via mechanical manipulation.

More specifically, our style of butchery—and the demand it has produced for more expensive middle cuts of the animal (parts like the belly, ribs, and loin)—is directly informed by the technology of band saws and other devices that aid in slicing through bones with ease. This technology surely made the American butchery process easier, faster, and more profitable for slaughterhouses and grocery stores, but it also damn near killed off what few butcher artisans remained in the United States during and after the '70s, '80s, and '90s.

BONE BUTCHERY VS. SEAM BUTCHERY

THIS TECHNOLOGICAL ADVANTAGE, of course, was unknown to our forebears in Europe. A century ago, even a handsaw was a rare luxury in breaking down the weekly slaughter; in most cases, a simple hatchet was more common.

Unlike our American system, the old-school European methodology of butchery was governed by an economic ideal: The idea was to use and sell as much of the animal as possible, with an eye to selling the animal as a whole as opposed to maximizing sales of only the choice middle cuts.

Back then, butchers—guys who built guilds around their craft[28]—worked from an ancient playbook where the seams between muscles guided their cuts. It was a time when cutting machines didn't exist, a butcher's speed depended solely on his skill with a blade, and the quality of his work depended on a deep understanding of animal anatomy and musculature. And on the other side of the counter, the populace was accustomed to and understood more about meat than just the standard steaks and chops. In short: The butcher of the past was more than just a guy who lined blocks of meat up on a band saw and ran them through. Rather, he was a craftsman who sculpted edible, meaty art from whole, hanging carcasses.

> The butcher of the past was more than just a guy who lined blocks of meat up on a band saw and ran them through. Rather, he was a craftsman who sculpted edible, meaty art from whole, hanging carcasses.

And that brings me to the unique differences between modern American butchery and the more traditional European style of butchery I learned about in Spain. A renaissance for the latter, I am happy to say, is currently in full swing thanks to the heavily tattooed and fabulously facial-haired hipster butchers in locales like Brooklyn, San Francisco, Austin, Portland, and beyond.

THE FIRST CUT IS THE DEEPEST

I'LL START WITH how Americans codify the butchery of our pigs into primal cuts, which are also known as wholesale cuts. These are the major muscles that most American butchers break pigs into before taking them down into the myriad of sub-primal, or retail, cuts—which we will discuss shortly.

In US butcher shops and slaughterhouses, a pig is broken down into four or five rectangular-ish primal sections, the true number dependent on whether you're a piggie romantic who includes the head and jowls in the accounting. It's a sad but true fact that pig head is an uncommon special-order item, even in the most exclusive and hipstery of joints, so let's just leave it as *four* common primal sections in American butchery:

➡ The shoulder section
➡ The loin and back rib section
➡ The belly and sparerib section
➡ The back leg (ham) section

Spanish butchers, however, see the pig differently. First, the head and jowl section is a big part of Spanish butchery and *charcuterie*. Depending on which region of Spain you happen to be in, you're likely to find *papada, lengua, orejas,* or even a whole *cabeza* on a whim. This is especially true in the butcher shops of the north, where a salted pig's head is a common sight and a welcome addition to many of the classic soups and stews of the region.

Second, Spanish butchers don't cut a pig into rectangular primal "sections" like their American counterparts. That is, butchery in Spain is generally muscle-by-muscle rather than section-by-section; it's a process governed by slicing along the seams separating muscles—as opposed to using band saws to cut through bones and to form blocks of meat.

Third, Spaniards consider the ribs (*costillas*) to be a separate piece from the prized loin (*lomo*) and even more prized *cabecero*. That's why the ribs are typically the first area to be separated from the middle section at any *matanza* or slaughterhouse. Conversely, Americans—given our Homer Simpson-esque penchant for bone-in pork chops—generally consider the entire loin with ribs as one primal cut.

Last, the Spanish don't place a high value on preserving the integrity of the belly (*panceta*) section like we bacon-loving Americans do. In fact, once the *lomo* and *costillas* are removed, the belly itself is dismantled into a number of sub-primal cuts or ground up for sausage.

All that said, here is what a typical primal* breakdown would look like on the Spanish side of things:

➡ Head (*cabeza*)
➡ Ribs (*costillas*)
➡ Loins (*lomos*) and *cabeceros*
➡ Back legs/hams (*jamones*)
➡ Shoulders (*paletillas*)
➡ Belly (*panceta*)

*Yes, I know I said that the Spaniards don't really do a primal breakdown per se. This list is just for the sake of a clear comparison to the US system...so let's just go with it.

THE SUB-PRIMALS: A MULTITUDE OF MEATY OPTIONS

AROUND THE WORLD, butchers have created a dizzying array of possibilities for breaking primal cuts of pork into sub-primal cuts, the ones found at the supermarket (hence their alias, "retail cuts"). But not only do these cuts vary from country to country, they also vary from region to region, depending on the traditions of the locale and the recipes that are popular there. What I'm getting at is this: No matter where you happen to be in the world, you're going to find different techniques for getting down to sub-primal cuts—and those sub-primals are going to be the ones that make the most sense to whoever is holding the knife and doing the cutting.

As Michael Ruhlman and Brian Polcyn succinctly put it in their book *Salumi,* "There is no one right way to break down a hog. The only wrong way is thoughtlessly or wastefully." That's a great point to bear in mind as we compare some of the more popular sub-primal cuts of the US and Spanish systems.

Head Section

IT'S NO SECRET that the Europeans are a little more into head meat than Americans—they simply have more of a culture for the sorts of preparations that come from this oft-discarded delicacy. Truth be told, if you ever are savvy enough to get your hands on a head in America, you'll probably break it down in a similar fashion to the Spaniards. There's really only so many ways to do it, and the muscles that everyone takes advantage of at the sub-primal level are all virtually the same:

➡ Jowls *(papadas):* Cured, smoked, braised, or ground for sausage
➡ Cheeks *(carrilleras):* Braised
➡ Ears *(orejas):* Braised and then sometimes fried
➡ Snout *(morro):* Braised or used in stocks for its high gelatin content
➡ Tongue *(lengua):* Braised, grilled or pickled *en escabeche*
➡ Whole head *(cabeza de cerdo):* Roasted or used in stocks for its high gelatin content

Loin and Rib Sections

IN PRETTY MUCH any culture, the loin and rib section is the most coveted part of the pig. This section has the most meat and generally requires little cooking, so these cuts are considered the more luxurious parts of the pig. Also, this is the area where you'll see significant differences in how the musculature of a pig is treated in seam and bone butchery.

In America, the loin and rib section is generally left intact, especially since bone-in pork chops are a big seller here. Likewise, the whole loin also gives us bone-in crown roasts or boneless pork roasts, both celebratory American holiday treats.

From left: *Solomillo, secreto, pluma,* and *presa*

Tenderloins are typically trimmed of their silverskin membrane before being roasted. And, of course, pork ribs know no higher calling than becoming smoky St. Louis or baby back-style ribs in the barbecue. Thus, the typical American subprimals from the loin section would look like this:

➡ Pork chops (bone-in or boneless)
➡ Ribs (St. Louis or baby back)
➡ Loin roast (bone-in or boneless)
➡ Tenderloin

For the Spaniards, seam butchery provides economic opportunities to stretch this valuable section a little further. The tenderloin *(solomillo)* is cleaned and set aside (as in America, it is typically roasted as is). Since the ribs *(costillas)* have already been separated away from the loin *(lomo)* and likely shoved into the hands of a visiting American cook who's been ordered to barbecue them for family meal, the *lomo* is cleaned of all fat and subdivided into a few specific cuts.

At the top of the *lomo* is Elvis the butcher's favorite grilled snack, the *pluma*. This very flavorful muscle, part of the *rhomboideus,* is removed by slicing along a natural seam at the head of the *lomo* that releases the *pluma* from the rest of the loin.

Next, the *cabecero* is removed. This muscle is located just past the head of the loin near the shoulder (hence its *nom de guerre* in most kitchens, *cabeza del lomo,* or "head of the loin"). In American butchery, this section is usually left on as part of the Boston Butt, and some Americans might recognize it by its nickname on the BBQ circuit (it's part of what's known as the "Money Muscle") or by its Italian names (*coppa* or *capicola*). In America, the *cabecero* is also sometimes included in a few different sub-primal cuts, including the blade roast, the CT Butt, or the blade steak.

The *cabecero* contains two other favorite grilled or roasted cuts of Spaniards: the striated collar *(aguja)* that appears in some sausage recipes and the boneless shoulder blade roast *(presa)*. Both are part of the American Boston Butt, which is part of the reason why it's so delicious when slowly braised or smoked, as Americans often do for Carolina pulled pork.

At this point the *lomo* is now clean, trimmed, boneless, and ready for either curing as a *lomo embuchado*, which involves marinating in a garlic and *pimentón*-spiked *adobo,* or—in the rare instance of the geographically protected (IGP) Riojano-Style *Chorizo* (see recipe on p. 302)—grinding for a sausage mix.

Thus, the typical Spanish sub-primal cuts from the *lomo* are as follows:

- Loin, boneless *(lomo)*
  - *Pluma*
  - *Cabecero*
    - *Aguja*
    - *Presa*
- Ribs *(costillas)*
- Tenderloin *(solomillo)*

Belly Section

MUCH LIKE THE head section, you can really only break the pig's belly *(panceta)* down a few ways, so there's not a lot to be said here. Whether you are in the United States or Spain, most bellies wind up being cured, smoked, braised, or ground into a sausage mix.

What's intriguing, however, is how relatively unimportant the belly is to the Spanish. At most *matanzas*, the *matanceros* bisect the *panceta* laterally to take advantage of what little lean meat they can find there and use the rest for sausage grind. As explained to me by Miguel Ullibarri, an *Ibérico* expert, historian, and former head of Spain's *Ibérico* consortium, the belly was traditionally more valuable to Spaniards as part of a sausage mix because "when a pig was the main resource for a family, distributing the 'eatability' of the different parts throughout the year was key; that's why the fattier cuts (like the belly) were often utilized for mixing in with the lean."

In America, however, pork bellies were so integral to our growth as a nation that they were traded as a staple commodity on the Chicago Mercantile Exchange for years. And these days, we're blessed with a multitude of unctuous braised, smoked, grilled, and cured preparations of the belly—not to mention God's own bacon made by the hands of artisans like Allan Benton.

In America, pork bellies were so integral to our growth as a nation that they were traded as a staple commodity on the Chicago Mercantile Exchange for years.

Shoulder Section

AS WITH THE belly, the art of smoking meats has given Americans a way with pork shoulder that is uniquely our own. In the American system of butchery, that has led us to breaking a pork shoulder into its two major constituent muscles—the shoulder butt and the picnic ham. Only occasionally are the trotter and hock used, and even then it's generally for smoking and adding to dishes like collard greens. So, a typical American sub-primal breakdown of the shoulder is as follows:

➡ Shoulders
   ◆ Picnic
   ◆ Butt
   ◆ Trotters
      ◆ Shanks/hocks

Ask a Spaniard what he'd do with a shoulder section *(paletilla),* however, and you'll find that the only thing most red-blooded Spanish men like better than two hams per pig is four hams per pig. That means leaving the shoulder intact (including the hoof or *pata)* and curing it exactly as you would for a ham. Or, if you happen to reside in Galicia, this would mean *lacón;* it's essentially the same thing, except *lacón* is a pork shoulder that is found in both cured and cooked forms and comes from only the local white pigs of the region.

Most importantly, by removing the *paletilla* and cleaning up the belly section a little, you will stumble upon one of Spain's greatest contributions to pork eatery—the *secreto.* Hidden just under the top layer of belly fat and located just behind the *paletilla,* the *secreto* is essentially a pork skirt steak comprised of the *latissimus dorsi* and *pectoralis profundi* muscles—a heavily marbled, perfect cut of grilling meat that most butchers traditionally kept for themselves and their families (hence the name *secreto,* or "secret").

Thus, a typical Spanish sub-primal breakdown would look like this:

➡ Shoulder *(paletilla)*
   ◆ Trotter *(pata)*
➡ *Secreto*

Ham Section

OTHER THAN THE great hams of our Southern states or the work of excellent artisan ham purveyors like Allan Benton, Bev Eggleston, or La Quercia, you rarely find Americans doing much in terms of whole cured hams. Instead, we typically break the back leg into more manageable parts for roasting or braising, such as the sirloin, top round, and a smaller version of a ham. As with the shoulder cuts, the shank and trotter are often removed in American butchery and smoked for use in pots of split pea soup or long-simmered greens.

That means that the typical American ham breakdown would include five muscles:

➡ Hams
   ◆ Sirloin
   ◆ Top round
   ◆ Ham
   ◆ Shank
   ◆ Trotter

In Spain, the story is much simpler: If there's a ham worth having, it's a ham worth curing. You rarely—if ever—see sub-primals coming from the back leg, since demand is so high for Spanish hams. This is especially true for *Ibérico* pigs, whose legs can fetch as much as $1,000 each—in some cases, the *jamones* are worth more than the rest of the pig combined.

THE PIGS OF SPAIN: A STORY OF RECLAMATION

SOMETHING IS HAPPENING in Spain these days. There's a new sound on the *dehesa,* a new sight amongst the sea of black-haired, black-hoofed *Ibéricos* that have dominated these lands. Changes, they are afoot.

From region to region, the people of Spain have slowly begun reclaiming and restoring native pig populations that had almost gone extinct. Mirroring our own swine surge here in the United States (and in other countries all over the world), all manner of rare and heritage breed pigs are finding popular support in Spain, from Asturias to the Basque country and all points in between—with local governments rising to support and promote their local pigs as emblematic symbols of regional pride and financial boon.

Although it's doubtful that these breeds will be readily available in your local market anytime soon, you should know about them. They represent a new wave of porcine perception that is sweeping not only the Spanish nation, but the entire world.

Ibérico Pig: King of the *Dehesa*

THE *CERDO IBÉRICO* is undisputed champ in a country of very proud pork eaters. Like Kim Kardashian, the *Ibérico* just has certain genetic advantages that make it desirable to those who prefer a little extra junk in the trunk. You see, the *Ibérico* has the capability to store a proportion of its fat intramuscularly, a genetic flaw that leads to a higher percentage of marbling in *Ibérico* meat than in any other type of pork.

And this, among other reasons, is why the *Ibérico* is one of the best *jamón*-producing breeds out there. Its meat is literally interlaced with striations of fat, which the enterprising Spaniards discovered can be made even more delicious through the natural process of the *montanera.*

The *montanera* is both a philosophy and a means of producing the best *Ibérico* meat possible. The months-long, controlled feeding program allows the *Ibéricos* to roam freely and happily through forests of acorn-providing holm and cork oak trees. Their diet is completely free range, the acorns supplemented only by grasses or fragrant herbs the pigs might find in the area.

What follows is nature's little miracle: Meat and fat are imbued not only with the natural nuttiness of the acorns (by happy coincidence, the pigs only consume the sweetest variety and spit out the bitter ones) but also with very high contents of oleic acid as a byproduct of the animal's acorn diet. This oleic acid content, one of the greatest assets of *jamón Ibérico,* is a big marketing point for the Spaniards since it provides the same LDL cholesterol-reducing qualities found in olive oil (hence the *Ibérico's* nickname: olive oil trees on legs). Also, a major reason why the fat of *jamón Ibérico* melts at room temperature is because of the meat's high levels of oleic acid—a characteristic that sets it apart from other cured hams.

But the Spaniards aren't done tinkering with the *Ibérico.* The latest trends include supplementing the pigs' diet with everything from pumpkin to dates to further enhance the sweeter qualities of the animal's already naturally sweet meat.

In the words of Mel Brooks: It's good to be the king.

White Pig: An Unsung Hero

IF YOU'VE EVER eaten a plate of *jamón Serrano* or a stew made with greens or beans, you probably have Spanish white pigs to thank for your meal. Generally speaking, Spain's white pigs are close relatives to Duroc, Pietrain, or Landrace hogs (all three are popular throughout the country), but commercially available *jamón Serrano* is typically made from only the Landrace breeds.

White pigs are a significant proportion of the pig population of Spain (and most of Europe as well). Also, they are a cheaper alternative for Spaniards looking to satiate their *jamón* cravings, since the relatively expensive feeding standard of the *montanera* makes *jamón Ibérico* pretty expensive even within Spain.

More specifically, strains of white pigs are sometimes used for cross-breeding to make a less expensive variant of *Ibérico jamón.* This is a source of constant consternation for *Ibérico* purists like Miguel Ullibarri, since regulations are only just starting to keep up with scientific and breeding advancements. Those advancements have produced an animal that can reach market weight much more quickly, but the process of doing so sacrifices some of the flavor and other characteristic qualities of purebred *Ibéricos.*

These cross-bred *jamones,* however, are typically easy to spot. The *pata* isn't always black, for one thing (myths abound about unscrupulous vendors allegedly resorting to painting *patas* for higher resale values); also, the ankle area of a purebred *Ibérico* is typically much smaller and daintier than the thicker white-pig hybrid. As a result of some cross-breeding chicanery, the Spaniards' grading systems for *jamones Ibéricos* have gotten much tougher, including using chromatography and product grading to ensure the sanctity of the *Ibérico* name.

Euskal Txerria: The Last of the Basque Pigs

THE ANCESTRAL PIG breeds of the Basque Country have sadly been whittled down to just a single survivor: the *euskal Txerria,* an adorably short-legged, multicolored pig whose population thankfully has been brought back from the brink of extinction to stable levels.

This pig—barely a blip on the Spanish map at present—features in many traditional Basque dishes and *charcuterie,* including the sausage *lakainka* (a *chorizo* variant either grilled fresh or cured and stored for the winter months). Since the *euskal* is fed a diet mainly consisting of acorns, chestnuts, and hazelnuts, the flavor of the fat picks up many of the same notes as that of the *Ibéricos.*

Gochu Asturcelta:
Smoke's Best Friend

EVERYTHING ABOUT THE *gochu Asturcelta,* a native Asturian pig, seems perfectly suited to the climate of northern Spain and the slow-cooking methodology of local stewpots. Big, floppy ears cover most of the pig's face, protecting it from the climate's cold and rain; its huge snout is perfect for rooting around in the wet earth; and its elephantine trotters and hooves are ideally suited to climbing the Asturian mountains in search of food. The animal's natural diet of hazelnuts and chestnuts gathered in the surrounding countryside (acorn trees don't grow in the north of Spain) add a wonderful quality to the flavor of the meat.

But this pig also has a very important secret. This secret is one of the reasons for skyrocketing interest in its breed: The *gochu* was destined by nature for the smokehouse. Jose Manuel Iglesias, president of the regional *Gochu Asturcelta* Association, explained to me that the meat of the *gochu* holds significantly more natural water than most other types of pork.[29] Given that the smoking process involves slow drying and/or cooking, the natural humidity of *gochu* meat keeps it perfectly moist and thus able to spend more time in the smokehouse than other types of pork. As a result, the *gochu* is slowly becoming one of northern Spain's most talked-about, and certainly one of the most delicious, four-legged foodstuffs.

CHAPTER

3

SALT, MEAT, LOVE, AND TIME

Decoding the secrets and
science of *charcutería*

VERYTHING NEEDS
water to survive. This axiomatic truth is at the very core
of understanding how to safely cure, ferment, dry, and eat
anything that falls under the banner of *charcuterie*.

It's the role of the *charcutier,* then, to manipulate *aqua
pura* at any given time in the curing process—to control
water activity by using ingredients that remove it (salt, for
example) and by controlling environmental factors to help
make a safe, delicious, and stable end product. This, in a
nutshell, is the whole point of *charcuterie,* and it's also the
focus of this chapter. Here is where you will learn how to
play God within your own little cured-meat universe.

CADA MAESTRILLO TIENE SU LIBRILLO.

EVERY MASTER HAS HIS METHODS.

I'll start by defining exactly what it means to cure meat, specifically as it pertains to the types and styles of *charcutería* that you can expect to find in Spain and throughout this book. But that gets us only halfway home, since you're eventually going to need to set up your very own meat-curing lair.

That's why I'll next talk strategy via a step-by-step guide of the entire *charcuterie* process, in which I will (1) discuss equipment options for home cooks and the professional crowd, (2) explain the ingredients involved in curing meat, and (3) introduce some of the techniques I picked up during my time working with the *sabias*.

THE CURE FOR WHAT AILS

CURING, WITH REGARD to either meat or health, is both etymologically and intrinsically linked to the same Latin root: *curare,* meaning "to take care of." Either way, you are "taking care of" the subject (in this case, meat) by removing nega-tive factors that could cause harm and installing in their stead a positive state of homeostasis.

In the case of *charcutería,* you have two means at your disposal for achieving homeostasis within a protein: a *salmuera* and a *salazón.*

A *salmuera*—aka Thanksgiving's favorite turkey helper, a brine—involves immersing the protein (typically whole muscles) for a period of time in a salt-based solution. Other aromatics and liquids can also come to the party and do some flavor mojo, but the point of the *salmuera* is to use the process of osmosis to affect the internal salinity of the protein, thereby creating an environment less hospitable to bacterial baddies that if left unchecked would otherwise spoil the meat.

A *salazón* involves the gradual removal of water from a protein through application of a salty rub called, not coincidentally, a cure. Like the *salmuera,* aromatic companions can join the cure—herbs, spices, sugars, and such are all fair game—but it's the salt content that really pulls water out of the protein and achieves preservation potential.

Once the *salmuera* or *salazón* has done its magic, there are a few different methods for finishing the job. The most common ways in Spain include fermenting, smoking, or dry curing the protein.

THE HUSTLE BEHIND YOUR CURED MEAT MUSCLE

NOW THAT YOU know what curing meat essentially means and how it works, the next step is to understand how to apply that knowledge to meats of different sizes and shapes.

More specifically, the manner in which you decide to preserve meat largely depends on exactly what you're curing and the end result you're working toward. Thus, this section is divided into the three major types of meat-based *charcutería* you'll encounter in Spain: (1) whole-muscle preparations, (2) sausages and other *embutido* preparations, and (3) *terrine, pâté,* and *confit* preparations.

Whole Muscles

WHOLE-MUSCLE CURING IS exactly what it sounds like: taking an entire piece of meat and applying either a *salmuera* or a *salazón* to it. In the case of a *salmuera,* that means immersing the muscle—say, a pork loin for making *lomo adobado*—in a brine for a period of time before moving forward with an *adobo* and then cooking the *lomo*.

At the *matanzas* that I worked, the *sabias* generally used a standard *salmuera* for everything that was brined. Since they knew the salinity of the brine, the weight of the meat, and whether the meat contained a bone, they also knew roughly how long the meat should stay in the brine. I've included their *salmuera* recipe in Chapter 4 (see p. 105), along with a timing chart for brining various types of proteins in it. As for a brining vessel, any foodsafe container will work, as long as the meat remains completely immersed in the liquid (in some cases, this may mean that you'll need to weigh the meat down with a few plates or cans).

In the case of a *salazón,* however, the meat is coated in a salt-based rub for an extended period of time—typically anywhere from half a day to a full day per kilogram of meat—and then weighted, rotated occasionally, and finally rinsed off when it feels sufficiently "bouncy" to the *salazón maestro's* learned touch.

For this application, a ziptop bag that just fits the meat is your best friend. For larger muscles, like *jamones, cecinas,* or *lacones,* however, you will probably need to invest in a large, foodsafe plastic tub that's dedicated to curing, since it will need to be sterile and should have holes in the bottom to drain any accumulated liquid.

Jamón Ibérico, for example, is made precisely in this manner. In Spain, the cleaned and trimmed *jamones* are rubbed and packed with a *salazón* before being stacked into gigantic mountains of meat. Every 8 to 10 days, the *jamones* are turned and repositioned in the pile—with each ham getting a turn as king of the hill, so the weight of the pile is distributed as evenly as possible amongst the *jamones.* This process continues until the *maestro* gives the OK and the *jamones* are rinsed and continue their journey to ham heaven.

Sausages and Other *Embutidos*

SETTLE IN: THIS is where the ride gets a little bumpy.

In the world of Spanish *charcuterie*, anything crammed into a sausage casing falls under the nebulous category *"embutidos,"* which comes from the verb *embutir,* meaning "to stuff." But although nearly all sausages are *embutidos,* not all *embutidos* are necessarily sausages. For example, you can also find *lomos* and other large-format cured muscles stuffed into casings and sold under the *embutido* title at *charcutería* shops around Spain. And just to make matters even more confusing, many sausages and other cured meats involve a variety of techniques to get them from raw product to something shelf stable—which tends to make classifying them as, for example, "smoked," "fermented," or "dry-cured" considerably more complicated since a number of techniques might be used to make the same product.

That's why I will describe and classify each style of *embutido* and exactly what it entails in this section. These styles include: (1) fresh *embutidos,* (2) semicured *embutidos,* (3) cooked *embutidos,* and (4) dry-cured *embutidos.*

➡ **FRESH *EMBUTIDOS:*** This family of *embutidos* is seasoned, mixed, and stuffed into a casing and held cold until being cooked.

The only deviation from this plan might come if, at the whim of the *charcutier,* the sausage meat is fermented for a day or two to give it a little bit of sour tang and depth. (I'll discuss fermentation a little later.) The most important thing to remember here is that that you'll have to cook fresh *embutidos* before you eat them; that's what I'm using to define this style of *charcuterie* as "fresh."

➡ **SEMICURED *EMBUTIDOS:*** Of all of *embutidos, semi-curados* are the most difficult to classify. They reside somewhere between raw and cured, since the whole point of making them is to ferment the sausage heavily and quickly, thus increasing the amount of acidity in the sausage. The resulting product is both deliciously sour and somewhat shelf stable; however, *semi-curados* still should be cooked before eating.

The *semi-curado's* fermentation period, which takes place over several days, is typically accompanied by a cold or hot smoking process. Smoking is a preservation technique typically found in the north of Spain, where the climate is too cold and wet to provide an adequate means of drying out *embutidos* in the normal fashion, so hybrid techniques like the *semi-curado* evolved over time.

Note, however, that smoking is a process used pretty much across the board for different types of *charcutería.* It's more of a means to an end than a category unto itself, so you can expect to see a slew of recipes in this book that call for an *ahumador* (smoking chamber); it's needed for everything from semicured preparations like *Chorizo Asturiano* (see recipe on p. 308) to dry-cured preparations like *Cecina* (see recipe on p. 115).

➡ **COOKED** *EMBUTIDOS:* These *embutidos* have a cooking step referred to in meat-industry parlance as a "kill step" as part of the preparation process. Thus in these recipes, you will make and stuff the sausage first and then cook it completely in some fashion. (Simmering the sausage typically gets the job done, though some recipes call for hot smoking.) Last, you'll chill the *embutido* to help it set and make it sliceable.

This style of *embutido* is edible cold or hot, but in most cases, eating it hot means throwing the sausage onto a grill or *plancha* for searing. That's my favorite way to eat many of the *morcilla* (blood sausage) preparations you'll find in this book.

➡ **DRY-CURED** *EMBUTIDOS:* Dry curing meat is the epitome of the *charcutier's* craft—it's a labor of love whereby the craftsman works hand in hand with nature herself to create something wholly, uniquely meaty and delicious.

Building on each of the other styles of *charcutería,* dry-cured *embutidos* involve mixing and stuffing an *embutido* like you would in the fresh style, fermenting it like the semicured style (though for a longer period of time and at different temperatures), in some cases smoking it for a period of time, and finally placing it into a drying chamber. There, nature takes its course until the cycle is complete.

This type of *charcuterie* will also require the most investment of your time and resources in terms of sourcing the right ingredients and equipment (something I'll get into shortly). Just bear in mind that if you are serious about going down the rabbit hole of crafting *charcutería* and the art of curing meat—a subject with incredibly delicious rewards—you must first understand the principles of dry curing.

Terrines, Pâtés, and *Confits*

THIS LAST CLASSIFICATION of *charcutería* involves all manner of spreadable delicacies, including *foie gras, rillettes, pâtés,* and other deceptively simple, yet labor intensive, recipes and techniques. I'm a fan of these recipes, because they require several techniques that really demand focus and attention—just the sort of passion for details that separates the rookies from the pros.

I'm talking about details like getting a perfect cure on your meat via a *salazón* or *salmuera,* cooking or curing to the right degree of doneness, and then serving your work with the right garnish to complement the dish. And when you nail it, paying such close attention to the minutiae leads to a very rewarding experience. And if something goes wrong? It'll inspire a deep-seated yearning to try again.

WEAPONS OF CHOICE

WHETHER YOU'RE A professional chef with some training or an aspiring *charcutier* trying to wrap your brain around a strange new world, this section will help you understand what you will need to safely make the sort of *charcuterie* recipes like the ones in the next chapters.

To start, we will discuss a few important points on safety and sanitation—I presume that most of what I say in this section is common knowledge so I'm not going to dive too deep; just play nice, work clean, and make good friends with your kitchen scale, *amigo*.

After that, we will break down each step of the *charcuterie* process, discussing within each step some equipment options, ingredients, and techniques from the *sabias* for making *charcutería* their way.

Practicing Safe Stuffing

THERE'S NO GETTING around the fact that the *charcutier's* art is one of precise measurements and exact temperatures. Since this isn't the sexiest of subjects, however, I'm going to limit what I have to say to three points: (1) selecting proteins, (2) working temperatures, and (3) measuring procedures.

After that, we can get back to the fun stuff... I promise.

SELECTING PROTEINS: If you are making *charcutería* at a *matanza* in Spain, then you (or the *matanceros)* more than likely personally took the life of a pig. The pig's owner—likely your friend or relative—knew the pig's name, what it ate, and where it slept. He or she saw to it that the animal lived a generally happy and healthy life and received an honorable and swift death. That is simply how things are done at the *matanza,* and it's especially true if that pig was a coveted *Ibérico*.

My point is that most *charcuterie* consists of a scant few ingredients: Most of the time,

it's little more than meat, fat, salt, and some spices. Your entire goal is to then methodically remove the water content from the meat and fat, thereby concentrating the flavors bestowed upon Piggie by God and whatever foods that your snouted friend ate. Thus, if your pig lived a shitty life, ate a shitty diet, and lived in shitty circumstances, guess what your *charcuterie* is probably going to taste like?

It's really pretty simple: Happy pigs make tasty *charcuterie*. So bear that in mind as you search out this main ingredient for your meat-curing projects: Befriend your butcher and/or farmer, know your pig and how it lived, and treat the animal's sacrifice with the respect it deserves.

At the risk of sounding like an episode of *Portlandia* or a douchey Olive Garden commercial, trust me...you really will taste the difference.

WORKING TEMPERATURES: Any time you are dealing with raw meat for *charcuterie* purposes, the colder, the better—to the point of partially freezing the meat when making sausage. That's because the process of making *charcutería* almost always involves heat transfer, whether it's from the warmth of your hands when rubbing a *jamón* with salt or the mechanical friction of grinding sausage meat. The potential of raising the meat's temperature is a real and constant danger.

And that's why your local health department rather dogmatically insists that all restaurants keep their refrigerators under 41°F (5°C): That's the highest temperature that will prevent bacteria and other foodborne baddies from multiplying and proliferating. If you fail to follow the techniques in the *charcutier's* playbook for storing raw, uncured meats at or below this temperature, you're rolling the dice every time you eat.

I'll get into some tricks for keeping everything chilled later in this chapter, so for now, just bear in mind that in terms of raw proteins, warm temperatures are like watching reality TV: best to be avoided in the interest of preserving your health and sanity.

MEASURING PROCEDURES: Do yourself a huge favor and go buy a kitchen scale right now if you are serious about endeavors like baking bread or curing meat. In these two surprisingly related culinary disciplines, a good scale will prove its weight to you in gold…and tasty treats.

Both baking and *charcuterie* are crafts of meticulousness and patience. They are both at their best when measuring ingredients in metric units down to the gram or kilogram—not only for the sake of precision, but also because that's how most of the world works when it comes to writing replicable recipes. By comparison, our US customary system of pounds and ounces simply isn't practical for measuring the small ingredient amounts used in *charcuterie,* and the last thing you want to do is use volume measurements (cups, etc.) for critical ingredients such as salt and curing salt. I have supplied equivalent measurements in the US customary system if you insist on sticking with what you already know, but note that while these equivalents will get the job done, they are not as exact as the metric measurements. For example, I've had to use the measurement ⅓ ounce for recipes calling for 8, 9, or 10 grams of an ingredient. Sorry, my fellow *Americanos,* but the metric system is the way to go for accuracy, and that's

how I wrote the *charcuterie* recipes in this book.

As a quick example, if you take a cup of kosher salt—say, Diamond Kosher brand—and compare it to a cup of another kosher salt, like Morton's, you'll wind up with almost twice as much with the Morton's salt (see the chart below). This would be disastrous to any curing project, and the reason is simple: Every salt is flaked differently, so they all take up different volumes in a measuring cup.

1 CUP OF SALT (IN GRAMS)

Morton's Kosher Salt	240 g
Diamond Kosher Salt	120 g
Supermarket-brand coarse sea salt	185 g
Maldon sea salt	125 g

That's why I'm only going to provide weight measurements for crucial ingredients, like salt and curing salts—the margin for error is so small with these ingredients that you really want to be as precise as possible. For other ingredients like spices, however, I'll provide the most convenient measure, since a little extra pepper in your *chorizo* won't mean the difference between success and failure.

Last, I'll almost exclusively be writing the *charcutería* recipes in such a way that they're scaled to what we call in *charcuterie* parlance the green weight—meaning the raw weight—of the main protein being used. Virtually every *sabía* I know uses this methodology, but for the purposes of this book I'm calling it the *Charcutier's Percentage* (see *The Charcutier's Percentage* on the next page).

THE *CHARCUTIER'S* PERCENTAGE

YOU'RE GOING TO notice something different about a lot of the *charcuterie* recipes in this book: I don't give you exact amounts for the various meats, spices, seasonings, and other ingredients. Instead, you'll see amounts weighted as a percentage against a standardized amount of the major ingredient in the recipe (since that's typically either meat or blood, the standardized amount is typically either 1 kilogram [kg] or 1 liter [L]).

That's because at all of the *matanzas* that I participated in, and with all of the *sabias* whom I've talked to and learned from, I realized that pretty much everyone works off of a percentage-based system. Doing so really makes life simple for *charcutiers,* since it allows us the freedom to adapt instead of being beholden to a static recipe.

Think about it: If you ask a butcher for exactly 7 pounds of lean pork meat and 3 pounds of fat, you're likely to get an amount of each that's either slightly over your target or slightly under it. If you're cooking a regular meal, this isn't an issue since you can just fudge the rest of the ingredients without worrying. But in *charcuterie,* precision is key, so if you're like me, you'll likely be frustrated with the recipe, stressing over whether the math is on point, or noticing your tolerance for calculating numbers dwindles as each glass of bourbon is consumed. By following the *sabias'* example, however, you can take any amount of meat and fat for a given recipe and apply percentages to all of the remaining ingredients based on some known rules of thumb.

The interesting thing is that the logic behind this way of thinking isn't groundbreaking—bakers use it every day. Specifically, in baking, all of the ingredients are based on a percentage against the total amount of flour, which is always 100 percent (it's a methodology called the *baker's percentage*). If you apply that concept to the world of *charcuterie,* you've got the means for ensuring that recipes will work without needing a degree in advanced calculus to scale recipes up or down.

On the next page, you'll see a chart with the rules of thumb for major ingredients in most *charcuterie* recipes. Bear in mind, however, that for every rule, there are exceptions. Some *chorizos,* for example, call for much more *pimentón* than the chart's maximum of 2 percent. I'm also not above rounding seasonings off to the nearest decimal, especially when I'm converting from weight to volume measures for less critical ingredients. Salt and curing salt numbers are more concrete, though, because those numbers are more a question of safety, whereas seasonings are more a matter of taste.[30]

THE *CHARCUTIER'S* PERCENTAGE

CHARCUTERIE TYPE	LEAN MEAT (%)	FAT (%)	SALT (%)	CARBOHYDRATE (%)	LIQUID (%)	PURE SODIUM NITRITE (%)	PURE SODIUM NITRATE (%)	TCM #1, INSTACURE #1, OR DQ #1 (%)	TCM #2, INSTACURE #2, OR DQ #2 (%)	FERMENTATION AGENT	BINDING AGENT (%)	SEASONINGS (%)
Whole muscle	n/a	n/a	6	2.50	n/a	.0625	.0218	.10	.05	n/a	n/a	n/a
Fresh *embutidos*	70–80	20–30	1.5–2.5	.5–1	1–10	n/a	n/a	n/a	n/a	n/a	Up to 10	.75–2
Cooked *embutidos*	70–80	20–30	1.5–2	.5–1	1–10	.0156	n/a	.25	n/a	Minimum of 10 grams	Up to 10	.75–2
Semicured *embutidos*	70–80	20–30	2.2–2.5	.5–1	1–10	.0156	n/a	.25	n/a	Minimum of 10 grams	Up to 10	.75–2
Dry-cured *embutidos*	70–80	20–30	2.2–2.5	.5–1	1–10	.0156	.0171	.25	.25	Minimum of 10 grams	Up to 10	.75–2

NOTES: All figures in this chart are percentages of the weight of the whole muscle (in the case of the whole-muscle figures) or the weight of the sausage mix/*masa* (in the case of the *embutido* figures). So, for example, 10 kg of *masa* for a fresh *embutido* like *Chorizo Fresco* (see recipe on p. 226) would use a maximum of 2.5% salt (250 g).

Note that I listed the nitrite/nitrate figures in their respective columns as the maximum allowable amount of curing salts in their *pure form* in the United States. Knowing these maximum percentages of curing salts is advantageous since, with a little math, you can be certain you aren't adding more than the maximum quantities.

I have also listed the maximum amounts for mixes like TCM #1 or TCM #2, which are recommended for use as a precautionary measure for most *charcuterie* producers. These mixes are made with specific percentages of nitrite and/or nitrate to safeguard against adding more than the legally allowable quantities of curing salts. If you are using another mix or using sodium nitrite/nitrate in their pure forms, you will need to calculate the parts per million (ppm) of the curing salts you're adding to ensure that they fall within safe guidelines for consumption.

For more information on this, check out the United States Department of Agriculture's Food Safety and Inspection Service website (www.fsis.usda.gov). It has lots of useful tips for figuring out all of this information.

THE TOOLS OF THE CHARCUTERIE TRADE

ONE OF THE biggest secrets about *charcuterie* is that a home cook can find and set up the right equipment for fermentation, smoking, and drying chambers with relative ease—provided that you can convince your significant other that a little funk in the basement air is a damn sexy thing.

My restaurant and butchery brethren out there, however, have a much tougher road to travel. Since much of the *charcutier's* equipment needs are designed to manipulate temperature, humidity, pH levels, and airflow, you'll often find yourself falling well outside the guidelines of the United States Department of Agriculture (USDA) and your local health department for raw protein storage. So, those looking to sell their *charcuterie* are legally obligated to answer to local and state regulators about operational goals, log temperature and hazard analysis and critical control point (HACCP) data, provide test sausages, and jump through myriad other required hoops that home cooks don't face.

Thus, as I discuss the options for setting up your curing rig, I'll offer a range of ingredients and techniques to help you accomplish your meat-curing goals. Your commitment, space availability, and budget will help you determine the right move for you within these options, but bear in mind this point: If you are a food professional or intend to sell your *charcuterie* to the masses, you're going to eventually need to spend some quality time with the fine folks of your local health department, as well as USDA representatives, if you want to get serious about *charcuterie* production in your facility.

Since we already covered sourcing meat, I'll presume that you've tracked down a reliable purveyor of delicious, happy pigs. Also, I'll presume that I've weaned you off volume measurements in favor of using a proper kitchen scale.

Next, I'll cover the equipment, ingredients, and technique for curing a typical *embutido* or whole muscle, all of which takes place during the following steps: (1) grinding, (2) seasoning and mixing, (3) stuffing, (4) linking and tying, (5) fermenting, (6) smoking, and (7) drying and storage.

Grinding

IN THE OLD days, you'd use a rotary grinder or sharp knife, along with some elbow grease, to chop and dice lean meat and fat into the proper sizes and shapes for *charcuterie*. These days, however, a lot of grinding options are available, as well as a slew of techniques for keeping equipment and ingredients as cold as possible throughout the grinding process.

EQUIPMENT: Modern times have brought all manner of electric grinding devices, including everything from attachments for stand mixers to gigantic, vacuum-sealed industrial grinders/tumblers for *charcuterie* batches reaching into the hundreds of kilograms.

You're going to want to have a dedicated grinder, since even a smallish, standalone grinder will really save you some calluses in the long run. They vary in price from just under $100 to much more, and if you're lucky, you might find

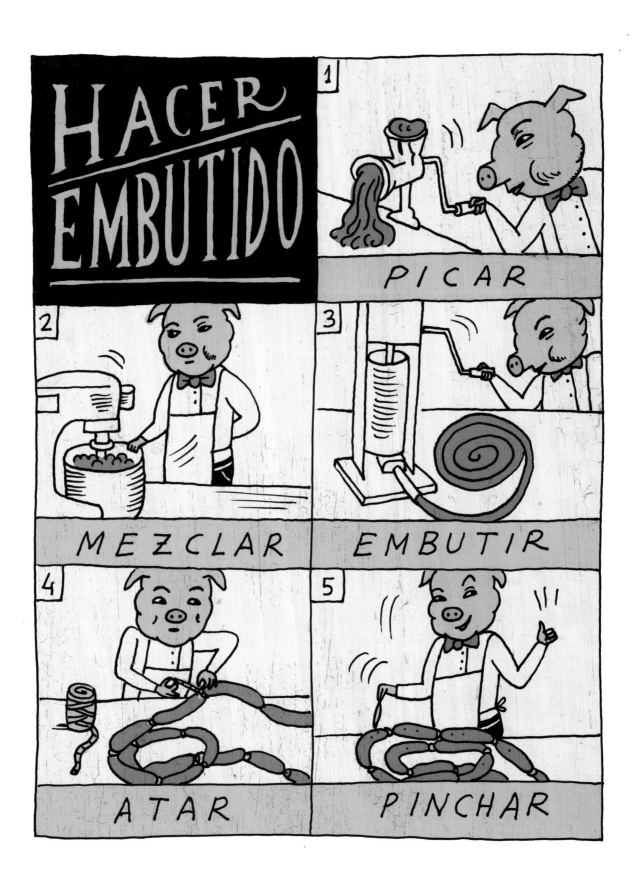

one on sale for less than $100. The one you want will have detachable, die-cast metal parts that are dishwasher safe and rustproof, a few die and stuffing tube choices, easily replaceable parts (having a few blades on hand are a big help), and a 350-watt (or more powerful) motor.

As for those stand-mixer grinding attachments: I've never been a big fan of them for grinding meat or stuffing sausage. The motors tend to wear down quickly when you're sending quasi-frozen pork fat through the machine, and most of the time the blades and dies either aren't sharp enough or aren't the right size to get the job done. But if you want to roll the dice and give it a shot, check your mixer's parts and user guide to see if they will suit your needs.

For the restaurant and butchery guys, I'm sure I don't need to tell you that if the name Hobart is stamped somewhere on your machine, you'll probably be fine.

INGREDIENTS: At most *matanzas* I've attended, we cut our meat into large chunks and mixed it with the salt, sugar, and any curing salts that we were using. Then, we'd send the chunks through the grinder so that the seasonings and curing agents were evenly distributed throughout the grind. We also used the finest die of a grinder to mince garlic for seasoning *embutidos*. Likewise, we ran cooked pumpkin, potato, and even rice through grinders for later use as binding ingredients for different sausages.

I'll explain this process further in the next section. For now, just remember that having an automated grinder handy obviously means it'll be a lot easier to grind piggie parts, and it can also help you prep a lot of the other ingredients for your *charcuterie* recipes.

TECHNIQUE: Pork back fat, the weapon of choice for most *charcuterie* applications due to its flavor and texture, melts at a relatively low temperature. Its melting point clocks in around 85°F (29°C), so a smart *charcutier* uses the following techniques to maintain the integrity of the sausage's fat:

➡ Chilling anything that's going to come into contact with the meat and fat in a freezer. This includes bowls, blades, dies, the hopper, and the auger.

➡ Chilling the meat and fat in a freezer. In their book *Charcuterie,* Michael Ruhlman and Brian Polcyn describe the right texture for the meat as being "crunchy."

➡ Making sure that the grinder blades are very sharp. If they aren't, a professional sharpener can touch them up for a small fee.

The whole point of going through these preparations is to avoid every sausage maker's worst nightmare: smearing. Smearing is a folly that can occur as the meat is being ground; if it happens, the ground meat looks less like individual specks of meat and fat and more like homogenized pink toothpaste. When it occurs, the fat is heated up beyond its melting point due to a grinding problem and is literally "smeared" into the meat—a situation that will ruin your batch and maybe even your entire day.

To avoid making pink meat goo, first make sure that you follow the three tips listed above. Next, stop the grinder as soon as you notice the grind looks odd and clear the die or auger of any blockage. Check your meat and fat temperature and chill again if necessary. Then, make sure that you have the blade installed in the grinder correctly, since some manufacturers sometimes sharpen one side of the blade and not the other.

Last, a trick the *sabias* often used was to step down their grind into phases—in *charcuterie* parlance, this is known as "progressive grinding." Basically, the idea behind progressive grinding is that it's a lot easier to go from big dice to small dice to smaller dice than it is to shove a big piece of meat through a die with tiny holes. So, as an example, if your goal is to get a very fine dice, you should start with a large grind and, step by step, take the grind down to the desired end result, re-freezing the meat in between steps if necessary. By doing so, the fat encounters less friction and the chance of smearing is reduced.

The following chart shows some typical progressive grinding plans.

TEXTURE REQUIRED	DIE PLATES USED	
	Size of first grind in inches (and mm)	Size of second grind in inches (and mm)
Coarse (second grind optional)	¾ (19)	⅜ (9.5)
Medium-coarse (second grind optional)	⅜ (9.5)	⅜ (9.5)
Medium	⅜ (9.5)	¼ (6)
Medium-fine	⅜ (9.5)	3⁄16 (5)
Fine	⅜ (9.5)	⅛ (3)

Obviously, it would be great if you had exact dies for each size of grinding meat, but that takes time and money, and you have to be pretty serious about making *embutidos* to make that kind of investment. If you have only the few dies your grinder came with, don't sweat it. Just shoot for the ballpark and step down as needed.

Seasoning and Mixing

IN MOST OF the *embutido* recipes that I learned with the *sabias,* we mixed salt, curing salts (if we used them at all), and sugar in with pre-cut muscle chunks prior to grinding. Doing so ensured that the salt got pushed around the ground meat evenly and served as another layer of protection by getting the salt dispersed.

Later in the mixing process, we added binding agents, fermentation helpers, and any appropriate seasonings to the mix. These ingredients all have a role to play depending on the type of sausage that you're making, so I discuss them below. The addition of salt, however, is especially important, so it warrants a quick explanation now.

For the science geeks: When you vigorously mix together salt and ground meat, you are in effect solubilizing two proteins inherent in meat, actin and myosin. These proteins are responsible for the sausage *masa's* capability to bind with water and stabilize the fat within it, thus keeping the fat from pooling together or floating to the top of the sausage when it's cooked.

For example, if you've ever mixed meatloaf into a big, homogenous ball, you get the concept of enabling actin and myosin. Conversely, when you're forming hamburger patties, you have the opposite goal, as overmixing and getting really dense patties through enabling actin and myosin is much less desirable.

EQUIPMENT: For the task of mixing, nothing beats a stand mixer with a paddle attachment. Just make sure that you chill the bowl and paddle in the freezer before getting started and that you start on a slow speed before stepping up to faster speeds.

If a mixer isn't in the cards for you, don't sweat it. At our *matanzas,* modern luxuries like this were generally unheard of and likely something the *sabias* would cluck their tongues at anyway. Instead, we used either a bowl and a sturdy wooden spoon or a large *lebrillo* and our two hands for mixing. Either way, the job was spurred along by shots of Chinchón to ease the arm cramps, since day-drinking is a perk of doing grunt work at a *matanza.*

INGREDIENTS: Both the grinding and mixing phases involve adding some of the most important ingredients for ensuring the success of your *charcutería*—ingredients like salt, curing salt, seasonings, binding agents, and fermentation agents. I'm going to go through these one by one, but bear in mind that for anything other than seasonings, you'll want to weigh everything out to the gram for reasons of precision and safety.

Salt—In Spain, sea salt is found in every *sabía's* larder. Nothing special here: It's either a coarse salt used for stacking and curing whole-muscle cuts or a fine salt for all other *charcuterie.* Here in the United States, however, sea salt isn't always easy to come by or is often quite expensive, so kosher salt is your next best option. In fact, the only type of salt you should stay away from is iodized (table) salt. (The iodine has a funny taste that can seem bitter.)

Curing salts—Curing salts are not the enemy you think they are...really.

The short story is that curing salts provide two major benefits to our meat-curing cause: (1) They give smoked and cured meats their telltale pink color, cured aromas, and piquant flavors, and (2) they protect you from contracting botulism and other foodborne illnesses that proliferate in high-temperature and low-oxygen environments.

What you need to know, however, is that curing salts come in two pure forms—sodium nitrite and sodium nitrate. They are NOT interchangeable, as they have very different usages and results:

➡ Sodium *nitrite,* which is mainly sold in a mix called Instacure #1, DQ #1, or TCM #1, is meant for anything that will cure for 14 days or less. Typically, it is used with meats that should be cured and acidified quickly or will eventually be cooked. Thus, this mix is used for a lot of cooked or quick-fermenting, semicured *embutidos.*

➡ Sodium *nitrate,* which is mainly sold in a mix called Instacure #2, DQ #2, or TCM #2, is meant for anything that will cure for more than 14 days. That's because over time, nitrate degrades into nitrite; think of nitrate as a time-release capsule for dosing nitrite into a long-curing product. That is why nitrate is used with meats that cure for a long time, such as dry-cured *jamones* or *chorizos.*

Here's the crux of it: When working with either of these ingredients, treat them with respect and keep them well away from your other spices to avoid confusion. A very small amount of either pure nitrite or nitrate can be lethal, which is why they are sold in diluted mixes and tinted pink.

Seasonings—During the mixing phase, you add all the other herbs and spices that contribute flavor to whatever you're making. In most recipes for Spanish *charcuterie,* this means that anything you want to give a red tint to will likely at this point get a good dose of ground fresh garlic, *pimentón,* oregano, pepper, or some combination thereof. Anything that isn't red, however, is probably a recipe that predates Columbus's discovery of the new world, so its spice profile will be more Medici-like (think: nutmeg, clove, allspice).

Sabias typically mix spices with whatever liquid is on hand to form a slurry, since the spices will mix in quicker in a dissolved format. In the case of *chorizos,* for example, the solvent was usually a few glugs of chilled white wine or water.

Binding agents—"Binding agent" is just a fancy name for something that makes a sausage stick together and provides a little texture. While modern *charcuterie* practices might call for things like soy protein isolate or nonfat dry milk powder to cheaply and quickly bind sausages, you'll never find these ingredients in a *sabía's* cupboard. Instead, they use what's on hand, which typically means cooked potatoes, rice, pumpkin, bread, or another starchy vegetable or grain, to bind and stretch their recipes.

To be clear here: Just because the *sabias* eschew the phosphates, milk powders, or other binders and extenders that the industrial *charcuterie* world favors doesn't mean that I'm completely against them too—in all fairness, curing salts and fermentation agents aren't typically used in family-run *matanzas* either. Some of these industrial binders have a logical place in the production of mass quantities of *charcuterie,* so if you decide to play with them, we can still be friends. I'm just not a fan of the texture and flavor they give to my *charcuterie,* so I generally avoid them.

Fermentation agents—I'll get into the fermentation phase of *charcuterie* in a bit, but the fact that fermentation products are typically added during the mixing phase necessitates a quick introduction to them now.

When fermentation occurs, certain bacteria multiply and consume any available food in the area. The food that these bacteria consume, mainly carbohydrates, forces them to release lactic acid. This acid lowers the meat's pH levels and, along with the salt you've added, helps the "good" bacteria suppress any "bad" bacteria that may be present.

Depending on where you are in the world, you'll find different bacterial strains that produce different flavors and acid levels in local sausages and yeast-risen breads (both rely on nature's "good" bacterial strains to help things along). Given our modern age of ingenuity, some companies have been able to cultivate those bacterial strains so that you can buy them, freeze them, and then introduce them into your *charcuterie* for a more authentic flavor and fermentation jumpstart. I'll give you more specifics on what to buy in the fermentation section, but the important thing to understand for now is that you will be adding these natural cultures at the end of the mixing phase.

TECHNIQUE: Mixing technique is less about mixing and more about knowing when to add the various ingredients and understanding when is the time to stop mixing. Here's how it works: Once you've ground your cut pieces of meat with salt or *ajosal* (a garlic and salt mixture—see *Ajosal* below), curing salt, and sugar, you place the ground meat into a chilled mixing bowl. Next, any binding agents and solubilized spices are added, and it's time to mix. During this first mixing phase, you are just looking to incorporate the ingredients together into a rough ball of meat.

Once the mix is relatively homogenous, the dissolved fermentation agent is added, and we get serious about mixing, stepping up the mixing velocity until the meat has a tacky quality and a white film forms on the outside of the bowl. (Science geeks: The actin and myosin have begun solubilizing.) At this point, you've achieved what charcutiers call the "primary bind" and can move on to the *prueba* (roughly translates to "taste test") and, later, the stuffing process.

For the *prueba,* you have the option to cook and taste your sausage *masa* and make any final additions to the recipe. While you really should avoid cooking and taste testing any *embutidos* that contain sodium nitrate for safety reasons (you want the nitrate to go through the conversion process to nitrite before consuming), you should definitely cook and taste a sample of any fresh or semicured *charcutería* to make sure you like the flavor before you go through the stuffing process. As for how to cook the *prueba,* the *sabias* normally just sear a piece in a skillet. While this gets the job done with little fuss, you should realize that frying the meat—as opposed to poaching it in what is sometimes called a "quenelle test"—will caramelize the sausage a little, which could alter what the *prueba* actually ends up tasting like. And it's also likely to lead to snacking on more meat than you intended, since the *prueba* is generally a delicious treat.

AJOSAL

IF A RECIPE includes garlic, *sabias* typically run the cloves through a grinder with the finest die. Then, they mix the ground garlic with all the sea salt required for that particular *embutido* and then push and grind the salt around the bowl, essentially pulverizing the garlic and forming a paste called an *ajosal,* or "garlic salt." This salt is used to season meat chunks with prior to grinding.

But whatever you do, don't mistake fresh *ajosal* for that nasty powdered garlic salt that's been sitting in your spice cabinet for years!

You'll see this technique in most recipes involving garlic, since an *ajosal* is a great way to make sure that garlic is incorporated across the entire sausage, as opposed to just in the pockets where larger pieces of garlic happen to fall.

Stuffing

TWO SCHOOLS OF thought exist regarding the appropriate time to stuff ground meat into sausage casings. Depending on what ingredients are involved and the type of sausage that is made, some people stuff their sausages immediately upon grinding, while others first wait for a period of time. Some *butifarra* recipes I have seen, for example, call for no less than five days between grinding and stuffing.

At our *matanzas,* we always let the ground meat set up overnight before stuffing to allow the meat to pre-ferment a little. We also wanted to let the salt do its magic on the actin and myosin, which allowed the *masa* to get a little denser and easier to stuff. If time is a factor, however, you can always stuff the sausages the same day you make the *masa.*

EQUIPMENT: Many standalone grinders offer a stuffing attachment. To me, these attachments aren't so great since, most of the time, when you're trying to stuff the *masa* into the machine you wind up with suction and air pockets that make it exceptionally difficult to form consistent sausage links.

If you can afford it, a better option is a standalone vertical stuffer. Vertical stuffers are available for anywhere from around $75 to $150 and, if you are doing decent-sized batches, they will make your *charcuterie* life much easier (and will keep you from throwing your stuffing attachment across the room in frustration). Go for one with cast metal parts that are dishwasher safe, rust free, and removable for easy loading.

INGREDIENTS: Casings come in all sorts of shapes and sizes, including natural casings from sheep, pigs, and cows and artificial casings that are made from either plastic or collagen. Almost all require thorough rinsing in water—both inside and outside—to clear any salt and/or to hydrate the casing.

I typically use natural casings that come packed in salt. The size depends on whatever size my *embutido* needs to be—*Chistorra* (see recipe on p. 231), for example, is thin and skinny so it goes in sheep casings; *Sobrasada* (see recipe on p. 304), however, is bigger, so it goes in a large pork casing or bung.

TECHNIQUE: You will come across four different types of *embutidos* in Spain, each requiring a little discussion on how they are stuffed: (1) regular *embutidos,* (2) large *embutidos,* (3) *morcillas* and other loose-filling *embutidos,* and (4) whole muscles.

Regular *embutidos*—Regular sausage stuffing really isn't rocket science, but you will need to make a few mistakes before getting it right. If you've got a freestanding stuffer, it's a pretty quick and easy process. Feed the casings on one end, add the meat in the other, and crank down on the plunger. Once meat comes out, tie off the casing in a bubble knot (see p. 90 for an example) and continue filling the casings while a friend coils them up for tying off and linking. Meanwhile, feel free to laugh at every phallic joke and reference that passes through your mind.

With time, this will become a one-person job: You'll learn how to crank down the plunger with one hand and feed the sausage with the other.

HACER MORCILLA

Just remember that the most important part of stuffing is knowing when to say when. Too much meat in the casing will always cause a blowout, so don't fill the casing to capacity before you have a chance to link and tie off the sausages. That said, even the *sabias* have a few casing blowouts every now and again, which is why they're smart enough to invite a bunch of their relatives over, get them drunk, and proceed with family therapy via *embutido* making.

Large *embutidos*—These big boys require a large bung cap, which is called a *ciego de cerdo* in Spain. Literally translated, the term means "blind part of the pig," as it refers to the fact that the bung is a somewhat obscured part of the intestinal tract. Typically, this type of *embutido* is stuffed by hand with chunks of meat, as with *Morcón,* the *Chosco* from Tineo, or the *Botillo* from Bierzo (see recipes on p. 329, 275, and 279).

Otherwise, it's filled like a typical sausage with some *chorizo masa* for something like the *Sobrasada* (see recipe on p. 304) from Mallorca. For such a sausage, you'd use the widest stuffing attachment you have and fill the bung to near-bursting capacity.

***Morcillas* and other loose-filling *embutidos*—**Sausage stuffers generally don't work for these, so I use a funnel to stuff my *morcillas* into casings. The process really isn't a lot different from the previous stuffing techniques—you need to run the casing onto the end of a funnel and secure it to the funnel with a string. Once you tie a bubble knot in the end of the casing, you simply ladle the sausage mixture into the funnel and, if necessary, use a wooden spoon or some other implement to plunge the mix down into the casing.

Whole muscles—Whole-muscle *embutidos* like *Lomo Embuchado* (see recipe on p. 313) are

MORCILLA

A WORD OF warning: Making *morcilla* can be a pretty messy process. Even if you ladle and fill your sausage with care, your apron and most other things in the immediate vicinity are bound to look like a scene from the TV show *CSI,* so just take some precautions and don't wear your Sunday best on sausage-stuffing day.

relatively straightforward: Stuff the cured muscle into a large casing and tie it off nice and tight with butterfly knots (see p. 90 for an example). If you want to be fancy, make several ties across the width of the *embutido* to evenly distribute the pressure.

Linking and Tying

STUFFING, LINKING, AND tying *embutidos* at a *matanza* normally happens toward the end of the second day, which means everyone is tired, buzzed, and verging on a food coma. That's why it's a family affair to get this final job done. It's the home stretch, so everyone pitches in.

Traditionally, a *matancero* cranks down on the stuffer's handle while a *sabía* forms up the links into a gigantic coiled sausage. She then ties off the open end, pierces the sausage all over with a pin, and passes it down the assembly line to another waiting *sabía* who methodically portions the large single sausage into smaller and smaller links. Each of those linked pieces is sent to the last *sabias* in line, who then tie off the ends, loop the sausage, and pass the completed *embutidos* to the men, who are typically hanging around smoking, drinking *Chinchón,* and occasionally helping when they are harangued into doing something useful.

The process for larger or whole-muscle *embutidos* is accomplished by the skilled hands of the head *sabía* who—fingers flying in a blur of motion—nimbly ties neat packages of meat with intricate linking knots that distribute pressure across the entire *embutido*. This is the sort of thing that takes a lot of practice to understand

NUDO BURBUJA

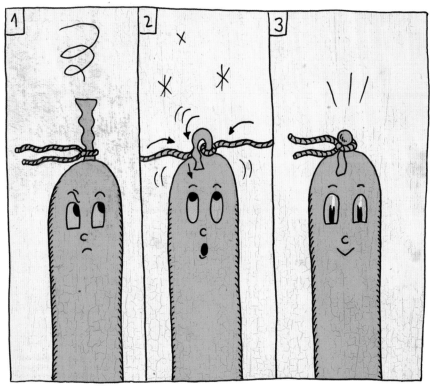

and master, but a little later I'll give you the basics for how to get the job done. Don't worry if you don't get the knots exactly right your first time, just keep your eyes on the goal: a tight, football-shaped *embutido*.

EQUIPMENT: You will only need two items here: butcher's twine and a sausage pricker. Butcher's twine is just heavy-gauge string; cotton is best, and it typically comes in a big spool. It's also great for tying up roasts and *pancetas*.

Other than twine, you will need something to prick the sausages with. You can buy a fancy-schmancy professional sausage pricker, but the *sabias* use a cork stuck with a needle. Whatever you use, just make sure to sterilize your poker before wielding it about.

TECHNIQUE: Once the stuffing is complete, the act of linking sausages at the *matanza* involves a few different jobs: (1) portioning, (2) tying off, and then (3) looping the sausages.

Portioning—A sausage is initially passed down the line in the form of a single, coiled sausage that needs to be cut into individual sizes and retied. This process is really pretty simple: The *sabía* measures out the appropriate size, depending on whether the sausage is meant to be a loop shape or not, and it's then cut and tied off in a fancy knot.

Naturally, some of the meat winds up being removed from each sausage so it can be tied. It's placed into a big bowl in the center of the table and then either fed back through the stuffer again later or fried off to make a *picadillo* snack.

Tying—*Sabias* tie off *embutidos* with three styles of knots. The knots help prevent the weight of the meat from dragging the sausage to the ground as it hangs. The knots are (1) a bubble knot, (2) a butterfly knot, or (3) a combination of knots.

The bubble knot *(nudo burbuja)* is really just a series of simple square knots (see illustration on p. 86). After tying the casing with a first knot, you fold the casing down and tie the second knot on top of the folded casing. This is followed by a third anchoring knot. This series creates a "bubble" in the casing and, once the bubble knot is formed, you can use the remaining string to hang the sausage or make a looped *embutido,* as how *chorizo* is often formed.

A butterfly knot *(nudo mariposa),* on the other hand, is typically used for large or whole-muscle *embutidos* or for slippery casings made from collagen or other synthetic materials. To tie a butterfly knot, make a slit down the middle of the very end of the casing. This bisects the casing's end and creates two "wings." Next, form a string into a pretzel shape and fit the two wings of the casing into the holes of the pretzel,

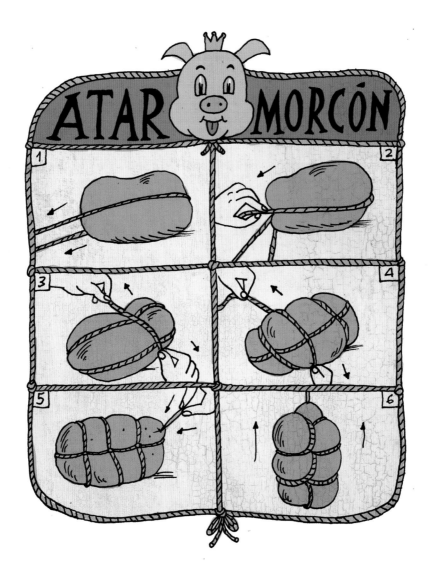

securing the ends of the string by tucking them underneath the base string. Next, wrap those strings around the sausage and tie them off with a simple square knot. Tie a second knot in the opposite direction to finish.

The *embutidos* that use a combination of knots are the largest style of sausages, including anything stuffed in a *ciego de cerdo,* like a *Morcón* (see recipe on p. 329). To tie off a *ciego*-stuffed *embutido,* the head *sabía* first closes off the end of the sausage with a very tight butterfly knot. Starting on the opposite end from her knot, she then brings two strings up the length of the sausage to the butterfly-knotted end and secures the strings in a double knot. This process is repeated with two more strings on the opposite side. In the end, the sides of the *embutido* are supported by four strings.

Next, she ties several square knots across the middle of the sausage, much like tying up a roast. This ensures that the meat is distributed evenly across the *embutido* and that the pressure is equal across the whole sausage.

Last, all the ties on the butterfly-knotted end are brought together in a few square knots, and the *sabía* forms a loop for hanging the *embutido.* After pricking, the sausage is hung up to ferment and dry.

Looping sausages—Sausages like *butifarras, morcillas,* and *chorizos* are often looped into a ring before being cured or cooked. The sausage ends are simply fastened together in a double knot with a long string. The extra string is used to form a loop to hang the sausage, which—over time—allows gravity to pull everything downward and give the *embutido* an oval shape as it dries.

Fermenting

FERMENTATION IS THE first step in the curing process where you need to hold temperature and humidity levels relatively high and relatively stable for two major reasons:

➡ First, chemical reactions that occur during the fermentation process add flavor to the final product. Fermentation imparts an acidic tang because it creates higher levels of acidity, and this sourness is common to many forms of Spanish cured meats.

➡ Second, fermenting meat enlists nature to help create a safe and edible product. Through fermentation, which essentially lowers pH levels by encouraging natural bacteria to proliferate and release lactic acid, we create an environment in which harmful bacteria cannot survive and replicate.

In this book, I discuss two types of *embutidos* that require fermentation: Semicured *embutidos* and dry-cured *embutidos*. The *semi-curados* aren't fully cooked, so your goal will be to ferment them rapidly and get a higher degree of acidity into the sausage over a shorter period of time. In this case, fermentation is more about the flavor of acidity than preservation, so you'll need to cook these *embutidos* before eating them. Fully dry-cured *embutidos* are fermented as a preservation technique rather than to develop high acidity, so your goal will be to slowly ferment these *embutidos* over time to avoid the strong acidic bite characteristic of *semi-curados*.

EQUIPMENT: You'll need an environment with a temperature range from 65°F to 90°F (18.3°C to 32.2°C) and humidity levels upward of 85 percent. A small thermometer/hygrometer combo, which shouldn't run more than $20, will help you gauge both variables. The temperature range will vary according to the type of fermentation agent that you use, while the humidity—depending on the size of the environment and the amount of fermenting meat—should stay naturally high due to the inherent moisture given off by the *embutidos*.

Unfortunately, there's nothing really on the market that's specifically made for the job of fermenting proteins. You'll have to improvise a bit, and an Internet search will give you a ton of ideas. For the home *charcutier*, some options include:

➡ A simple plywood chamber fitted with a non-porous lining and a heat source. A light bulb works well to bring the temperature into range, but beware that humidity can make certain bulbs shatter.

➡ A warming drawer, such as a bread warming drawer or an electric crisper. Note that the capacity of most drawers is pretty small, and crispers tend to use hot air to remove humidity from the environment. That means that you'll have to put the meat into a ziptop bag to keep the humidity high.

➡ A foodsafe plastic tub with a heat source and possibly a small humidifier. Again, a light bulb should work (see first item above).

➡ A closed oven with the light on. Depending on the size of the oven and the amount of meat being cured, humidity levels may not be high enough. To remedy this, place a bowl of hot, heavily salted water in the oven with the meat (the salt in the water kills any potential bacteria that could start growing in the water).

➡ A small refrigerator that is either not turned on or modified to hold an exact temperature and humidity level (a small humidifier should get your humidity in line). This configuration works really well and provides the best control, but just realize that it also requires more of an investment in both time and resources.

For chefs and culinary professionals (or the very ambitious home cook), an electric proofing box for bread is your best option. Most proofing box models include both temperature and humidity controls, so they make perfect fermentation chambers that your health department should have no problem with.

INGREDIENTS: In the section about the mixing phase, I talked a bit about how fermentation agents help lower pH levels. Remember your high school chemistry? Since pH works on an inverse scale, a lower pH means a higher level of acidity. In the fermentation chamber, these agents get to work, so now you'll need to know specifics about brands and types of cultures.

The cultures I recommend are the same as everyone else recommends: The brand is Bactoferm, made by CHR Hansen, and its products are pretty much charcuterie industry standards. CHR Hansen makes a variety of strains you can get your hands on, but the two I recommend for the recipes in this book are (1) Bactoferm F-RM-52 and (2) Bactoferm T-SPX.

➡ F-RM-52 provides fast acidification. The culture likes to ferment between 70°F and 90°F (21.1°C and 32.2°C) and prefers dextrose as a carbohydrate source. You'll use this agent for the semicured *embutidos* in this book, since they need to acidify quickly.

➡ T-SPX provides a slower, milder acidification and is typically used in southern Europe, including Spain. It's important to note that this particular strain likes to ferment between 65°F and 80°F (18.3°C and 26.6°C) and prefers table sugar (sucrose) as a carbohydrate source. This product is best with *embutidos* that will cure for a while—say, a month or more.

For either culture to work, you'll need to activate them, much as you would activate yeast for making bread. Just add a little distilled water (the chlorine in tap water could kill the bacteria), mix to dissolve, and let everything come alive for 20–30 minutes. Then, add the bacterial culture as the last step in making your *masa*. See the mixing section on p. 85 for more details.

Note that even for the smallest recipe batch, you'll need to use a minimum dosage of at least ¼ of the packet of fermentation agent (it normally works out to ¼ to ⅓ ounce [around 10 g] of starter). As with all of your ingredients, make sure to read the product labels carefully, because they will fill you in on the details for getting the most out of each one.

TECHNIQUE: Since I have already discussed the techniques for mixing fermentation agents into your *masa* and for setting up your fermentation chamber, the only thing left to cover is the technique for testing whether your *embutidos* are fermenting properly and how to tell when they are ready.

My guideline for checking fermentation, which has never steered me wrong in my own meat curing, is to ensure that the pH drops below 5.3 within 48 hours of hanging the sausages in the fermentation chamber. This isn't a hard and fast rule, but with time and a little practice, you'll notice that the correct drop in pH is consistent with a change in the meat—it toughens up a bit when the acidity is right.

For the home *charcutier,* I really like the method Paul Bertolli discusses in his book *Cooking by Hand.* Bertolli suggests wrapping a sample of the prepared *masa* in plastic wrap and holding it in the fermentation chamber alongside your other *embutidos.* You'll then use this test sausage to periodically check the pH level of all the sausages.

When it comes time to check the pH, use the same pH test strips marketed for testing the acidity of wine, beer, and other liquids. Just dilute 1 part *masa* with 2 parts distilled water and blend to combine. After dipping the test strip in the liquid, compare the color change of the strip with the key on the container. This will give you a reasonably accurate assessment of your *embutido's* pH.

Professionals, however, should understand the larger truth: Fermentation occurs on a sliding scale according to the fermentation agent used and the temperature in the chamber. These factors, coupled with the number of hours the meat is held at temperatures above 60°F (15.6°C) (the critical temperature at and beyond which *Staphylococcus* bacteria grows) will govern your HACCP plan as well as how often the local health department will come knocking at your door. Thus, the pros out there will probably want to invest in an electronic pH meter and perhaps even a water activity (Aw) meter. Both are hefty investments, but they will also help ensure the quality and consistency of your product and may even be mandated by your local health department.

FERMENTATION

HOWEVER YOU CHOOSE to ferment your meat, I encourage you to do it in a different environment from your drying room. Since these two chambers have different temperature and humidity requirements, you will likely get to a point where you want to ferment in one chamber while you dry in the other... especially once you catch the meat-curing bug.

Smoking

WHEN IT COMES to smoking *charcuterie,* the Spanish don't mess around, especially in the northern area of Asturias. Up in the Asturian mountains, an on-property smoking shed is sometimes more common than modern extravagances like consistent running water, indoor bathrooms, or Internet access.

For most northern Spanish *charcutería* recipes, smoking times range into the days or even weeks in chambers the size of walk-in refrigerators, so you'll need to figure out your own smoking strategy for hitting the sweet spot of either a little smoke over a lot of time or more smoke over less time.

EQUIPMENT: Planning your smoking rig means strongly considering the space you have available, how much money you want to invest in the rig, and how amenable your neighbors are to the sweet smells of smoking meats (or how open they are to some meaty bribery!). Here are a number of options that range from under $200 into the thousands of dollars for more professional setups:

Most Asturian homes have backyards (and by backyards, I mean acreage in the vast Asturian mountains) that include a smallish shed for smoking meat. Those sheds are typically simple wooden rooms with hooks or poles embedded in the ceiling and a smoldering fire built either directly on the floor or in a rolling fire pit. The fire is placed in the center of the room, directly under the hanging *embutidos,* if hotter smoking temperatures are required. Otherwise, it is placed in a corner, and colder smoke works its magic over many days.

If you don't have acres of space or are a little concerned about leaving a smoldering fire burning for days at a time near your home, the next best alternative is purchasing a home smoker that uses either charcoal or an electric heating element. Ideally, you should choose a smoker that can be used (or adapted to be used) for either cold or hot smoking, but a lot of brands out there can get the job done with a variety of configurations. If you happen to have a hot smoker already, there are conversion kits that can convert them to also be used for cold smoking with little more effort than drilling a hole.

As for smoking times, remember that every unit differs, so you'll have to take into consideration the size of your unit when gauging the level of smoke you want for a recipe. Bear in mind that the smoking times I provide in this book are for large chambers with a lot of airflow and low-to-medium levels of smoke. If you're using a smaller smoking chamber with high levels of smoke, your smoking time will be much shorter than I recommend in the recipe—perhaps as little as half of the prescribed time.

For the professional crowd, there's a number of smoking and holding cabinets on the market. I've been a fan of the Alto-Shaam smokers for years, but they aren't cheap. Shop around and be sure to get one with the capacity to support your production.

INGREDIENTS: Finding the right wood for smoking *embutidos* is the easy part. Spanish *charcuterie* is smoked mainly over oak, which is really

easy to find in the United States. The only other wood I have ever seen used is in Galicia, when sometimes laurel wood (from the same tree that provides bay leaves) is used for smoking *chorizo*.

TECHNIQUE: When you are smoking anything, there are two ways that you can do it: (1) cold smoking and (2) hot smoking.

Cold smoking occurs in an environment with temperatures anywhere from 52°F to 71°F (12°C to 22°C) and over a time frame of 1 to 14 days. The amount of time to let anything cold smoke is proportional to the amount of smoke in the room, so you can either opt for less smoke over more time or more smoke over less time. Obviously, too much smoke all at once imparts much more smokiness to the finished product—which

is why this method is pretty hard to get right consistently. That's why a smaller quantity of smoke over a longer period of time is the best option as you're starting out.

Hot smoking, on the other hand, occurs at 105°F to 140°F (41°C to 60°C) over ½ to 2 hours, with the temperature gradually rising the whole time. Hot smoking is governed by the internal temperature of the meat or *embutido* you're smoking. Thus, when the product you're smoking reaches the right degree of doneness, it's either ready to be eaten or to be placed in the drying room for the next phase of the curing process.

It's important to note that both smoking methods are antibacterial in nature, so they prevent mold development on the surface of whatever you are smoking. Thus, you aren't likely to

THE PELLICLE

ANYTHING YOU INTEND to hot smoke should be left, uncovered, in a cool environment for at least 2 hours prior to smoking. This process forms a "pellicle," which is a tacky layer on the outside of the meat that allows the smoke to adhere more readily during the short smoking time.

see the normal white bloom on smoked foods you expect on the surface of cured meats that move directly from the fermentation chamber to the drying room.

Drying

THE DRYING CHAMBER is the last stop for your *charcutería* in its journey from raw hunks of meat to cured, preserved deliciousness. This is where the fruits of your labor will spend quite a bit of their time. A lot can go wrong, so it's worth planning carefully to control the three major ambient variables: (1) temperature, (2) humidity, and (3) airflow.

That's why most families who put on *matanzas* or cure their own meats have a *bodega,* or drying room, where they can store meats in a temperature- and humidity-controlled environment. Sometimes those controls are electronic and automated, and sometimes—like in the mountain town of Jabugo—it involves simply opening a window to keep the air fresh and circulating and to let the microflora do their job.

EQUIPMENT: As with the smoking and fermentation chambers, you can create your own perfect drying environment by repurposing something you already have or can acquire easily and cheaply. You should select a space that can provide a temperature range of 40°F–80°F (4°C–26.6°C) and relative humidity levels between 65 and 90 percent. As with the fermentation chamber, use a thermometer/hygrometer to gauge the levels.

Clearly, these ranges require a more flexible environment, and you'll find that they also differ from what other books about curing meat recommend. That's because for some of my *charcutería* recipes (the one for *jamón,* for example), I've provided variable temperatures and humidity levels that mimic the fluctuations of nature. These fluctuations are one of the quirkier aspects of old-school *charcuterie* rooms, where opening and closing windows was the only way to influence the naturally variable environment and allow the *jamones* to "breathe" during the fall drying phase and "sweat" during the summer. In modern Spanish *charcutería* facilities, open windows are a hazard, so tweaking the curing room's environment like I recommend in some recipes is an accepted "cheat" used by some Spanish producers.

For the amateur crowd, an Internet search can be your best friend in getting ideas for building out your drying space. Some of the ways I've done it include:

⇒ If you have a decently insulated basement located in the right part of the country (parts of the San Francisco Bay Area, for example), you might already have an environment that naturally maintains the right ambient temperatures and humidity for curing meat. I converted a friend's basement just like this by sectioning off a foodsafe area (meaning no chemicals or other hazardous items were stored in it) where I installed a rack system for hanging the meat and covered the entire area with a large tarp. In less humid months, we placed either a humidifier in the area or a large pan of heavily salted water on the bottom rack. Obviously, in this environment you won't have a lot of control over ambient variables, but it got the job done and made for some happy Italians.

➡ I have also converted full refrigerators, wine refrigerators, proofing boxes, and other small refrigerators into curing cabinets. It requires a little work and an understanding of refrigeration technology, but in the end, you'll have much more control over the product's quality. I would recommend this DIY project to most of the home-curing crowd, because it's low cost, relatively easy to accomplish, and will provide an environment that allows for all of the right controls.

That said, actually putting a drying space together is a chapter unto itself, and there's enough information already out there on the Internet for you to draw up some plans (all you need to do is search for "meat-curing chamber" to get a dozen blog entries with photos and even schematics). Just take care to do your due diligence, making sure that everything is not only safe but also runs consistently. The last thing you want is to get electrocuted trying to wire a light bulb into a converted fridge or to put a bunch of meat into your cabinet and then have the system fail.

For professionals or super-meathead amateurs, there is one final option: A couple of companies exist that make professional meat-curing cabinets...but they aren't cheap. Italian companies like Staginello and Cerato Fratelli make beautiful curing cabinets that do everything discussed here, and domestic companies like Enviro-Pak will design custom curing rooms. You'll find their information in the purveyors section on p. 441. Check it out, and join me in drooling over the possibilities.

INGREDIENTS: There's one final ingredient that makes the *jamón* from Jabugo or the *chorizo* in Asturias taste different from the ones you'll make (which is not to say you shouldn't still try to make your own spin on them!). That difference is Mother Nature, and she provides microflora unique to every area that, hopefully, will come to populate your curing cabinet as well.

As your creations begin to dry and ripen, you'll hopefully notice a white mold starting to form on the *charcuterie's* surfaces. This mold—if it is the correct chalky white one and not something nefariously green or black—is perfectly edible. It's related to forms of *penicillium* like the ones found in blue cheese, and it'll give your *charcutería* a characteristic taste and aroma.

You can introduce this flora into your chamber a number of ways. Some *charcutiers* open a window and let nature takes its course. Some hang a sausage or two that already has this desirable mold on it in their chamber and hope migration occurs. Still others take matters into their own hands by spraying a mold culture into the chamber. This culture, Bactoferm 600 sausage mold, is available from CHR Hansen in freeze-dried form. Much like the company's fermentation agents, all you'll need to do is hydrate according to the package's directions and spritz away.

TECHNIQUE: Obviously, if you're new to meat curing or are building a new rig, you'll need to monitor things carefully and baby everything a little at first. Constant fluctuations in temperature and humidity will always occur—especially when you first introduce new *embutidos* and their inherent moisture into the chamber.

Also, you'll need to keep an eye out for a few warning signs as your meat slowly ripens and cures. If you catch problems early enough, you'll be able to salvage your projects and won't have to throw everything away. These signals are specific to different types of *charcutería,* so I'll break them up into troubleshooting for either *embutidos* or whole muscles.

Embutido troubleshooting—With nonsmoked *embutidos,* you'll want to keep an eye on the color changes and especially the mold growth on the casing for signs that the environment is on target. For example, growth of a white, chalky mold on the outside of your *embutidos* is fine, provided it doesn't happen too rapidly. This might mean that the environment is too humid.

Other signs of an excessively humid environment and/or insufficient airflow include condensation on the roof of the chamber and an unusual "off" smell in the chamber. This may lead to formation of black mold on the *embutidos,* since black mold often occurs when moisture drips on meat. Green mold, or mold that's really any color other than white, is generally not good either. If you catch it early enough, however, a quick wipe-down of the meat with vinegar will stop the bad guys in their tracks.

Conversely, an under-humidified environment will lead to something called case hardening. This occurs when the sausage's casing doesn't receive sufficient moisture to stay hydrated, so it dries out and traps moisture inside the sausage, leading to spoilage from the inside out. To avoid case hardening, make sure that your humidity levels stay in line, and if you happen to see the casings drying out, spray them with a little purified water and adjust the humidity levels accordingly.

Whole-muscle troubleshooting—Since *jamones* and *lacones* tend to cure for a long time (in some cases, more than a year), a lot can go wrong. Aside from the same mold issues that exist with *embutidos,* hams are especially susceptible to winged invaders and other creepy crawlers that may try to make a home in your meats. To prevent this, most *jamón* producers fastidiously keep track of their investment by dipping the hams in very hot sunflower or vegetable oil every three to four months. Since the ancient *bodegas* where these *jamones* hang are generally loaded with cultures, a bloom of good mold is usually back on the meats within a week.

Whatever you hang in your drying environment, you are generally looking for a weight loss of around 30 percent as compared to the green weight of that item. The only difference is in the semicured *embutidos,* in which case you are looking for a 10 to 15 percent loss against the green weight since this isn't a fully cured product.

As the final word, remember the prevailing rule of the dry-curing arts, one that holds especially true in the early stages of your curing career, or when you're working with a new curing environment for the first time:

When in Doubt,
Throw It Out!

NITRATE SHENANIGANS

VIRTUALLY EVERY *CHARCUTERIE* nerd I know rolls his or her eyes and shakes a fist at the marketing ploy known as "nitrite-free" cured meats. The truth is that The Man is messing with your world and playing games with your meat, since most of those super-expensive frou-frou bacon brands at your organic market are made "nitrite-free" by giving the pork bellies a bath in celery juice.

Well, guess what celery juice is super high in? Yup...nitrites.

Thanks to a heavily contested study from the 1970s that posited a link between cooking nitrite-laced foods and the formation of carcinogenic nitrosamines, nitrites and nitrates have been stained with a black mark for longer than I've been alive and learning the meat-curing arts.

Here's the kicker: Nitrates are a naturally occurring chemical compound we ingest every day in enormous quantities in doctor-recommended foods like leafy greens, celery, carrots, fruits, grains, and pretty much everything else that's grown in the ground and has access to nitrogen. When we ingest those healthy foods, our bodies naturally convert the sodium nitrate to sodium nitrite, which of course means that humans are designed to consume certain levels of the compound.

Even the American Medical Association contradicted the commonly held stigma about nitrate consumption when it reported that "given the current FDA and USDA regulations on the use of nitrites, the risk of developing cancer as a result of consumption of nitrites-containing food is negligible."[31]

Listen, I encourage you to do your own research on the subject and decide if the benefits of adding nitrates to your *charcuterie* outweigh the risks of not adding them. For me, I'm okay with a few grams of pink salt if it can allay my fears of botulism and give me peace of mind.

But do me one favor: Stop paying more for something that should be delicious and simple, like bacon, because The Suits are trying to pull a fast one on you. Even better: Make your own bacon or support great Southern artisans like Allan Benton who make the sort of bacon that God himself wants to eat with his eggs every morning.

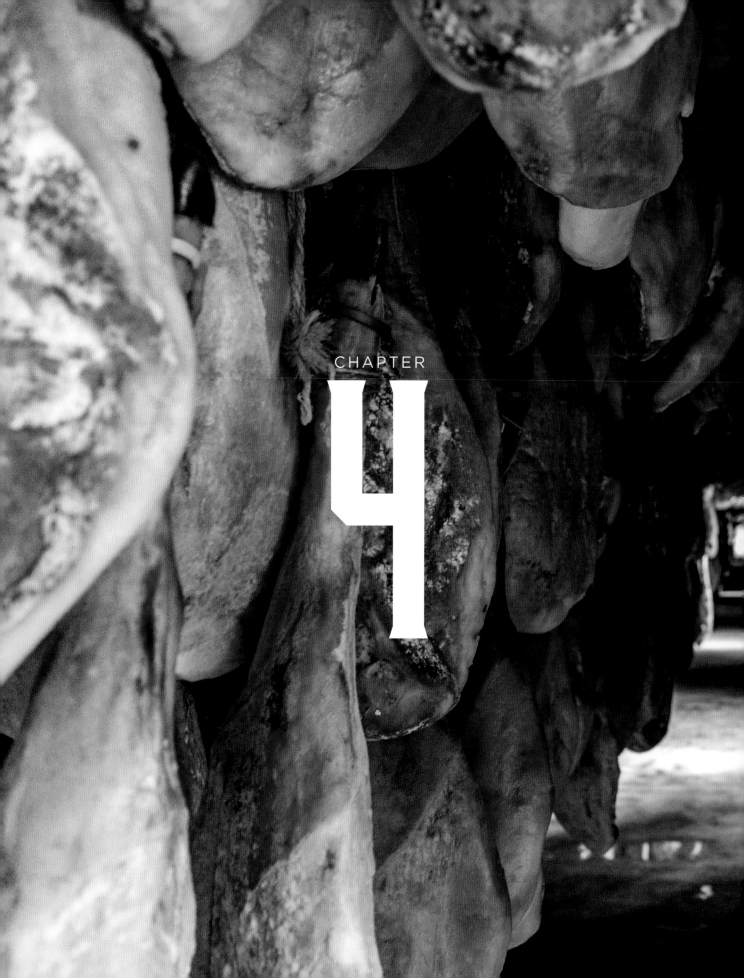

CHAPTER

4

SALMUERAS Y SALAZONES

IN THE WORLD of *charcutería,* it doesn't get much more basic than the foundation techniques of *salmueras* (brines) and *salazones* (salt cures).

Much has been written about salt's importance throughout the history of civilization. It's fundamental to our very existence as a major component of our body chemistry, has helped build societies as a means of trade and commerce, and has been the impetus for wars and conquest.

In the *charcutier's* kitchen, salt is a fundamentally imperative element for drying and preservation. In this role, salt ties up and binds water—the life-giving element for all things—thus depriving spoilage bacteria and microbes of what they need to survive. It also adds salinity to and enhances the flavor of whatever is being cured, a welcome byproduct of the preservation process.

Since food basically goes bad because of spoilage caused by microscopic invaders, the historical reason for the principles of *charcuterie* has been to kill off bacteria when storing proteins and vegetables for extended periods of time. Now, of course, refrigeration is able to help us stave off spoilage for longer periods of time by slowing bacterial growth, but traditions and the fact that cured stuff is delicious keeps us happily pursuing God's work in the name of great preserved foods.

In this section, you will find recipes that use *salmueras* and *salazones* to dry and cure whole muscles, fish, bones, and other large-format *charcutería.* I'll start with *salmuera* recipes, which are foods brined with different flavoring components. Then, you'll learn about the larger group of *salazón*-cured *charcuterie,* which includes salt rubbing everything from bones to fish, as well as the recipes and techniques used to make the famous *jamones* of Spain.

Always remember that the *Charcutier's* Percentage is there to help you calculate your ingredient needs. Since the recipes are written per 2.2 pounds (1 kg) of meat, use the percentages to calculate how much of each major ingredient you will need to safely cure the quantity of meat that you have. The other ingredients, which are usually things like spices and seasonings, can of course be safely altered to reflect your individual tastes.

SALT: STEP 1

NOTE THAT FOR many of these *salazón* recipes, salting is only the first step. That is, once you've salt cured a protein you will find that many of the recipes call for further drying, smoking, or cooking to complete the curing process. For that reason, make sure that you read and understand Chapter 3, which discusses fermentation, smoking, and dry-curing environments, before embarking on any of these recipes.

BASIC *SALMUERA*

A *salmuera* is a brine, which is probably nothing new to you if you drop your Thanksgiving turkey into salt water every November. *Salmueras* are routinely used during America's favorite holiday of gluttony to extract blood, kill off unwanted bacteria, and help salvage many an overcooked bird by adding moisture to the meat.

What might be new to you, however, is the egg test, which *sabias* use to check the salinity of their brine. Think about it: Back when measuring water and salt by weight wasn't so easy, most housewives or *charcutiers* needed a simple method to make sure that they weren't wasting their precious salt. To do so, they'd place a fresh egg (for curing anchovies, an anchovy was used) in a container of water and then whisk in salt until the egg ascended to a specific height.

In my experience, this method is reliable with one caveat: The egg needs to be *really* fresh, which is something that—unless you have a few chickens out back like the *sabias*—is hard to tell given our modern supermarket eggs of dubious provenance. That's why even though I mention the egg test as a means of double checking your brine if you want, a trusty scale for measuring everything is the way to go.

Note that this recipe will make enough brine for 2.2 pounds (1 kg) of meat. For more information on the math here, see the sidebar on brine percentages on p. 107.

1   To a large foodsafe container, add the water. Whisk in the salt and sugar.

2   Seal the container. Refrigerate the resulting *salmuera* until ready to use.

▶ **NOTE:** Sugar substitutes are fair game here—consider sweeteners such as honey, maple syrup, birch syrup, sorghum, or barley malt. The list is endless, but just make sure to measure by weight since volume measurements vary and will throw off the brine.

YIELD

17 ounces (500 g/500 mL) brine, or enough brine for 2.2 pounds (1 kg) of meat

INGREDIENTS	CHARCUTIER'S PERCENTAGE
17 ounces (500 g) lukewarm tap water	
1 ounce (31 g) kosher salt	**6.2%**
½ ounce (12 g) granulated sugar	**2.4%**

TRUCOS DE LA COCINA: BRINE PERCENTAGES

STANDARD BRINES ARE often based on an amount of water that's equal to 40 to 50 percent of the protein's weight, so that's what I used for basing the *salmuera* recipe on 2.2 pounds (1 kg) of protein. The quantity of salt will yield a 6.2 percent brine, which is the perfect solution strength for most proteins. If you need a more aggressive brine for a different recipe, you simply add more salt to the mix on a recipe-by-recipe basis to raise the base *salmuera's* potency.

Make sure to check out the brine table below for recommended brining times per protein in this base *salmuera*. The amount of time that your meat spends in its brine is really important, and mainly depends on the meat's thickness and its intended final usage.

Brine Table

OSMOSIS—THE MAGIC that allows salt to enter a meat submerged in brine—is a fickle mistress governed by the protein's thickness and length of exposure to the brine. Too little brining time, and the meat won't be fully seasoned and juicy. Too much brining time, and the meat could become unbearably salty.

So, here's a rule of thumb: If you will be roasting your brined meat, err on the side of the lower brine times in this table. That's because I presume you will be salting the surface liberally before roasting since salty, crackly skin is one of life's great pleasures. For everything else, including *pâté* or terrine garnish, the lower limits of the given brining window will do.

PROTEIN FOR BRINING	SIZE/WEIGHT	BRINING TIME (HOURS)
Chicken breasts (boneless and skinless)	8 ounces per breast	1½–2
Whole chicken (medium to large)	2–4 pounds	6–10
Whole turkey (small to medium)	Up to 15 pounds	24
Whole turkey (large to extra large)	15 pounds or larger	36–48
Pork chops, bone-in	Per 1 inch of thickness each	1
Whole pork loin, boneless	6–10 pounds	24–36
Whole pork tenderloin	1–3 pounds	6–12
Pork or lamb tongue	1–3 pounds per tongue	12–24
Calf or beef tongue	1–3 pounds per tongue	48–72
Pig ears and tails	1–3 pounds per ear or tail	48
Split pig head	8–10 pounds per half head	48–72
Pork trotters	3–7 pounds per trotter	48–72

CODILLO EN SALMUERA

I first encountered *codillo en salmuera*—a brine-preserved pork shank—while cooking for one of the best-known chefs in Andalucía (and a really awesome guy), Chef Daní Garcia. I was tasked with making a luxuriously rich stock for *puchero,* which is the Andalucían version of the famous Spanish Sunday stew/hangover cure called *Cocido Madrileño* (see recipe on p. 332). I accompanied a senior cook to the walk-in refrigerator, where he plopped different things into my pot for the *puchero*: a chicken, some sausages, some salted bones, and a gnarled, wet, slightly pink hunk of meat that he pulled from a vat of brine. I later learned that hunk of meat was the *codillo,* which gives the broth a serious punch of porky flavor and gelatin.

Codillos are typically used as part of a soup or stew, but can also be used to add flavor to braised greens—similar to how smoked ham hocks lend a helping hand in flavoring Southern-style braised greens or bean dishes.

1 To a medium foodsafe container, add the Basic *Salmuera*. While the *salmuera* is still warm, whisk in the kosher and curing salts until fully combined. Add the juniper berries, bay leaves, thyme, rosemary, and garlic.

2 Warm a small, dry skillet over medium–low heat. Add the peppercorns, coriander seeds, and mustard seeds to the skillet and toast for 2 to 3 minutes, until they are fragrant. Remove from the heat. Add the toasted spices to the *salmuera.*

3 Seal the container and refrigerate the brine until ready to use.

4 Transfer the *salmuera* to a foodsafe container large enough to accommodate the *codillos.* Add the *codillos* to the chilled *salmuera* and weigh them down with a few plates to keep them submerged. Seal the container.

5 Refrigerate the container for at least 4 days. Though the brined *codillos* can be preserved in a sterilized jar for a good amount of time, I would use them within 14 days.

6 Use the *codillos* as a flavoring ingredient for soups and stews or for braising greens.

NOTE: If the *codillos* have been in the brine longer than 1 week, make sure to soak them in cold water overnight prior to using and change the water once or twice during the soaking period. Otherwise, they may add too much salt to the dish you are preparing.

INGREDIENTS	CHARCUTIER'S PERCENTAGE
per 2.2 pounds (1 kg) of *codillos* (ham hocks) **100%**	
1 recipe Basic *Salmuera* (see recipe on p. 105) **50%**	
1 ounce (25 g) kosher salt **2.5%**	
⅛ ounce (3 g) TCM #1 or DQ #1 curing salt mix (see p. 83) ....**.3%**	
2 tablespoons smashed juniper berries	
2 dried bay leaves	
10 fresh sprigs thyme	
1 fresh sprig rosemary	
5 cloves garlic, lightly crushed	
1 tablespoon (5 g) whole black peppercorns	
1 tablespoon (5 g) whole coriander seeds	
1 tablespoon (5 g) whole mustard seeds	

TRUCOS DE LA COCINA: BRINING SPICES

THE SPICES AND herbs used in this brine are also perfect for virtually any other cut of pork you can think of. You'll want to toast the spices lightly in the pan, as it will bring out their natural aromatic oils.

Some other good candidates for this brine include grilled *chuletas* (pork chops), a whole roasted *aguja* (pork collar), or a *lomo* (loin).

BOQUERONES

Every Spaniard has an opinion about *boquerones*—that ubiquitous, vinegar-cured anchovy staple found in *tapas* bars all over Spain. Some prefer to start the anchovies in a *salazón,* while others prefer a quick bath in a *salmuera.* Everyone seems to have a preference for how long the anchovies should "cook" in the vinegar. The longer they stay in the vinegar, the more firm they become, so it's all a matter of texture and acidity preference.

This recipe yields *boquerones* with a little kick from the red pepper flakes and a hint of citrus from the lemon zest. These tasty treats are the perfect companions for a summer day, some toasted bread, and a cold *caña.* Make a good amount at one time and keep them covered with olive oil in your refrigerator; they can keep for quite a while in the oil (known as the *aliño),* but they do tend to disappear quickly.

Just make sure that you get the freshest anchovies that you can find for this recipe. Ideally, they shouldn't be more than a day or so out of the water.

INGREDIENTS	CHARCUTIER'S PERCENTAGE
per 2.2 pounds (1 kg) of very fresh anchovies **100%**	
1 recipe Basic *Salmuera* (see recipe on p. 105) **50%**	
2 medium lemons	
1½ cups (350 mL) white wine vinegar	
2 cloves garlic, minced	
2 medium shallots, minced	
2 ounces (60 g) minced fresh flat-leaf parsley	
1 pinch red pepper flakes	
½ cup (120 mL) extra virgin olive oil	

1 Descale, decapitate, and eviscerate the anchovies, rinsing off any surface blood. For each fish, you should be left with 2 clean filets attached to the backbone.

2 To a small nonreactive container, add the Basic *Salmuera.* Stack the anchovies, skin-side down, in the *salmuera.* Seal the container. Refrigerate for 2 to 3 hours (see Note).

3 Remove the anchovies from the *salmuera.* Rinse them gently and place them into another nonreactive container. Discard the *salmuera.*

4 Remove the zest from the lemons. Slice the lemons in half and juice them. Reserve the zest and juice in separate containers. Discard the pulp.

5 In a nonreactive measuring cup, combine the vinegar and lemon juice, creating an acid bath. Pour the mixture over the anchovies to cover. If they are not completely submerged, add more vinegar to cover. Seal the container.

6 Place the container with the anchovies in the refrigerator. They should remain in the acid bath until they are white in color and reach your desired level of firmness, which could mean anywhere from 3 hours to overnight.

7 Remove the anchovies and discard the acid bath. Remove the spines from the anchovies, taking care not to tear the flesh. Place the deboned anchovies on a large plate or baking dish.

8 Layer the reserved lemon zest, garlic, shallots, parsley, and red pepper flakes on top of the filets. Cover with the olive oil, seal the container, and refrigerate for at least 2 hours.

9 Serve the *Boquerones* at room temperature directly from the marinade.

NOTE: Some *sabias* I know test the *salmuera* for proper salinity by placing 1 anchovy in the water. If the anchovy floats on the top, there is too much salt. If it sinks, the water needs more salt. If the anchovy floats right in the middle, the salinity is correct.

BASIC *SALAZÓN*

Using a *salazón* is the most basic trick in the charcutier's playbook: It's just a salt rub + meat + time. Pure salt can be a little harsh, though. That's why I use a little sugar in my *salazón*—for balance.

Much like the *salmuera,* the *salazón* serves as a preserving agent, extracting water or blood while killing off not-so-friendly bacteria found on the surface of most meats. In addition, the *salazón* can play a vital role in changing the texture of some meat and fish, thereby improving flavor while also extending shelf life.

For example, in Andalucía, we often placed fresh loins of *merluza* (called hake in the United States) in a *salazón* of salt and some toasted spices for exactly 18 minutes. While the amount of time might seem inconsequential, this precise curing time delivered fish with denser, meatier flesh that required no additional seasoning. And when you have very fresh fish like that Andalucían *merluza*—fish that came off the boat literally hours before we cooked it—the best culinary rule to follow is that the less you screw with your product, the better.

1   In a small mixing bowl, combine the salt and sugar.

2   Transfer the resulting *salazón* to a foodsafe ziptop bag or container and store at room temperature until ready to use.

NOTES: Toasted spices, fruit zest, and/or fresh herbs are fair game to add to a *salazón*. I'm a big fan of fruit zest in particular, as the oils of the citrus really permeate the cure into the fish. It's particularly great for *crudo* (raw) preparations.

If you are planning to add TCM #1 (sodium nitrite mix) or TCM #2 (sodium nitrate mix) to your *salazón* for making a cure mix, I recommend adding it only after you know the weight of the meat you are using; don't premeasure it into your *salazón*. Not only is it safer this way (since you can control the dosage), but it keeps your *salazón* flexible for other recipes that don't require curing salts.

YIELD

10 ounces (280 g) dry cure

INGREDIENTS	CHARCUTIER'S PERCENTAGE
7½ ounces (210 g) coarse sea salt	**75%**
2½ ounces (70 g) granulated sugar	**25%**

YEMAS CURADAS (CURED EGG YOLKS)

The idea of preserving eggs isn't a new concept. It's been around in Chinese cuisine for centuries in the form of Century Eggs, but the idea of dry curing just the egg's yolk is a relatively new concept on the culinary scene. Curing yolks was an idea inspired by the modernist cuisine movement, whose epicenter is in Spain, where manipulating the texture of eggs has been a longtime culinary fascination.

There's really not much to the technique: You just bury fresh egg yolks in a *salazón* (I like to lace mine with maple sugar). When the yolks are dry enough to handle, hang them up for further curing. Eventually, you'll be able to grate their deliciousness over pasta or salads. This is a very versatile weapon in your curing arsenal, since once you master the technique, you can play with the cure however you want. Some of my favorite substitutions include brown sugar, soy salt, or salts flavored with herbs and spices.

1   In a measuring cup, thoroughly whisk together the maple sugar and salt. Pour half of the resulting *salazón* into a medium baking dish.

2   Using an egg shell, make 12 depressions in the *salazón*. Gently place 1 egg yolk into each depression. Cover all the yolks with the remainder of the *salazón*. Cover the baking dish with plastic wrap.

3   Store the dish in the refrigerator for 7 days.

4   Remove the yolks from the dish, brushing off any *salazón* that clings to the eggs. (Doing so may require a light spritz of water to help dislodge any remaining salt.)

5   Wrap each yolk individually in cheesecloth, and tie the cheesecloth with some butcher's twine. Using the ends of the twine, hang the yolks in the refrigerator for 5 to 7 days.

6   For serving, either grate or finely chop the yolks. (Depending on how long you cure the yolks, they can be quite potent.)

INGREDIENTS

per 12 large egg yolks

3½ ounces (100 g) maple sugar

3½ ounces (100 g) kosher salt

HUESOS SALADOS

Huesos salados, another discovery I encountered while cooking in Andalucía, are a great example of how preservation techniques from before refrigeration still have a purpose in today's modern culinary world. Historically, killing a pig in Spain meant using every conceivable—and inconceivable—part of the animal. Thus, even the pig's bones were preserved in salt until they had completely dried out; they were then typically used to flavor soups, stocks, or broth.

What's really interesting, however, is what occurs to the bones while they are immersed in a salty slumber. Immersing fresh meats or seafood in salt changes the resulting product's flavor compounds, and so too does salt change the flavor of bones. I discussed this phenomenon with our modern-day culinary prophet Harold McGee, who says: "If the bones are cured for weeks or months, then those proteins (in or on the bones) will be partly broken down into flavorful amino acids."

To me, this explains why many *sabias* (and also crafty chefs like Daní Garcia) specifically seek out salted bones or salt their own. It's a winning proposition: The bones have an indefinite shelf life without fear of spoilage, and they make soups and stocks even more flavorful than their fresh counterparts.

INGREDIENTS	CHARCUTIER'S PERCENTAGE
per 2.2 pounds (1 kg) of cleaned lamb, pork, or beef *huesos* (bones)	**100%**
17½ ounces (500 g) Basic *Salazón* (see recipe on p. 111)	**50%**

1   Place the *huesos* in a large foodsafe container filled with water. Seal the container and soak the *huesos* for a minimum of 2 hours or overnight, depending on the size of the bones, in the refrigerator. Change the water several times during the soaking period, as you are trying to get as much blood out of the bones as possible.

2   Remove the *huesos* from the water and dry them thoroughly. Drain.

3   Pour half of the *salazón* into a large ziptop bag. Lay the bones on top of the *salazón,* and then cover them with the rest of the *salazón.* Make sure the bones are completely submerged in the *salazón* to prevent spoilage. Seal the bag.

4   Place the ziptop bag in the refrigerator and allow the *huesos* to cure for a minimum of 7 days (after this point, they will last pretty much indefinitely in the cure).

5   When you are ready to use the *huesos,* rinse them thoroughly to remove all traces of the *salazón.*

6   Use your *Huesos Salados* to flavor any kind of stock, soup, or stew.

CECINA

Cecina has a sketchy history as Spain's version of mystery meat. Donkeys, horses, cows, or almost any other farm animal—or nonfarm animal, in some extreme cases—have historically been considered fair game when making *cecina,* mostly due to the need for protein during lean times of war and scarcity.

Today, *cecina* is often referred to colloquially as *jamón de vaca* to distance it from its dubious past. It's a justifiably famous export with IGP-protected status in the area of Castilla y León, so the stuff we make here in America can't technically be called *"cecina de León"* even though we have the means for making awesome *cecina*. The technique is really no different than making a ham: All you need is a source for great beef (if you want to make horse or donkey cecina, you're on your own, *amigo)* and the space to smoke and dry it.

Spaniards know that the best *cecina* comes from the oldest, gnarliest cow or ox they can find—ones that worked the fields for years and are unfit for butchery into regular steaks and chops. These animals make the most flavorful *cecina* because their meat is very tough and flavorful, but here in America, older meat is a tall order since the age of animals fit for consumption is regulated. Something in the quality neighborhood of American Kobe or Black Angus would be ideal here, but really any animal treated with love and as old as you can find will give you a great final product.

Last, remember that Spaniards break a cow down differently than Americans. You are looking for one of these four large cuts from the back leg of a cow:

- ⇒ *Tapa:* Inside (top) round, normally weighing in around 18 pounds (8 kg)
- ⇒ *Babilla:* Knuckle or round tip, normally weighing in around 15 pounds (7 kg)
- ⇒ *Contra:* Outside (bottom) round, normally weighing in around 22 pounds (10 kg)
- ⇒ *Cadera:* Sirloin, normally weighing in around 13 pounds (6 kg)

INGREDIENTS	*CHARCUTIER'S* PERCENTAGE
per 2.2 pounds (1 kg) of *tapa* (beef top round), *babilla* (beef round tip), *contra* (beef bottom round), or *cadera* (beef sirloin), cleaned of all silverskin	**100%**
2 ounces (60 g) Basic *Salazón* (see recipe on p. 111)	**6%**
Coarse sea salt, to cover	
OPTIONAL	
⅛ ounce (3 g) DQ #2 or Instacure #2 curing salt mix	**.3%**

CONTINUED ON NEXT PAGE

CECINA CONTINUED FROM PREVIOUS PAGE

1   Weigh the meat to obtain its green weight; record the weight for later.

2   Place the Basic *Salazón* in a small bowl. If using, whisk in the curing salt. Set aside.

3   Place the beef on a work surface, such as a baking sheet. Rub the curing mixture into the meat, pressing hard and distributing it evenly over its entire surface. Place the meat and any leftover curing mixture into a large foodsafe container.

4   Cover the meat with some plastic wrap and weigh it down with a few cans. Close the container and refrigerate the meat overnight.

5   Remove the cans and plastic wrap and pour the coarse salt onto the meat until it is completely covered with salt. Re-cover the meat with the plastic wrap, replace the cans, and return the closed container to the refrigerator for ½ day per 2.2 pounds (1 kg) of meat, flipping the meat every day to redistribute the cure and then covering it again with the salt.

6   Remove the meat from the cure and rinse it thoroughly under cold water. Pat the meat dry and set it aside on a clean plate.

7   Tie one end of the meat securely with butcher's twine and secure a loop in order to hang it.

8   Hang the *cecina* in a drying chamber set at 54°F to 60°F (12°C to 16°C) and 80% to 85% relative humidity. The *cecina* should remain in the chamber for 15 to 45 days (this will depend on the green weight of the meat). You will know when the meat is ready when it loses around 8% to 10% of its green weight (see p. 76 for details).

9   Stock a large cold smoking chamber with oak wood and bring the temperature to 90°F (30°C) and the relative humidity to 80% to 85%. Hang the *cecina* in the smoking chamber and smoke for 7 to 10 days, or break it up into 4-hour-long sessions for 10 to 12 days, returning the *cecina* to the drying chamber set at 54°F to 60°F (12°C to 16°C) and 80% to 85% relative humidity in between sessions (see Notes).

10 Return the *cecina* to the drying chamber set at 54°F to 60°F (12°C to 16°C) and 80% to 85% relative humidity for 3 to 4 months, until it has lost about 35% of its green weight.

11 Serve the *cecina* as you would a *jamón*—sliced runway-model thin and laid out on a plate.

NOTES: These smoking times are for large chambers with a lot of airflow and low-to-medium levels of smoke. For more information on smoking meat, see Chapter 3.

Some restaurants in León use a *jamón* stand to show off their enormous legs of *cecina.* These bars typically use the finest legs of *cecina* from old Galician cattle—definitely the sorts of places you will want to track down and try a *ración.*

A special shout-out goes to the Animal Science departments of Texas A&M University and Cal State Chico for your help with the translations of the cuts of meat in this recipe.

UNTO GALLEGO

"El caldo sin unto no está en su punto" goes a Gallego truism. The saying implies that any soup, stew, or broth meant to contain *unto* (a Galician version of cured leaf lard) will never reach its fullest potential without it.

To a native Gallego, nothing else will do as a cooking medium, so you can find *unto* in most Galician markets as a commercially made product that is nearly pearl white in color. Homemade *sabía*-endorsed varieties, however, have a yellowish tinge as a result of their longer curing times. They often even have a bit of intentional rancidity prior to curing.

To make *unto,* the leaf lard is folded and pressed, salted for a period of time, and then cured in a dark space, since light can cause rancidity in pork fat. The *sabias* usually then roll the *unto* in newspaper and hang it in the fireplace for a few days to smoke, but your smoking rig will get the job done just the same. As a final step, the *unto* is then typically dried until nearly crumbly.

Unto is perfect for any dish that calls for smoky, porky goodness, but use it sparingly, since a little carries a pretty strong porky punch.

INGREDIENTS	CHARCUTIER'S PERCENTAGE
per 2.2 pounds (1 kg) of thoroughly rinsed and drained *unto* (leaf lard)	**100%**
21 ounces (600 g) Basic *Salazón* (see recipe on p. 111)	**60%**

1 Weigh each piece of *unto* to obtain a green weight; record the weights for later.

2 Pour ⅓ of the Basic *Salazón* into a large foodsafe container. Set aside.

3 To a large mixing bowl, add ⅓ of the Basic *Salazón.* Add the *unto* and toss together until the lard is completely coated.

4 Turn out the *unto* onto a work surface. Flatten it out and roll it up tightly, just as you would a jelly roll. Pour any *salazón* that remains in the mixing bowl into the foodsafe container containing the Basic *Salazón.*

5 Place the tightly rolled *unto* into the container. Pour the remaining ⅓ of the Basic *Salazón* on top of the roll, pressing down hard to compact the cure into the fat. Seal the container.

6 Place the foodsafe container inside a large, dark plastic bag to block out all light (the bottom half of a garbage bag works well here, but so will any dark foodsafe bag). Place the wrapped container in a cool (39°F to 54°F [3°C to 12°C]), dark place to cure for at least 1 month.

7 Rinse the *unto* roll thoroughly, until all the cure has been removed, and pat dry. Tie the *unto* roll securely with butcher's twine, securing a loop on one end in order to hang it.

8 Stock a large cold smoking chamber with oak wood and bring the temperature to 90°F (30°C) and the relative humidity to 80% to 85%. Hang the *unto* in the smoking chamber and smoke for 24 hours, or break it up into 6-hour-long sessions for 2 to 4 days, returning the *unto* to a drying chamber set at 54°F to 60°F (12°C to 16°C) and 80% to 85% relative humidity in between sessions (see Note).

9 Place the *unto* in a drying chamber set at 54°F to 60°F (12°C to 16°C) and 80% to 85% relative humidity for 3 to 4 weeks, or until it has lost about 35% of its green weight and is dry and crumbly.

10 Use as a cooking medium or as a flavoring component in soups, stews, or braised greens, such as *grelos.*

NOTE: These smoking times are for large chambers with a lot of airflow and low-to-medium levels of smoke. For more information on smoking meat, see Chapter 3.

TOCINO SALADO

Italians are rightfully famous for their *lardo*. It's the Big Daddy in the world of cured fats, especially the *lardo* from Colonnata, which is cured in marble vats. Spaniards, however, are famous for the acorn-tinged fat of their *Ibérico* pigs, so it only makes sense that some innovative Spanish *charcutiers* have decided to use Italy's *lardo* techniques to cure their superior *Ibérico* fat.

These days, you can get frozen *Ibérico* back fat from some specialty purveyors like Wagshal's Delicatessen in Washington, DC. Or you could be a fancy-pants locavore, if you want, and track down some fat from Mangalitsa hogs (also known as wooly pigs for their unique pelt). Really, you could use the back fat of any well-raised, well-treated piggie that was finished on an acorn diet. As any proud Spaniard will tell you—whether you ask for an opinion or not—acorn feeding makes a big difference in the flavor of the meat and fat.

INGREDIENTS	CHARCUTIER'S PERCENTAGE
per 2.2 pounds (1 kg) of *tocino* (cured pork back fat), skin on and cut into large squares	**100%**
21 ounces (600 g) Basic *Salazón* (see recipe on p. 111)	**60%**

1   Weigh each of the *tocino* squares to obtain a green weight; record the weights for later.

2   Pour ½ of the Basic *Salazón* into a large foodsafe container.

3   Place the *tocino* squares on top of the *salazón*.

4   Cover the *tocino* squares with the remaining ½ of the Basic *Salazón,* pressing down hard to compact the cure into the fat. Seal the container.

5   Place the foodsafe container inside a large, dark plastic bag to block out all light (the bottom half of a garbage bag works well here, but so will any dark foodsafe bag). Place the wrapped container in a cool (39°F to 54°F [3°C to 12°C]), dark place to cure for at least 6 months.

6   Rinse the *tocino* thoroughly, until all the cure has been removed, and pat dry. Tie the *tocino* securely with butcher's twine, securing a loop on one end in order to hang it.

7   Place the *tocino* in a drying chamber set at 54°F to 60°F (12°C to 16°C) and 80% to 85% relative humidity for 3 to 4 weeks, or until it has lost about 35% of its green weight.

8   Use in a hearty stew or thinly sliced and spread on hot, grilled bread.

NOTE: *Tocino Salado* is a key ingredient in recipes like *Cocido Madrileño* (see recipe on p. 332) or other hearty stews where fat is emulsified into the broth (see p. 335 for more on this technique).

PANCETA CURADA

The bond between meat-curing traditions in Italy and Spain is pretty obvious, especially in this recipe. To anyone with some *charcuterie* experience, this should read very much like a basic Italian *pancetta* recipe. *Pancetta* is the ubiquitous rolled, cured pork belly that's become an American favorite over the years, though many Italians would call this version *tesa* since the belly is left flat instead of rolled up.

You will most likely find this *panceta curada* hanging right next to *chorizos* and *morcillas* in the *charcutería* shops of Madrid since Madrileños, given their cosmopolitan nature, are less likely than other Spaniards to thumb their noses at something so uniquely Italian.

I really like to let my *panceta* hang for a while once it's cured. Doing so concentrates it with winey and garlicky flavors, making it a great addition to pasta dishes or along with some eggs in the morning.

1 Weigh each piece of the *panceta* to obtain a green weight; record the weights for later.

2 In a nonreactive bowl, add the Basic *Salazón*. Whisk in the curing salt until fully combined. Set aside.

3 Warm a small, dry skillet over medium–low heat. Add the peppercorns, allspice, and cloves to the skillet and toast for 2 to 3 minutes, until they are fragrant. Remove from the heat.

4 Using a spice mill or a mortar and pestle, grind the toasted spices with the juniper berries into a fine powder. Whisk the nutmeg and *salazón* into the ground spices, creating a cure. Set aside.

5 Line a baking sheet with parchment paper. Place the *panceta* squares on the baking sheet. Sprinkle the squares with the cure, pressing down hard to compact the cure into the *panceta*. Lightly sprinkle the squares with the wine, and then press the garlic into each square.

6 Place the prepared squares into a large ziptop bag. Crumble the bay leaf and scatter it over the squares and seal the bag.

INGREDIENTS	CHARCUTIER'S PERCENTAGE
per 2.2 pounds (1 kg) of *panceta* (pork belly), skin on and cut into large squares	**100%**
1.4 ounces (40 g) Basic *Salazón* (see recipe on p. 111)	**4%**
⅛ ounce (3 g) TCM #1 or DQ #1 curing salt mix	**.3%**
1 tablespoon (5 g) whole black peppercorns	**.5%**
2 whole allspice berries	
2 whole cloves	
3 juniper berries, smashed	
1 teaspoon (.5 g) freshly grated nutmeg	
1 tablespoon (20 mL) Spanish red wine, such as Rioja	
2 cloves garlic, peeled and lightly crushed	
1 fresh bay leaf	

OPTIONAL

3 tablespoons (15 g) whole black peppercorns

3 tablespoons (20 g) *pimentón dulce* or *picante*

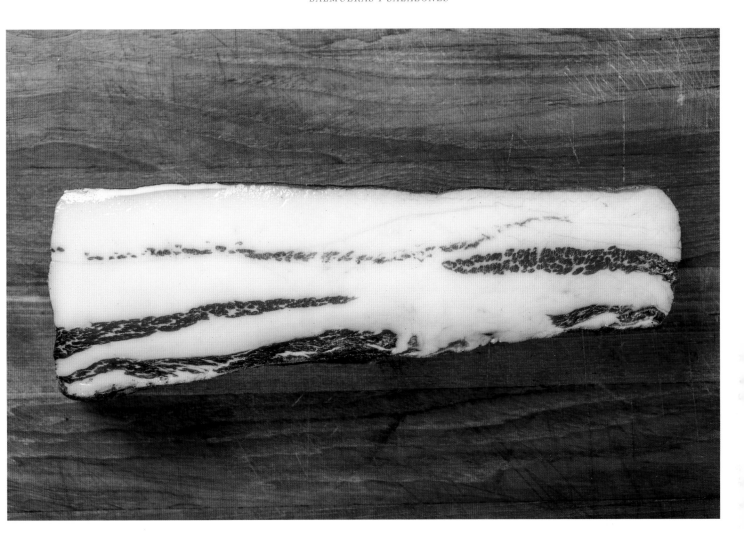

7   Place the ziptop bag into a large baking dish and weigh it down with a few cans. Refrigerate the dish at least overnight, and up to 2 weeks depending on when the meat is sufficiently cured.

8   Remove the cans, open the ziptop bag, and check the meat for firmness (see Note).

9   Remove the *panceta* from the cure and rinse it thoroughly under cold water. Pat the meat dry and set it aside on a clean plate. If using, press the remaining peppercorns or *pimentón* into the meat at this time.

10 Tie one end of the *panceta* securely with butcher's twine and secure a loop in order to hang it. (I typically leave the meat flat and tie it up like a Christmas present with crossing ties in the middle, but you can also roll it up like traditional *pancetta* if you want to be fancier.)

11 Hang the *panceta* in a drying chamber set at 54°F to 60°F (12°C to 16°C) and 80% to 85% relative humidity for 3 to 4 weeks, or until it has lost about 35% of its green weight.

12 Use the *panceta* as you would bacon, such as fried and served with breakfast or crumbled over pasta. Its rendered fat is excellent as a cooking medium.

NOTE: When you remove the *panceta* from the ziptop bag after its overnight refrigeration, you should notice a change in the feel of the meat. It should firm up a bit as it cures.

BACALAO

If there's one cured food that can give the *Ibérico* pig a run for its money in a Spanish popularity contest, it's *bacalao*. Salted cod has a long history in Spain, dating back almost 500 years to a time when Basque whalers, who traditionally preserved whale fat and flesh with salt, decided to try the same process with cod. The process worked out well, and a Spanish culinary star was born.

Bacalao's place on Spanish dining tables was later solidified when the Catholic Church declared it a favorite protein substitute during meat fasting days and Lent. Since the number of days that Spaniards were forbidden from eating meat totaled almost half the calendar year, the *bacalao* trade boomed on those days and the humble cod eventually became a religious icon.

Not only was *bacalao* essential to trade and the sustenance of nations, but the act of salting cod actually makes the fish taste better by compacting it and changing its flavor profile—a technique that also works well for other fish. As Harold McGee points out in his book *On Food and Cooking:* "*Micrococcus* bacteria generate flavor [in the salted fish] by producing free amino acids and TMA; and oxygen breaks up to half the very small amount of fatty substances into free fatty acids and then into a range of smaller molecules that also contribute to aroma."

Sadly, according to the Seafood Watch Program at the Monterey Bay Aquarium, stocks of Northwest Atlantic cod (specifically Georges Bank cod) have been drastically overfished, almost to the point of collapse. Fortunately, Gulf of Maine and Northeast Atlantic cod stocks are still fairly strong. Look for them, and specifically make sure that the fish was caught with a hook and line, as opposed to trawling. Trawling is a bad business that messes with marine habitats.

Lesson of the day: Protect the planet, *amigos*...safe fishing means more delicious fishies for years to come.

INGREDIENTS	*CHARCUTIER'S* PERCENTAGE
per 2.2 pounds (1 kg) of rinsed, fresh cod loins, bones, spine, and skin removed (see Note)	**100%**
17½ ounces (500 g) Basic *Salazón* (see recipe on p. 111)	**50%**

1   Cut the cod loins into 1-inch-thick (2.5-cm-thick) pieces. Weigh each piece to obtain a green weight; record the weights for later.

2   Place the Basic *Salazón* in a small mixing bowl. Set aside.

3   Line a baking sheet with parchment paper. Place the cod loin pieces on the baking sheet. Sprinkle both sides of the pieces with about ½ of the *salazón,* pressing down hard to compact the cure into the cod. Cover the bowl containing the remainder of the *salazón* and reserve.

4   Wrap the cod loin pieces in a layer of cheesecloth. Either attach a sanitized hook to one end of each piece or tie butcher's twine around each piece and make a loop for hanging.

5   Hang the cod loin pieces in a drying chamber with excellent airflow. The temperature should be set at 54°F to 60°F (12°C to 16°C) and the relative humidity should be 80% to 85%. The loins should hang over a large bowl to catch any water that drips from the fish, and the water that accumulates in the bowl should be drained daily. Dry the loins in the chamber for 2 days.

6   Unwrap the cheesecloth from the loins and apply the remaining *salazón* to them. Rewrap the loins and rehang them in the drying chamber for 1 day, until they have lost about 50% of their green weight (with good airflow in the chamber, this will take about 1 day per inch (2.5 cm) of thickness on a loin).

7   Before using the *Bacalao,* you must soak the pieces in a few changes of water for 24 to 36 hours. Serve in salads such as *Esqueixada* (see recipe on p. 173), or use in the Basque *Bacalao* recipes on pp. 124, 163, 165, and 167.

NOTE: Consider reserving the bones, spine, and skin of the cod for another use, such as *pil pil* sauce (see recipe on p. 163).

MOJAMA

The tuna trade is a major industry in southern Spain. In the province of Cádiz, for example, a well known tuna-harvesting city named Barbate has an entire restaurant culture devoted to serving its native *atún* in various guises.

Several years ago, I won a Spanish Institute for Foreign Trade (*Instituto Español de Comercio Exterior;* ICEX) scholarship. The scholarship included travels throughout Spain with fellow chefs who'd also won, and our travels included a trip to Barbate to sample a *degustacion de atún* at a small mom-and-pop joint. The restaurant served us everything: petite slices off the back of the fish's head, *tataki* preparations of the loin, grilled heart, and even a traditional dish of braised tuna testicles.

Whatever the fishie equivalent of nose-to-tail-to-balls is, these guys were *all* about it. But it's the thin-sliced, cured tuna loin called *mojama* that I remember best through some very foggy *jerez* goggles.

The process of making *mojama*—the loins are salted and hung to dry in the sea breeze until they are solid as wooden planks—was taught to early Iberian tribes by the Phoenicians as a means of trade. The people of the area took the idea and ran with it and, centuries later, *mojama* is still served around Barbate and most southern cities in the traditional fashion: sliced thinly, like *jamón,* and served simply with a good olive oil and some toasted Marcona almonds.

Outside of Andalucía, however, more progressive chefs use *mojama* as a deliciously briny garnish on dishes like grilled asparagus topped with a fried egg (see recipe on p. 157). But this is frowned upon by old-school Andalucíans, for whom even a little lemon is considered the height of culinary blasphemy.

If you intend to make *mojama,* know that—like most *charcuterie*—the quality of the end result depends greatly on the quality of the tuna you start with. For that reason, only fresh, grade-A tuna loin—the kind used for sushi or sashimi, trimmed of the bloodline—will make the best *mojama.*

INGREDIENTS	CHARCUTIER'S PERCENTAGE
per 2.2 pounds (1 kg) of rinsed, fresh, sushi-grade tuna loins, bloodline trimmed (see Note)	**100%**
17½ ounces (500 g) Basic *Salazón* (see recipe on p. 111)	**50%**

CONTINUED ON NEXT PAGE

MOJAMA CONTINUED FROM PREVIOUS PAGE

1   Cut the tuna loins into 1-inch-thick (2.5-cm-thick) pieces. Weigh each piece of the tuna to obtain a green weight; record the weights for later.

2   Place the Basic *Salazón* in a small mixing bowl. Set aside.

3   Line a baking sheet with parchment paper. Place the tuna loin pieces on the baking sheet.

4   Pour a ½-inch (13-mm) layer of the *salazón* into a deep, nonreactive baking dish. Lay the tuna loin pieces on the cure, leaving space between each piece. (If there isn't enough room, you can just stack the pieces with a layer of cure in between.) Cover the tuna loin pieces with another ½-inch (13-mm) layer of the *salazón*. Cover the baking dish with plastic wrap.

5   Weigh the tuna loin pieces down with a few cans. Refrigerate the dish overnight.

6   Remove the cans and the plastic wrap. Remove the tuna pieces from the cure and rinse them thoroughly under cold water until all the cure is removed. Pat the tuna pieces dry and set them on a perforated rack.

7   Place the rack over a dish or bowl to catch any water and refrigerate for 2 days. During this time, the cure will redistribute within the tuna.

8   Wrap the tuna pieces in a layer of cheesecloth. Either attach a sanitized hook to one end of each piece or tie butcher's twine around each piece and make a loop for hanging.

9   Hang the tuna pieces in a drying chamber with excellent airflow. The temperature should be set at 54°F to 60°F (12°C to 16°C) and the relative humidity should be 80% to 85%. The loins should hang over a large bowl to catch any water that drips from the fish, and the water that accumulates in the bowl should be drained daily. Dry the loins in the chamber for 10 to 21 days, until they have lost about 50% of their green weight and the color has darkened significantly.

10 Serve sliced very thinly, plated with a few Marcona almonds and a drizzle of good olive oil. It is also delicious grated over seafood dishes, like you would use a salted roe like Italian *bottarga*.

NOTE: Traditionally, *mojama* is cured in salt only. I like using the Basic *Salazón* recipe to cure it, however, because I believe the sugar helps balance the harshness of the salt.

ANCHOAS EN SALAZÓN (CURED ANCHOVIES)

The Spanish author Josep Pla once wrote, "You will remember the anchovies of l'Escala for quite a while, though not forever, since in time you might confuse them with your first love." I don't know how Josep's first love felt about being compared to an anchovy, but hopefully she took comfort in knowing he was talking about the finest, most amazing anchovies on the planet.

Anchovy curing has been going on since the ancient Greeks arrived in Catalonia and taught the native population their preservation techniques. Since that time, curing anchovies has been a thriving business on the Catalan coast, especially considering that the quality of Catalan anchovies is among the best in the world. In fact, Catalan anchovies are so well known that a 50-mile stretch of land running from Girona (where you can find an anchovy museum) into the south of France along the Vermilion Coast is also known as the *Costa de l'Anxova* (the Anchovy Coast).

As with the recipe for *Boquerones* (see recipe on p. 110), the key here is finding anchovies that are as close to "right off the boat" as possible. The sooner you can get them from boat to *salazón,* the better your final product will be.

NOTES: Don't be tempted to substitute extra virgin olive oil for the vegetable oil or olive oil blend specified here. *Sabias* rightfully believe that pure extra virgin olive oil muddles the anchovies' flavor.

These cured anchovies are great to use as an ingredient in sauces or as a topping for *tapas* like *Montadito de Matrimonio* (see recipe on p. 155).

INGREDIENTS	CHARCUTIER'S PERCENTAGE
per 2.2 pounds (1 kg) of very fresh anchovies, heads removed, gutted, and rinsed	**100%**
8¾ ounces (250 g) coarse sea salt	25%
Vegetable oil or olive oil blend, as needed	

CONTINUED ON NEXT PAGE

ANCHOAS EN SALAZÓN (CURED ANCHOVIES)

CONTINUED FROM PREVIOUS PAGE

1   Line the bottom of a large foodsafe storage container with a layer of anchovies.

2   Place a layer of the salt on top of the anchovies. Top the salt with another layer of anchovies, and then a layer of salt. Continue until you finish with a layer of salt.

3   Cover the top layer with some plastic wrap, and then weigh the anchovies down with something heavy, such as a piece of wood. Cover the container and place something heavy, such as a brick, on top. Refrigerate the container for 3 weeks to 2 months, until the anchovies' flesh is rosy in color and they give off a rich aroma.

4   Remove the weights and uncover the container. Remove the anchovies from the cure and rinse them thoroughly under cold water until all the cure is removed. Pat dry and set the anchovies on a cutting board.

5   Filet the anchovies by removing the spines and pinbones.

6   Place the anchovies in a medium baking dish. Cover the anchovies with the vegetable oil (see Notes), cover the dish with plastic wrap, and place in the refrigerator. After an overnight chill, the anchovies are ready to serve. They will keep for about 1 month in the refrigerator.

TRUCOS DE LA COCINA: CANNING

MANY OF THE recipes in this book can be eaten immediately or stored indefinitely if canned. If you don't want to take the time to can your foods, of course, you don't have to. These are all techniques for preservation, so you can keep most of these dishes for at least 1 month in the refrigerator before having to worry. If, however, you decide to stock up *Little House on the Prairie*-style, here's how to go about it.

Canning jars typically come in three parts: the jar, the cap, and the band seal. The first step is to sterilize the jars and caps. Thoroughly wash and rinse each part in hot, soapy water. Meanwhile, get water boiling in a pot deep enough to ensure that the jars will be completely covered by at least 2 inches (5 cm).

Once the water is boiling, immerse the ends of any tongs or jar grippers you'll be using, since you'll need to grab the jars with something sterilized. Next, submerge each jar for 5 minutes and each cap for 2 minutes. Don't worry about the seals.

Once you've sterilized all the parts, place them on a rack to drain with a towel underneath. While the jars are still hot, fill them with whatever you're canning, but make sure to leave about ½ inch (13 mm) of space for expansion. Cover with the cap and seal with the band. Finally, return the jars (covered again by 2 inches [5 cm] of water) to the boiling water for 10 minutes each.

Remove the jars and let them come to room temperature. If all went well, you should hear lots of hissing and popping as the jars equalize. Test their seals by flipping the jars upside down. If the cap stays depressed, you are good to go.

Generally speaking, canned food will keep in a cool, dark place for a very long time, no refrigeration required, until you pop the seal.

This is what happens when you bring
American bourbon and rye to a *matanza.*

CARLOS TRISTANCHO'S
TATAKI OF *IBÉRICO* PORK

Carlos Tristancho is a man of many talents: actor, dancer, businessman, *maestro* of all things *Ibérico,* and a good friend. Aside from hanging out with Carlos and introducing him to the wonders of American bourbon, one of my favorite things about his Rocamador *matanzas* are the goodies that come off of Carlos's grill—especially his *secreto Ibérico,* a cut of meat between the belly and shoulder that is essentially a pork skirt steak.

Carlos typically salts the meat overnight to intensify its flavor and then cooks it *tataki*-style for an awesome Day 2 lunch. You munch on the barely-more-than-rare *secreto* with some charred bread, take a swig of Mahou (the official matanza beer of choice), and rinse, wash, and repeat until you succumb to the sweet, sweet embrace of the meat sweats.

A few years ago, I was asked to showcase some Ibérico Fresco meat at the StarChefs International Congress in New York and I knew instantly that I had to include Carlos's preparation. Truth be told, this technique works really well with the *pluma, presa,* or even a *solomillo* if you can't get a *secreto.*

Just know that if you plan to cook your meat as rare as Carlos does, you'll want to make sure the pork comes from someone you trust—for *Ibérico* meat in the United States, I'm a fan of either Ibérico Fresco or Fermín. The white miso (fermented soybeans with rice) is optional, but white miso and pork go together like drunk cooks, live fire, and liquor. I heartily recommend all of the above for a good time.

INGREDIENTS	CHARCUTIER'S PERCENTAGE
per 2.2 pounds (1 kg) of pork *secreto,* preferably *Ibérico,* cleaned of all silverskin	100%
⅓ ounce (10 g) Basic *Salazón* (see recipe on p. 111)	1%
2 tablespoons (25 g) white miso	2.5%

1   Place a rack inside a baking sheet. Place the pork on the rack.

2   Sprinkle the Basic *Salazón* on both sides of the pork. Next, smear a layer of the white miso on each side. Place the baking sheet in the refrigerator for 1 day. (You should see some water in the baking sheet the next day; discard the water.)

3   Light a charcoal grill or warm a cast-iron skillet over medium–high heat. Remove the pork from the refrigerator and pat it dry.

4   Place the pork on the grill or in the skillet and sear the outside of the pork for 2 to 3 minutes on each side, until a crust forms.

5   Allow the meat to rest for 10 minutes. Slice into thin slices across the grain and serve.

CHAPTER FEATURE

JAMÓN IBÉRICO

MY *JAMÓN IBÉRICO* guru, Miguel Ullibarri, said it best: "You can tell everything about a man by the *jamón* that he makes."

Formerly the general manager of the *Ibérico* ham consortium Real Ibérico, Miguel is a man who takes his pork products very seriously—a fact that I learned the hard way when I tried calling him once when he happened to be in the midst of slicing a leg for some friends. "I need to call you back tomorrow—I am focusing on the *jamón*!"

These days, Miguel is the resident *Ibérico* expert for A Taste of Spain, a group that runs gastronomic tours for foodies from all over the world. It's the perfect gig for someone with a passion for sharing great *jamón* and the people who make it.

And that's why, when he learned that I was writing a book about Spanish *charcuterie,* Miguel introduced me to 200 very special hams that have forever spoiled me for the rest of my *jamón*-eating life...

ARMANDO & LOLA'S STORY

TWO HUNDRED IS the exact number of hams that Armando and Lola Lopez Sanchez (aided by their children Helena, Natalia, and Armando, Jr.) cure in a given season at their family's sprawling farm, Finca Montefrío, in Huelva. Located in the Sierra de Aracena Nature Reserve, the *finca* is a 20-minute drive from the famous *jamón*-producing town of Jabugo, home to famous *jamón* companies like Sánchez Romero Carvajal (the company that makes the 5J line of *jamones*, a favorite of the king of Spain).

By comparison, Armando and Lola's operation is miniscule. They put great focus on each year's stock of 100 purebred *Ibérico* pigs that freely roam the *dehesa,* as well as the coming years' stocks of incrementally smaller animals in a manner that Miguel passionately calls "near-*Ibérico* perfection." Slated to start in 2013, all *jamones* from *Ibérico* pigs will state the percentage of pure breeding in the animal. Animals with 100 percent *Ibérico* pedigrees will be labeled as *Puro Ibérico*. Anything else, down to 50 percent *Ibérico* lineage, will be labeled *jamón Ibérico*.

Armando's *Ibéricos* are enormous titans of jiggling, black-haired jelly. These mountains of meat strut without a care in the world as they munch acorns in their domain. But don't be fooled: When Armando calls to them, they run like a fat kid running to a cupcake, and will bowl over anything in the way of their 300-plus pounds of momentum.

Later in the year, after the *montanera* is over and the supply of young, sweet acorns has been exhausted (it takes 26 pounds [12 kg] of acorns for a pig to put on 2.2 pounds [1 kg] of bodyweight), the pigs will become some of the rarest and best *jamones* you can find in Spain.

JAMÓN FROM THE GROUND UP

MOST OF THE world's high-quality hams go through similar processes that turn perishable raw meat into long-term luxury. That is, many of the same steps that are used to make *jamón Ibérico* are also used to make *prosciutto di Parma* from Italy, a Jinhua ham from China, Southern hams from the United States, and so on. They are all trimmed back legs of pork that have been salted and matured in some form or another. However, the major difference in making the *Ibéricos*—and it's a point that is generally overlooked except by *jamón* romantics like Miguel—is that all *Ibéricos* are born, bred, and live their lives in Spain's *dehesas.*

These pigs, and the *jamones* made from them, are a natural phenomenon of the *dehesa* ecosystem. There, thanks to evolution and Mother Nature, the pigs forage across many acres of ideal conditions—a minimum of 1 to 2 hectares is required per pig (that's almost 90 trees per animal!)—to fatten and flourish on a diet of herbs, grasses, and holm oak acorns.

This combination of acorn diet, good exercise, and the *dehesa* itself is unique to the system of raising *Ibéricos;* not surprisingly, it's also why true *jamón Ibérico* is such a rare and expensive thing, and why there's a few different grades of Ibérico ham in Spain. It's not that higher or lower grades are planned—logically, any farmer wants the best-quality animal. It's just that Mother Nature only makes so many acorns in a given season. Therefore, the pigs are segregated by which ones will spend the most time eating acorns in the dehesa, with the highest class getting to spend the most time foraging, and lower qualities of ham coming from animals fed a diet supplemented with cheaper feed to offset a poorer season's acorn yield.

The *Ibérico* grading system is often in a state of flux, but the current system goes like this:

➡ *Ibérico de bellota:* This, the highest quality of *Ibérico,* is bestowed upon a pig that has gained at least 50 percent of its weight during the *montanera,* reaching a final market weight of of around 330 pounds (150 kg) without the aid of any cereals to supplement the animal's diet. These pigs must be slaughtered between December 15th and April 15th, the traditional time in Spain for the *matanza* to occur.

➡ *Ibérico de recebo:* This grade of *Ibérico* follows similar regulations as the *cerdo Ibérico de bellota,* but its *montanera* period can be supplemented with grains and cereals to help fatten it up. This animal needs cereals for fattening because the number of acorns for the year was insufficient to bring it to market weight.

➡ *Ibérico de cebo de campo:* This grade of animal puts on the majority of its weight via cereals and grains as a result of a poor acorn harvest. While it has also spent its entire life in the free range, there simply weren't enough acorns to go around—for example, strong winds may have knocked a majority of the acorns to the ground too early in the season.

➡ *Ibérico de cebo:* This fourth category of *Ibérico* is a result of the boom in interest in the *Ibérico* name around 15 years ago (back when there was no legal protection for the qualities we see today). The classification was the result of a desire on the part of Spain's pork lobby to introduce a cross-bred pig into the *Ibérico* grading system. Thus, *Ibérico de cebo* ham is a product that comes from an animal that is not a purebred *Ibérico*—something

traditionalists like Miguel hope to one day see removed from the *Ibérico* grading system.

Since a pig's quality in death is so closely linked to the capabilities of its *dehesa* and the farmer, to experts like Miguel the *dehesa* is the most important ingredient in cultivating a great *jamón Ibérico.* To guys like that, the environment and curing process yields a porky sum that is greater than its constituent parts of pigs and acorns. Or, as Miguel puts it,

The more I speak with people linked to the world of *Ibérico,* the more I conclude that the true treasure behind *Ibérico* is the *dehesa,* and the ways the *Ibérico* people know how to handle it to the benefit of the *Ibérico* pigs. It's a true miracle that this awesome ecosystem and the rural wisdom linked to it have survived.

THE MAKING OF A JAMÓN

THE CURING PROCESS, which I'll describe here using the example of Armando's amazing pigs, involves five major phases: (1) butchering/shaping, (2) salting, (3) equalizing/resting, (4) maturing, and (5) finalizing. Even though every individual and company has its own variation on each step, the steps themselves are pretty universal for making a properly cured ham.

Butchering and Shaping the *Jamón*

FIRST, THE LOCAL *abattoir* slaughters and removes the hair from Armando's 100 pigs, and the meat is then broken down and segregated for various uses. (The family has been doing nose-

to-tail since way before it was cool, so they also make delicious *chorizos, morcillas,* and other *Ibérico* cured meats at the *finca.*)

Once separated from the carcass, the *jamón* is then shaped. This process involves shaving the skin off the outside of the leg in clean, single strokes and trimming the leg to remove any soft or excess fat on the outside of the ham. At this point, the aitchbone may also be boned out, though most *matanzas* and small-scale operations don't bother with this step.

This entire trimming process, which results in a teardrop shape often referred to as a sierra cut, typically leaves a decorative V-shaped marking of skin on the outside of the leg that extends down from the trotter. Ultimately, the goal is to create an oval ham with very clean cuts and no puncture marks into the flesh, since this ensures that there are very few places for bacteria to hide and grow as the *jamón* cures and matures.

For the final and most important step, a *maestro jamonero,* the person responsible for the *jamón's* journey throughout the curing process, runs his forearm along the femur, expelling any last vestiges of blood from the femoral artery. Any blood that remains within the artery can sour a *jamón* from the inside out, essentially wasting thousands of potential sales dollars, as well as the 5 years of work and sacrifice it took to raise the pig in the *dehesa*.

Salting the *Jamón*

AS MIGUEL POINTS out, the objective of the salting phase is not to make a salty ham or even to season the meat. These are just byproducts of the process. Rather, the idea of salting the ham is to create homeostasis and make the *jamón* a stable product that can survive its coming years of maturation.

A *jamón's* green weight is an essential piece of data for the process, so it is usually attached to each *jamón* for safekeeping via barcode. Therefore, the first step of the salting process involves weighing the *jamones*.

Next, the *jamones* are entombed in a salt grave, a process you'd think would require much more care and precision than just stacking hams with coarse sea salt shoveled in between. (You'd be wrong.)

Before getting stacked, the hams get a serious rub down, first with some fine salt to make sure that all of the nicks and crevices are protected, and then with coarse sea salt. Then, the hams are stacked up like a Great Wall of Pig, with shovels of more coarse sea salt placed on and around each piece to make sure that every square inch is happily buried. The meatier end of each *jamón* is always cast at a slight downward angle from the hock-end, as gravity helps pull the salt into the meatier parts of the ham.

In this salting room, the piggie parts sit for some time at a temperature hovering around 37°F (3°C) and at relative humidity levels around 85% to 90%. Everyone in the *jamón* business claims that how long their *jamones* sit is a trade secret, but the general rule of thumb is thus: 1 to 2 days per 2.2 pounds (1 kg) of ham, with the upper end of this time frame making for a saltier end product. Most *jamón* makers flip the hams around every so often—somewhere around every 8 to 10 days—and they also flip the orientation of the *jamones* each time, so that the top hams get cycled to the bottom and the weight is evenly distributed.

Through all of this, only one person really knows what's going on: The salting room *maestro*. By personally pushing and prodding each part of the muscle on each ham every day, the *maestro* discovers which areas on each ham are more or less cured and also when it's time to pull each ham from the salt. In this world, the *maestro* is the king of Salt Mountain.

But what about curing salts, you ask? What about some kind of mandated preservation agent to keep us all safe?

It's probably true that unrefined sea salts like the Spaniards use contain trace elements of curing salts. But a more important fact to recognize is that Armando (and most every purveyor of high-end hams, to be fair) knows the name of each pig from whence his hams came. He knows what they ate, when they were killed, how they were killed, and that—because of this knowledge—the meat will be perfectly safe to eat without the use of curing salts. A real *jamón Ibérico* is a time-honored tradition of meat + salt + time. Thus, many artisanal *jamón* producers don't bother with modern additives like sodium nitrate.

Equalizing and Resting the *Jamón*

ONCE THE *JAMÓN* feels ready to the *maestro's* learned touch, it is quickly rinsed and dried, removing any excess salt clinging to the surface. From there, the hams are moved to a resting room, the first in a series of drying rooms where the *jamón* will hang for a while in order to equalize the salinity within the muscles and allow any internal water to drip out. The resting room is normally kept at around 50°F (10°C) and 80% to 85% relative humidity, and the *jamones* are hung hoof-side up to ensure that gravity pulls all the interior salt content downward and into the thicker muscles of the ham.

This whole process takes a good long time, around 30 to 45 days, until around 10 percent of the ham's green weight drips away.

Maturing the *Jamón*

THE PENULTIMATE STEP in the process, maturation, means for Armando a beautiful *bodega* full of *jamón*. In that small room, his 200 coveted hams hang near open windows for as long as 3 years, quietly drying, aging, and stinking up the place with a smell akin to God's own funk as they lose 30 to 35 percent of their green weight.

That's because here on Armando's little *finca*, natural mountain breezes and the microclimate of the Jabugo area mean that every season brings its own ebb and flow to temperature, humidity, natural bacteria, and air flow—the life-giving elements of any curing room.

In other facilities—the sorts of places that the USDA might approve for export to the United

States—producers have found a way to mimic these variables of Mother Nature. They do so by mechanically fluctuating temperatures and humidity in the curing room, thus allowing some producers to create artificial environments where *jamones* flourish.

Either environment, however, necessitates defensive measures against over-eager molds or would-be winged invaders attracted to the smells of decaying protein and fat (especially in the *au natural* environs of open-windowed *bodegas).* So every 3 to 4 months, the *jamones* get a quick dip in moderately hot sunflower or vegetable oil—any fat, really, other than olive oil, as its flavor is far too aggressive for the delicate *jamones.* Really small operations, like family homes or tiny *bodegas,* might simply use an oilcloth to remove the surface mold from the *jamones,* followed by application of a paste made from *manteca* and a little *pimentón.*

Patience in this step is a delicious, multiyear virtue...

Finalizing the *Jamón*

AFTER ANYWHERE FROM 1 to 3 years in the *bodega,* Armando might release the Kraken if it's ready—a true gift for anyone lucky enough to reserve a guest cottage at his *finca* and experience a plate of his wares.

To determine whether the time has come, Armando will employ his *cala,* a bone that traditionally comes from a horse, to puncture the *jamón* in three different areas. Interestingly, depending on its proximity to the bone, each area of a *jamón* has a different intensity of smell, and one sniff of each puncture point will tell Armando if his *jamón* is ready.

A JAMÓN WORTH 1,000 WORDS

AND WHEN IT *is* ready, dear reader, believe me when I tell you: There are few things on this planet that compare to the smells from one of Armando's *jamones* being freshly cut and the taste of those transparent slices. We're talking about something like true love or your first blowjob, something that transcends the limitations of mere language.

Which brings me back to Miguel's words: "You can tell everything about a man by the *jamón* that he makes." Every one of Armando's *jamones*—every smell, every taste—tells the story of his passion for *Ibéricos,* for treating them right, for feeding them right, for taking their lives right, and for honoring their sacrifice by curing them right. Words can't do such an experience justice. You really need to go to the *finca* yourself and live it...Armando is waiting.

JAMÓN

Miguel Ullibarri once wisely pointed out to me that the quality of the pork used to make a *jamón* is magnified by the dehydration and maturation processes of curing—a simple but very important fact. So, let that be a guiding principle for you: If the ham that you choose to cure is of low quality, you will definitely taste it in your finished *jamón*.

That is why, in Spain, many people pay extra to have their *jamón* cut by a *maestro cortadero*, a person certified as a master *jamón* slicer. Such a person stakes his or her reputation on selecting only the highest-quality *jamones* for showcasing their skill. It's part novelty, part luxury, and part insurance policy to be confident that the *jamón* you're about to eat has been closely examined and given a thumbs up from a certified *jamón* buff.

All I'm saying is that you should bear these facts in mind if you plan to invest the time and effort in making your own *jamón*. Even though *Ibéricos* aren't available here, there's plenty of delicious pork in the United States that's worthy of your *jamón*-curing efforts. In particular, I recommend the Ossabow and Berkshire breeds for some really great hams.

Once you have acquired your pork, realize that dry curing it into a delicious *jamón* is no small undertaking. It's not that terribly difficult, but it does require planning, space, and a significant investment of time and money to create a reliable curing rig that will keep your meat happy for well over a year. Before you undertake this effort, be sure you read Chapter 3 thoroughly to figure out the right approach for your own dry-curing environment.

Please keep in mind that this recipe, like most of the *charcuterie* recipes in this book, is written for ease of scaling up or down. Your ham will likely have a weight range anywhere from 4 pounds (1.8 kg) to more than 20 pounds (9.1 kg), so remember to use the *Charcutier's* Percentage as your guide to find the right quantities for each ingredient for the weight of your *jamón*.

INGREDIENTS	*CHARCUTIER'S* PERCENTAGE
per 2.2 pounds (1 kg) of bone-in hind leg of pork, skin on	**100%**
1¾ ounces (50 g) Basic *Salazón* (see recipe on p. 111)	**5%**
Coarse sea or rock salt, as needed	

OPTIONAL

⅛ ounce (3 g) DQ #2 or Instacure #2 curing salt mix	**.3%**
¼ cup (25 g) *pimentón dulce*	
3½ ounces (100 g) *manteca* (pork lard)	
Sunflower oil, as needed	

CONTINUED ON NEXT PAGE

JAMÓN CONTINUED FROM PREVIOUS PAGE

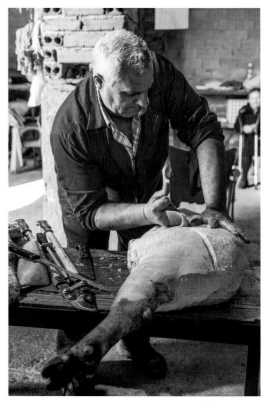

1   At most *matanzas,* we didn't bother to remove the aitchbone (an oblong pelvic bone that covers the hip socket joint). Instead, we removed it after the ham was cured. If you choose to remove the aitchbone from your ham, now is the time to do so by cleanly cutting around the aitchbone, thereby separating it from the meat and tendons holding it in place at the hip ball socket. (See Note.)

2   On a large cutting board or block, clean and sculpt the ham into what's known as a sierra cut by cutting away the skin, glands, and meat on either side to form a teardrop shape. Make sure your cuts are clean and leave smooth tracks. If desired, once you have trimmed the skin down to the fat layer, you can cut a "V" shape into the fat, as they do in Spain for some *jamones.*

3   Weigh the pork and record its green weight for later.

4   Run your forearm down the hind leg's femur bone to ensure that all of the blood has been removed from the femoral artery. Continue to massage it until blood stops emerging from the hip joint area.

5   If using, place the curing salt into a small bowl with the Basic *Salazón* and stir to combine.

6   Transfer the pork to a large tub or platter and cover the pork in the Basic *Salazón*, making sure to rub it into every crack and crevice of exposed flesh. Take care to make sure the femur, ball joint (if exposed), and/or aitchbone are well covered.

7   Transfer the rubbed *jamón* to a large plastic tub with holes cut in the bottom or a similar large container with a rack. (The ham must be able to lay flat and above any water that will drain from it, and there must also be space for weight to be placed on top.) Cover the ham completely with the salt (how much you use isn't really important, as long as it's completely covered).

8   Place something heavy (around 15 to 25 pounds [6.8 to 11.4 kg]) on top of the salt-covered pork and refrigerate it at 37°F to 41°F (3°C to 5°C) for 1 to 2 days per 2.2 pounds (1 kg) of weight. During the refrigeration process, carefully monitor the liquid that drains from the meat and make sure to drain the liquid every so often. If the ham ends up touching the liquid, make sure to cover any exposed areas again with the salt. During this period, the meat should get much firmer to the touch and lose 10% to 15% of its green weight.

9   Remove the *jamón* from the cure and rinse it under cold water, using a stiff brush to remove any surface salt. Pat the *jamón* dry.

10 Tie the *jamón* securely with butcher's twine on the hock end, securing a loop at one end to hang it. Hang the *jamón* in a drying chamber set at 54°F to 60°F (12°C to 16°C) and 80% to 85% relative humidity for 30 to 45 days. During this period, the cure will equalize within the meat and the meat will lose 10% of its green weight.

11   For drying, choose one of the following two options: (A) Hang the *jamón* in a drying chamber set at 54°F to 60°F (12°C to 16°C) and 80% to 85% relative humidity. (B) Mimic the traditional seasonal European method of drying by altering the curing environment as follows:

➡ 43°F to 60°F (6°C to 16°C) and 60% to 80% relative humidity for a minimum of 90 days.

➡ 68°F to 79°F (20°C to 26°C) and 70% to 85% relative humidity for a minimum of 90 days.

➡ 54°F to 72°F (12°C to 22°C) and 70% to 90% relative humidity for a minimum of 115 days.

If you are concerned about flies or other insects, you can make a *pasta* (spread) by mixing together the *pimentón* and *manteca* until it forms a thick paste. Slather the paste onto the *jamón,* especially on all exposed areas of flesh. Otherwise, consider the option of bathing the *jamón* quickly in *very* hot sunflower oil every 3 to 4 months during the maturation process. If your drying chamber has a good supply of mold, it should bloom again about 1 week after each oil bath.

After 22 to 30 days per 2.2 pounds (1 kg) of green weight, the meat should lose 35% of its green weight.

12   Review the *Trucos de la Cocina* on the next page for the proper cutting technique for your *jamón.*

NOTE: Removing the aitchbone is something you should do with cleanliness and precision, since the more cuts and nicks you leave in the flesh means the more opportunities for bacteria to enter the ham. It's really not a difficult process, but words don't really do it justice when there are so many videos on the Internet already. For that reason, I recommend you check a few of those out before attempting the removal.

Cortar Jamón

CORVEJÓN

CONTRAMAZA

MAZA

PUNTA

TRUCOS DE LA COCINA: SLICING A JAMÓN

JAMÓN SLICING IS an art in and of itself. The *maestro cortadero's* goal is to get the thinnest cuts possible so that the *Ibérico's* inherent oleic acid within each slice starts melting on your tongue before you even begin to chew. Thus, a prescribed course of action and cutting steps will ensure that you work with, and not against, the bones of the *jamón* in order to trim it—bit by bit—until those bones are bare.

The same also holds true for a *paletilla* (shoulder ham). In fact, since the bones of the *paletilla* almost exactly mirror those of the *jamón* (with the exception of the hip joint bones), the cutting steps are almost the same. Those steps are (1) selecting equipment, (2) racking, (3) marking, (4) skinning, (5) cutting, (6) flipping, and (7) final trimming.

Selecting equipment: To cut a *jamón* properly, you will need something to hold the ham in place and something to cut it with. A solid rack called a *jamonero* clamps down on the trotter side of the ham, pinning everything in place as you cut. It can hold the ham either horizontally or vertically, but vertical types are harder to find since they require a little more skill to work with. Prices for *jamoneros* run from fairly cheap all-wood types to more intricate types that can cost thousands of dollars.

You need three basic knives to take down a *jamón:* a skinny boning knife for close to the bones, a paring knife for fine work around the bones, and a long, skinny knife called a *cuchillo jamonero* for slicing the ham into thin *lonchas.*

Racking: The ham should be placed in the *jamonero* with the sole of the hoof pointing upward. If you aren't allowed to have one with the hoof attached (which is unfortunately mandated by law in the United States), then simply position it so the kneecap is facing downward. The top of the *jamón,* known as the *maza,* will receive the first series of cuts. The point at the lowest point of the stand (opposite the hoof side), the *punta,* is the second trimming area. The side opposite the *maza,* known as the *contramaza* (or sometimes the *babilla)* is the third trimming area.

But all that said, this style of cutting is appropriate only if you will be consuming the *jamón* quickly, since it takes advantage of the fattier nature of the *maza.* Conversely, since the *contramaza* has less fat and dries out quicker, start on that side if you'll be consuming the *jamón* over a longer period of time.

Marking: The first cut on the *jamón* should be a hunk out of the shank area that's ideal for dicing and chopping up for *croquetas* or in soups. This leaves a sort of "notch" in the *jamón* to start slicing from. Make a 45 degree cut with the boning knife aimed down into the bone at the point where the shank begins. This area is called the *corvejón,* which is also the name of the cut. Once you make the inward *corvejón* cut, slice the top layer of skin and fat inward toward the cut, thus creating the notch.

CONTINUED ON NEXT PAGE

 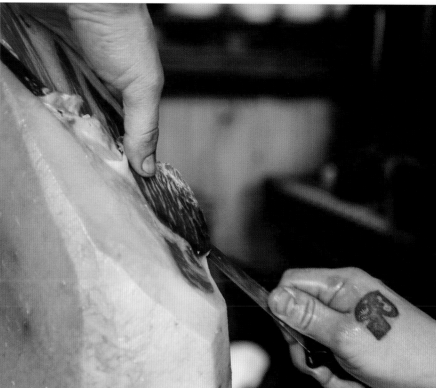

TRUCOS DE LA COCINA: SLICING A JAMÓN
CONTINUED FROM PREVIOUS PAGE

Skinning: Now that the *corvejón* cut has given you a place to start, start skinning the *jamón* with the *cuchillo jamonero*. Remove all the rancid yellow fat from the areas that you plan to cut and consume, but don't remove all of the fat from the ham. If you do, the *jamón* will dry out over time. Once you reach the layers of pearly white fat, save those slices for either cooking with or for keeping the *jamón* lubed up in its own natural fats for storage.

Cutting: Start cutting the *jamón* at the roundest, highest point on the *maza* side, coming down toward the *punta*. As you make thin, downward slices, you will eventually encounter the coxal bone, which gets in the way of slicing. Use the boning knife to cut around and mark it. This allows you to keep as flat a cutting surface as possible as you continually trim down the *maza* side to the femur. If you do it right, you will continually end your cuts at the coxal bone until you hit the femur and can cut no further.

Flipping: At this point, turn the *jamón* around so that the *contramaza* side is now on top. Your goal here is to continue cutting from the highest point down to the *punta*. You will find quickly that the tip of the hip bone will get in the way. Just as before, mark your way around the hip bone. This will allow you to continually make slices that terminate and release at the *punta*. The other bone you'll encounter as you slice your way down the *contramaza* side of the *jamón* is the knee cap. Just mark around it and continue cutting.

Final trimming: If all went according to plan, you should now have a relatively clean *jamón* from the top and bottom. The remaining meat will mostly be on the sides, which is not in a position to be cut into slices. This meat is typically trimmed off in hunks and cut into small dice called *taquitos* or *tacos* for adding to soups, stews, or *croquetas*.

PALETILLA

If curing a whole *jamón* seems a little too daunting, you can start with something a little smaller and less needy of your attention. This bone-in shoulder ham, for example, is a great gateway curing project since it cures much more quickly than a *jamón*.

First, let's quickly get some confusing definitions out of the way: The term *paletilla* or *paleta* can refer to the entire bone-in, cured shoulder from an *Ibérico* or white pig. But don't confuse this term with the next recipe, *Lacón*, since *lacón* also refers to the exact same cut. The difference is that *lacón* is a Galician specialty and can come only from a white pig like a Duroc or Berkshire hog—in the rest of Spain outside of Galicia; however, the term *"lacón"* can also refer to a boneless, cured, cooked, or smoked shoulder ham.

As for curing the *paletilla,* it follows a very similar process as the *jamón*, so it's great practice to get a feel for what it will take to cure a much bigger muscle.

INGREDIENTS	CHARCUTIER'S PERCENTAGE
per 2.2 pounds (1 kg) of bone-in shoulder of pork, skin on and trotter attached..........**100%**	
1¾ ounces (50 g) Basic *Salazón* (see recipe on p. 111)......................**5%**	
Coarse sea or rock salt, as needed	

OPTIONAL

⅛ ounce (3 g) DQ #2 or Instacure #2 curing salt mix..........**3%**	
¼ cup (25 g) *pimentón dulce*	
3½ ounces (100 g) *manteca* (pork lard)	

1   Weigh the pork shoulder and record its green weight for later.

2   If using, place the curing salt into a small bowl with the Basic *Salazón* and stir to combine.

3   Transfer the pork to a large tub or platter and cover the pork in the Basic *Salazón,* making sure to rub it into every crack and crevice of exposed flesh.

4   Transfer the rubbed *paletilla* to a large plastic tub with holes cut in the bottom or a similar large container with a rack. (The shoulder must be able to lay flat and above any water that will drain from it, and there must also be space for weight to be placed on top.) Cover the shoulder completely with the salt (how much you use isn't really important, as long as it's completely covered).

5   Place something heavy (around 10 pounds [4.5 kg]) on top of the salt-covered pork and refrigerate it at 37°F to 41°F (3°C to 5°C) for 1 day per 2.2 pounds (1 kg) of weight. During the refrigeration process, carefully monitor the liquid that drains from the meat and make sure to drain the liquid every so often. If the pork ends up touching the liquid at all, make sure to cover any exposed areas again with the salt. During this period, the meat should get much firmer to the touch and lose 10% to 15% of its green weight.

6   Remove the pork shoulder from the cure and rinse it under cold water, using a stiff brush to remove any surface salt. Pat the pork dry.

7   Tie the pork securely with butcher's twine on the hock end, securing a loop at one end to hang it. Hang the pork in a drying chamber set at 54°F to 60°F (12°C to 16°C) and 80% to 85% relative humidity for 7 days. During this period, the cure will equalize within the meat and the meat will lose 8% to 10% of its green weight.

8   If using the *pasta*, mix together the *pimentón* and *manteca* until it forms a thick paste. Slather the paste onto the *paletilla,* especially on all exposed areas of flesh.

9   Hang the *paletilla* in a drying chamber set at 54°F to 60°F (12°C to 16°C) and 80% to 85% relative humidity for about 15 to 25 days, until it has lost 35% of its green weight.

10 Slice the *paletilla* in a similar fashion to a *jamón* and serve as a *tapa* with bread.

LACÓN COCIDO

Lacón, at least for the time being, has a bit of a personality disorder. If you ask the Indicación Geográfica Protegida (IGP) labeling authorities, *lacón* is essentially the same shoulder-cut ham as a *paletilla,* except that a *paletilla* might be the shoulder of an *Ibérico* pig. A *lacón,* on the other hand, can come only from a white, non-*Ibérico* pig like a Duroc or Berkshire. The *lacón* is then cured, using basically the same process in the *paletilla* recipe, until the shoulder is a fully cured and shelf-stable product.

But that's only half of *lacón's* confusing story: In many Galician *charcuterie* shops, you will also find boneless cooked and smoked shoulder hams called *lacón,* as well as hams made from just the bottom half of the shoulder (which American butchers call the picnic ham) by the same name as well. In essence, then, the word *"lacón"* is just a placeholder for a cured/cooked/smoked ham from the shoulder region, which is what this recipe is all about.

Here, you will see how the Spaniards brine and then cook a shoulder ham. This recipe is traditionally prepared in the northern area of Galicia, where *lacón* hails from, though these days you can find cooked *lacón* in just about every supermarket in Spain.

The *salmuera* used here is pretty basic, just salt, sugar, and curing salt, so you should add any of the optional flavorings if the mood strikes you (some Italian *prosciutto cotto* recipes, for example, use a similar brine recipe based on lots of rosemary and garlic).

Last, this recipe calls for a special piece of equipment: a brine injector. This giant hypodermic needle looks like it belongs in Dr. Frankenstein's laboratory, but it's actually important for this style of *charcuterie.* By using the injector to pump brine into the ham, you can cut the recipe's curing time in half and ensure that all parts of the ham get brined.

INGREDIENTS	CHARCUTIER'S PERCENTAGE
per 2.2 pounds (1 kg) of bone-in shoulder of pork, skin on	**100%**
17 ounces (500 g) Basic *Salmuera* (see recipe on p. 105)	**50%**
1¼ ounces (35 g) kosher salt	**3.5%**
.14 ounce (4.5 g) DQ #1 or Instacure #1 curing salt mix	**.45%**

OPTIONAL

2	sprigs fresh rosemary
2	sprigs fresh thyme
1	fresh bay leaf
1	clove, ground and toasted
½	teaspoon (1 g) ground coriander, toasted
½	teaspoon (1 g) ground allspice
½	teaspoon (1 g) garlic powder
½	teaspoon (1 g) freshly ground black pepper
½	teaspoon (1 g) fennel seed
¼	teaspoon (.5 g) ground cinnamon
2	juniper berries, smashed

CONTINUED ON NEXT PAGE

LACÓN COCIDO CONTINUED FROM PREVIOUS PAGE

1 To a medium foodsafe container, add the Basic *Salmuera*. While the *salmuera* is still warm, whisk in the kosher and curing salts until fully combined. If using the optional ingredients, toast the clove and coriander at this time and add them, along with all of the other optional ingredients you choose to use, to the finished *salmuera*.

2 Weigh the pork shoulder and record its green weight for later.

3 In a medium mixing bowl, place enough *salmuera* to equal 10% of the weight of the meat. For example, if your meat weighs 11 pounds (5 kg), use 1.1 pounds (500 g) of brine. Set the bowl aside at room temperature and place the container with the rest of the *salmuera* in the refrigerator.

4 Place the pork shoulder on a baking sheet. Dip the brine injector into the mixing bowl of brine. Pull in the plunger and fill the injector. Inject the meat with the brine, starting at the hock. Continue injecting the meat with the brine, focusing on the thickest areas near the ball joints and trotter, until all of the brine has been injected. If any brine spills out, re-inject it into the ham.

5 Place the meat in a foodsafe container that is just large enough to hold it and a few ceramic plates to weigh it down. Pour the reserved, chilled *salmuera* over the meat and weigh it down until it is completely submerged beneath the level of the liquid.

6 Place the *lacón* in the refrigerator to cure for 1 day per 2.2 pounds (1 kg) of green weight. Each day, flip the meat over in order to ensure that all sides continue to be submerged in the brine.

7 Remove the meat from the brine and rinse it under cold water. Pat it dry.

8 You can cook the *lacón* by either roasting it in an oven or hot smoking it.

If oven roasting: Preheat the oven to 375°F (190°C). Roast the *lacón* for 20 minutes per 1 pound (450 g) of green weight, until it reaches an internal temperature of 140°F (60°C). Cool the *lacón* to room temperature for 30 minutes before either slicing and serving or chilling it in the refrigerator.

If hot smoking: Stock a large hot smoking chamber with oak wood and bring the temperature to 100°F (38°C) and the relative humidity to 70%. Meanwhile, dry the *lacón* for at least 2 hours in the refrigerator, until a pellicle forms on the surface. Apply smoke to the *lacón* for 20 minutes per 1 pound (450 g) of green weight, gradually increasing the temperature by 10°F (5.6°C) every hour, until the internal temperature of the ham reaches 158°F (70°C) and the *lacón* is firm to the touch. Cool the *lacón* to room temperature for 30 minutes before either slicing and serving or chilling it in the refrigerator (see Note).

9 After it has chilled completely, *lacón* is an awesome sliced cold cut for sandwiches or *charcuterie* plates. Gallegos also use the cured form of *lacón* with almost every type of stewed vegetable that they cook, but especially with *grelos* (see recipe on p. 289).

NOTE: These smoking times are for large chambers with a lot of airflow and low-to-medium levels of smoke. For more information on smoking meat, see Chapter 3.

SALMUERA AND SALAZÓN

TRADITIONAL AND MODERN RECIPES

MONTADITO DE MATRIMONIO

Any time you top a little piece of bread with something delicious, you have what the Spaniards call a *montadito*. And, in a world of infinite *montadito* possibilities, the first among equals is the *Matrimonio*.

Matrimonio is one of the most popular *montaditos* you'll find in Madrid. And you don't need to be Shakespeare to understand the metaphor at work here. A black, salt-cured *anchoa* is paired with a white, vinegar-cured *boqueron:* The bride is in white, and the groom in black, and both are topped with a citrus–herb mix called a *gremolata*. An edible yin and yang of anchovy-curing techniques, it's like this is the universe deliciously balanced on a piece of crunchy grilled bread.

TO MAKE THE *GREMOLATA*:

1   In a small mixing bowl, combine the parsley, garlic, and lemon zest. Mix well. Season with the salt and black pepper and set aside.

TO MAKE THE *MONTADITO*:

1   In a grill pan on medium–high heat, toast the bread slices for 1 minute on each side, until warm and just a little charred. Remove from the heat.

2   Rub the bread slices with the garlic and the two cut sides of tomato, grating the tomato into the bread.

3   Drizzle the bread liberally with the oil and sprinkle well with the salt. Place 1 piece of *anchoa* and *boqueron* on each slice of bread. Top with the *Gremolata*.

YIELD

4 servings

FOR THE *GREMOLATA*:

3   tablespoons (10 g) coarsely chopped fresh flat-leaf parsley

1   clove garlic, grated on a Microplane

Zest of 1 lemon

Kosher salt, to taste

Freshly ground black pepper, to taste

FOR THE *MONTADITO*:

4   slices good grilling bread, such as ciabatta or baguette

1   whole clove garlic, peeled

1   ripe tomato, halved

3   tablespoons (50 mL) extra virgin olive oil

Flaky sea salt, such as Maldon, to taste

4   *Anchoas en Salazón,* rinsed (see recipe on p. 129)

4   *Boquerones* (see recipe on p. 110)

ESPÁRRAGOS CON MOJAMA

Here in the United States, asparagus is available almost year-round in stalk sizes ranging from pencil thin to Ron Jeremy-esque levels of thickness. But no matter how large the asparagus here might be, they're almost always the usual shade of green. In Spain, however, the late-spring asparagus harvest yields not just the cultivated green variety but two others as well:

➡ Very tender white asparagus, beloved by Spaniards, with a color borne from mounding dirt each day over the growing stalk, which prevents both green pigment and toughness.

➡ A wild, thin, and long asparagus known as *espárragos trigueros.*

For this recipe, any type of asparagus works well except for that insipid canned garbage the Spaniards love so much for some reason. Canning fresh asparagus is common throughout the growing season in Spain, where canned goods are viewed as a means to year-round enjoyment of off-season vegetables. In fact, canned asparagus was such a popular *tapa* in bygone days that it is still standard fare at some old-school *tapas* bars. These dishes often includes some variation of white asparagus, typically right from the jar or can served with a little mayonnaise or *Alioli* (see recipe on p. 413).

I'll stick to the fresh stuff, *gracias*—especially in this dish, which I found at a few progressive restaurants in Barcelona during asparagus season. The *Mojama* does a beautiful job of adding salty, briny flavor to the dish, but I'm sure the Andalucíans are whipped up into a frothing rage over their proudest tuna export being used as a modern culinary garnish.

YIELD

4 servings

FOR THE ASPARAGUS:

Water, to cover

2.2 pounds (1 kg) green or white fresh asparagus, stemmed and peeled

Kosher salt, to taste

FOR THE ALMOND *PICADA*:

⅓ cup (50 g) ground almonds, toasted

2 slices toasted bread, coarsely ground

1½ tablespoons (10 g) *pimentón dulce*

½ cup (125 mL) extra virgin olive oil

FOR THE FINISHED DISH:

Extra virgin olive oil, as needed

4 large whole eggs

Kosher salt, to taste

Alioli (see recipe on p. 413), as needed, for serving

1¾ ounces (50 g) freshly grated *Mojama* (see recipe on p. 127), for sprinkling

TO BLANCH THE ASPARAGUS:

1 In a large saucepan, bring the water to a rolling boil. Season with the salt until the water tastes like the ocean.

2 Fill a large mixing bowl with enough cold water and ice to completely cover the asparagus. Set aside.

3 Place the asparagus in the boiling water and cook for 1 to 2 minutes, until just barely done (the amount of time will depend on the thickness of the asparagus). Remove from the heat.

4 Remove the asparagus from the boiling water and immediately plunge it into the ice bath. Chill for 5 minutes, until it is sufficiently cold. Once chilled, pat the asparagus dry with paper towels and set aside.

TO MAKE THE ALMOND *PICADA:*

1 In a small mixing bowl, combine all the Almond *Picada* ingredients. Mix well and set aside.

CONTINUED ON NEXT PAGE

ESPÁRRAGOS CON MOJAMA CONTINUED FROM PREVIOUS PAGE

TO MAKE THE FINISHED DISH:

1   Warm a small skillet over medium–high heat. Add about ¼ inch (6 mm) of the oil and heat until the oil is almost smoking.

2   Line a baking sheet with paper towels. Set aside.

3   In a small bowl, crack 1 of the eggs. Season it with the salt. Gently slip the egg into the oil, which will sputter violently at first. As the egg cooks, baste it with the oil. Cook for around 2 minutes, until it is crispy and becomes slightly golden at the edges but still has a runny yolk. (If using a small egg, you can "cheat" by separating the yolk from the white, cooking the white first, and adding the yolk to the pan for the last 20 seconds of cooking time. Baste to finish. (See the *Trucos de la Cocina* on this page for details.)

4   Remove the egg and place it on the baking sheet to drain. Repeat with the remaining eggs.

5   Light a charcoal grill and let the coals burn until they become white, or heat a broiler on high heat (and prepare your smoke detectors for a workout). Place the asparagus on a baking sheet and drizzle with the oil and salt. Grill or broil the asparagus for 2 minutes, until lightly charred on one side.

6   Smear a little of the *Alioli* on a serving platter (if you want to be fancy, do the ol' fine-dining "spoon swipe"). Place the charred asparagus across the *Alioli*. Top with the fried eggs and Almond *Picada*. Sprinkle with the grated *Mojama*. Serve.

TRUCOS DE LA COCINA: SPANISH FRIED EGGS

WHEN I FIRST began training with my mentor, Chef Steve Chan, he insisted that he could tell everything about a cook by the way he made eggs. Steve is not alone. As Chef Daniel Boulud points out in his book *Letters to a Young Chef*, egg cookery has always been a favorite test by old-school French chefs for gauging the skill of prospective hires. You see, you can tell exactly how much care a cook puts into his cooking by watching his technique in handling an important ingredient like an egg.

What's interesting is that I saw this truth yet again while cooking in Spain: Spanish chefs absolutely revere a properly fried egg, since it is literally the technique that has launched a thousand of their recipes. It's all in the method: Most people just throw a splodge of oil into a pan, heat it up, and then fry the egg. The results are often uneven at best or—heaven forbid—result in an overcooked yolk.

By comparison, Spaniards put a bunch of good olive oil in a warm pan and then slip an egg into the hot oil. This causes the egg white to bubble up and form a protective coating around the yolk. As the egg cooks, you spoon some of the hot oil over the top—the Frenchies and fine-dining crowd call this basting technique *arroser*—until the whole thing starts to get puffy and deliciously golden on the bottom and sides. That crunchy exterior, which is known as *puntillas* or "lace" for its similarity to the fringe on a matador's costume, is the stuff dreams are made of, especially when sprinkled with a little coarse salt.

Serve your sexy, runny, crispy egg with some grilled bread and sliced *jamón Ibérico.* Now *that's* the breakfast of *campeónes.*

CALDO BLANCO

True confession time: *Caldo Blanco* is a pretty obscure find even in Spain. More than likely, you'll encounter it only as a component of another dish, like *Sopa Blanca* (see recipe on p. 161).

As a matter of fact, there's really not a lot of information out there about this very old-school broth since it's a holdover from leaner times that most people would rather forget. Back in those days, the daily meal (notice I said "meal," not "meals") was mostly just bones + water + whatever veggies and meat you could find. The cook's goal, then, was to somehow turn these ingredients into a sort of "stone soup" that could provide warmth, comfort, and a source of nutrition.

I have no doubt you'll be surprised by how the bones and marrow in this dish create a pearly white and super-meaty broth. In some ways, it's similar to Japan's Tonkotsu style of ramen, which involves emulsifying marrow into the broth to add rich flavor. This technique of boiling fat and emulsifying it into broth in order to create a creamy consistency is used quite a bit throughout Spanish cookery, especially in heartier dishes like *Cocido Madrileño* (see recipe on p. 332).

1   In a large stockpot over medium–high heat, cover the beef shin with the water. Bring the water to a boil for 5 minutes. Remove from the heat. Drain and reserve the shin in the stockpot.

2   Add the *huesos* and vegetables to the stockpot. Cover again with cold water. Bring the water to a boil. Reduce the heat to medium and simmer the *caldo* for at least 4 hours, skimming impurities as needed but leaving the fat in the broth to emulsify.

3   Place the thyme sprigs, bay leaves, peppercorns, and cloves in a small piece of cheesecloth tied with kitchen twine. Add the sachet to the stockpot. (You may need to add some water to the pot keep the bones covered.)

4   Simmer the *caldo* for another 1 to 1½ hours, until it is a milky white color and the herbs have permeated the broth. Remove from the heat.

5   Taste and adjust the seasoning of the broth. Strain it into a bowl.

6   Either serve the broth immediately (to make *Sopa Blanca* [see recipe on p. 161], for example) or chill the broth if you plan to use it later.

YIELD

1 gallon (3.75 L)

2.2   pounds (1 kg) beef shin

    Cold water, to cover

4.4   pounds (2 kg) *Huesos Salados* (see recipe on p. 114), rinsed of excess salt

1   bunch celery, ribs only, cleaned and cut into large dice

4   leeks, cleaned and cut into large dice

3   medium white onions, peeled, destemmed, and cut into large dice

3   parsnips or large white turnips, peeled and cut into large dice

10   sprigs fresh thyme

2   fresh bay leaves

20   whole black peppercorns

2   whole cloves

SOPA BLANCA

The logical progression from making a *Caldo Blanco* is to make a *Sopa Blanca,* which is a white soup made with the white broth. The *sopa* is a homage to the color white: ingredients like white beans, potatoes, parsnips, and even the cabbage all lend some flavor, but the soup is really all about the pearly white color and meaty flavor of the stock.

1 Place the beans in a large bowl. Cover with cold water and leave to soak overnight.

2 Drain the beans and place them in a large stockpot over medium–high heat.

3 Add the *Caldo Blanco* to the stockpot. Bring to a boil and skim off the foam. Reduce the heat to medium. (From this point forward, do not stir the pot. Doing so will break up the beans. Instead, just give the pot a shake from time to time.) Simmer for 30 to 40 minutes, until the beans are just tender.

4 Add the vegetables and simmer for 20 minutes, until the beans and potatoes are cooked through and the soup has thickened. Remove from the heat.

5 Taste and season with the salt, white pepper, and nutmeg as needed.

YIELD

2 quarts (2 L)

½ pound (225 g) dried white cannellini beans

Water, to cover

1 recipe *Caldo Blanco* (see recipe on p. 159)

1 pound (450 g) Idaho potatoes, peeled and "cracked" into medium dice (see *Trucos de la Cocina* on this page)

4 parsnips or white turnips, peeled and cut into medium dice

1 head green cabbage, cored and shredded

4 leeks, cleaned and cut into large dice

Kosher salt, to taste

Freshly ground white pepper, to taste

Freshly grated nutmeg, to taste

TRUCOS DE LA COCINA: POTATO CRACKING

WHENEVER *SABIAS* ADD potatoes to a stew, they have a special technique for getting the starch to leech out of the potato and thicken the broth. That technique involves inserting a paring knife into the potato and "cracking" off bite-sized pieces. It's a great technique since you can vary the size of the resulting pieces, and by cutting the potato in this fashion—as opposed to a more anal-retentive Frenchie dice—you introduce more of the potato's cellular starch into the liquid.

BASQUE MOTHER SAUCES

ANY COOK WHO takes his or her craft seriously can probably rattle off to you the five basic mother sauces of the French kitchen: hollandaise, mayonnaise, *espagnol, velouté,* and *tomate.* Learning each was a prerequisite for my time cooking under my mentor, Chef Steve Chan, and it's also something most culinary-school kids have to recite ad nauseam.

When I got to Spain, however, I soon learned that the culinary traditions of other countries have their own criteria for what is deemed a "mother sauce." In the Basque region, for example, three major mother sauces—*pil pil, verde,* and *Vizcaina*—are the foundation for much of their cooking.

Those sauces, often served with fish like *bacalao,* as I recommend in the following recipes, are an integral part of both the old-school Basque culinary vernacular, as well as the *Nuevo Vasco* movement.

BACALAO AL PIL PIL

Adding chilies and garlic to hot olive oil makes a unique-sounding sizzle and pop to Basque ears. The sound—known in the local dialect as *"pil pil"*—has a significant meaning directly related to one of the cornerstone mother sauces for Basque cuisine.

Pil pil sauce is a singularly unique emulsification that starts with spicy, garlic-imbued olive oil. You then poach desalinated *Bacalao* filets in the oil, gently swirling the fish in the oil to release gelatin, fat, and natural juices. The resulting sauce that forms from continually agitating the cooling, gelatin-imbued oil is silky and yellow, like a hollandaise, but is in fact a uniquely Basque derivative. That is why *pil pil* has become a point of national pride for Basque cooks and is one of the most popular techniques for fish dishes you will find in País Vasco.

Getting the sauce right, however, requires a little finesse. The trick lies in getting enough gelatin into the sauce (this is why gelatin-rich, skin-on *Bacalao* is used) and forming the emulsification when the olive oil isn't too hot or too cold. It needs to be lukewarm to achieve a smooth sauce.

But that said, here are a few cheats that I picked up from my Basque pals to help you get the sauce right every time:

If you have any leftover bones or heads from butchering cod (a great reason to salt your own!), you can *confit* them in olive oil ahead of time before poaching the cod filets. This will give you a very rich and flavorful *pil pil* sauce and will also make emulsification a little easier, since the *confit* process essentially laces the olive oil with lots of gelatin.

In a worst-case scenario, use an immersion blender to get things started with the sauce—it works like a charm. Just don't do it around an old-school Basque cook, or you'll be mocked mercilessly.

In this dish, the quality of the olive oil really makes a difference in both color and flavor. For that reason, shoot for an oil that's as yellow and flavorful as you can get—something like Piqual or Arbequina will serve you well.

YIELD

4 servings

2.2	pounds (1 kg) *Bacalao* (see recipe on p. 124), skin on and cut into 4 large squares
	Water, to cover
1	cup (215 mL) good extra virgin olive oil, such as Arbequina or Piqual
	Water, as needed
4	cloves elephant garlic, peeled, destemmed, and sliced into thin rounds
2	Fresno or other red chile peppers, cut into thin rounds

CONTINUED ON NEXT PAGE

BACALAO AL PIL PIL CONTINUED FROM PREVIOUS PAGE

1   In a large mixing bowl, cover the *Bacalao* with the water. Soak in the refrigerator for 1 day, changing the water at least twice during the soaking period.

2   Remove the *Bacalao* from the water and pat it dry. Set aside.

3   In a skillet over medium–high heat, warm the oil for 4 to 5 minutes, until just rippling. Place 2 to 3 paper towels on a plate and set aside.

4   Place the garlic and chilies in the skillet and fry lightly for 2 minutes, until the garlic is golden brown and the chilies have darkened in color. Remove the garlic and chilies and place them on the lined plate to drain. Reserve the oil in the skillet and reduce the heat to medium.

5   Place the *Bacalao* in the skillet skin-side up. Cook for 8 to 10 minutes, until cooked through. Remove from the heat.

6   Place the *Bacalao* on the plate containing the garlic and chilies and cover with foil. Set aside.

7   Pour most of the oil in the skillet into a bowl, leaving behind ¼ cup (60 mL) of oil and any cooking residue in the pan (this residue is full of gelatin and protein, which will set the *pil pil* sauce). Allow the oil in the skillet and in the bowl to cool for 10 minutes before proceeding—the oil should be warm to the touch.

8   Begin whisking the oil in the skillet, while at the same time shaking and swirling the pan, until a creamy emulsion begins to form. Sprinkle a little water into the pan (this will help the sauce to bind) and swirl it into the sauce.

9   While continuing to whisk, slowly drizzle the reserved olive oil back into the pan, just as you would when making a mayonnaise, until the emulsion reaches the desired thickness.

10 Transfer the *Bacalao* to a serving platter. Pour the *pil pil* sauce over the *Bacalao* and garnish with the fried garlic and chile rounds. Serve warm.

BACALAO EN SALSA VERDE

Of all of the Basque preparations for *bacalao,* this version is hands down my favorite since it's a great example of the New Basque cuisine that swept through Basque kitchens in the 1970s and 1980s. Back then, great chefs like Juan Mari Arzak and Pedro Subijana—guys who were influenced by the *nouvelle cuisine* movement of France—put on events akin to today's pop-up restaurants with the goal of steering Basque cooking away from its heavy, butter-fortified sauces in favor of lighter preparations that were more ingredient driven.

This dish is a great example of what they were getting at. The original *salsa verde* recipe for this dish calls for the same technique used to make a classic *velouté,* essentially thickening fish *fumet* with a heavy roux before finishing everything with chopped herbs.

By comparison, the updated version I learned from some Basque friends features a cleaner and lighter sauce that lets the ingredients do the talking. In this recipe, a little flour lightly thickens the sauce. The flour binds the oil and water with the fish's natural gelatin, which is released using a swirling technique like the one used to make *pil pil* sauce. As a result, the dish is really light and fresh, especially with the addition of herbs and lemon.

TO REHYDRATE AND COOK THE *BACALAO:*

1   In a large mixing bowl, cover the *Bacalao* with the water. Soak in the refrigerator for 1 day, changing the water at least twice during the soaking period.

2   Remove the *Bacalao* from the water and pat it dry. Reserve ½ cup (120 mL) of the soaking liquid. Set aside.

3   In a medium skillet over medium–high heat, warm the oil for 4 to 5 minutes, until just rippling. Line a plate with 2 to 3 paper towels and set aside.

4   In a shallow bowl, season the flour with the salt and black pepper. Dredge each of the pieces of *Bacalao* in the flour, shaking off any excess flour.

5   Sear the *Bacalao* in the skillet for 2 to 3 minutes on each side, until golden brown. Remove from the heat. Remove the *Bacalao* pieces from the skillet and place them on the lined plate to drain. Cover with foil and set aside. Reserve the oil in the skillet.

YIELD

4 servings

FOR REHYDRATING AND COOKING THE *BACALAO:*

2.2   pounds (1 kg) *Bacalao* (see recipe on p. 124), skin on and cut into 4 large squares

Water, to cover

¼   cup (60 mL) extra virgin olive oil

All-purpose flour, as needed

Kosher salt, to taste

Freshly ground black pepper, to taste

FOR THE HERB SAUCE:

4   whole cloves garlic, peeled, destemmed, and minced

Generous pinch kosher salt

¼   cup (15 g) minced fresh flat-leaf parsley

2   tablespoons (5 g) minced fresh chives

2   tablespoons (5 g) minced fresh chervil

FOR THE SALSA VERDE:

1   medium yellow onion, destemmed and minced

Kosher salt, to taste

Pinch red pepper flakes

20   littleneck clams, scrubbed

½   cup (120 mL) Spanish white wine, such as Verdejo

Juice and zest of 1 lemon

Freshly ground black pepper, to taste

CONTINUED ON NEXT PAGE

BACALAO EN SALSA VERDE CONTINUED FROM PREVIOUS PAGE

TO MAKE THE HERB SAUCE:

1   In a mortar and pestle, crush together the garlic and salt to form an *ajosal*. If desired, you can finish the *ajosal* in a food processor fitted with the "S" blade.

2   Add the parsley, chives, and chervil to the *ajosal* and continue crushing or processing until a rough green paste forms. Drizzle in the reserved *Bacalao* soaking water and continue crushing or processing until the paste becomes a somewhat thick sauce. Set aside.

TO MAKE THE *SALSA VERDE* AND THE FINISHED DISH:

1   Add the onions to the skillet and season them with the salt. Place the skillet over medium–high heat and sweat the onions in the oil for 10 minutes, until they are soft but have not taken on color.

2   Add the red pepper flakes, clams, and wine to the skillet. Cover the pan and bring the liquid to a boil. Reduce the heat to medium and simmer, covered, for 6 to 10 minutes, until the clams are just ready to open. Remove from the heat.

3   Add the *Bacalao* and the Herb Sauce to the skillet. Begin whisking the sauce in the skillet, while at the same time shaking and swirling the pan, until a creamy emulsion begins to form as with a *pil pil* sauce.

4   Transfer the clams and *Bacalao* to a serving platter. Add the lemon juice to the skillet. Taste and reseason the sauce as necessary with the salt.

5   Pour the sauce over the clams and *Bacalao*. Top with the black pepper and lemon zest. Serve warm.

BACALAO EN SALSA VIZCAINA

There's a number of surefire ways to start a bar fight in Spain: Wear a Real Madrid jersey in Barcelona; overcook your seafood in Galicia; discuss the virtues of adding onion to a *tortilla española* in Madrid; or ask a group of Basque cooks about *Salsa Vizcaina*.

Salsa Vizcaina—one of the Basque people's oldest and most cherished sauces—is a real powder keg of a conversational topic. No two recipes are exactly alike, and, as is normally the case with old recipes and people who are highly opinionated about their cooking, everyone thinks his or her recipe is the best.

Some recipes call for the Spanish version of tomato sauce, called *Tomate Frito* (see recipe on p. 407). Others say that *jamón* is an indispensable ingredient. If you add some *galletas Marias* (Spanish cookies similar in taste to a graham cracker that help bind things up) to your sauce, you'll either be praised for your authenticity or derided as a heretic.

The way I see it, this recipe is going to make some Basques happy, and the rest will just be pissed at me. I may not be able to please everybody, but in the end, the sauce is delicious.

1   In a large mixing bowl, cover the *Bacalao* with the water. Soak in the refrigerator for 1 day, changing the water at least twice during the soaking period.

2   Remove the *Bacalao* from the water and pat it dry. Set aside.

3   In a medium skillet over medium–high heat, warm the oil for 4 to 5 minutes, until just rippling. Place 2 to 3 paper towels on a plate and set aside.

4   Place the garlic and bread in the skillet and fry lightly for 4 minutes, until both are golden brown. Place both on the lined plate to drain. (See Notes.) Reserve the oil in the skillet and reduce the heat to medium.

5   Place the *Bacalao* in the skillet skin-side up. Cook for 8 to 10 minutes, until cooked through.

6   Place the *Bacalao* on the plate containing the garlic and bread and cover with foil. Set aside.

7   Add the onions to the skillet and season them with the salt. Sweat the onions in the oil for 10 minutes, until they are soft but have not taken on color.

YIELD

4 servings

2.2   pounds (1 kg) *Bacalao* (see recipe on p. 124), skin on and cut into 4 large squares

Water, to cover

½   cup (120 mL) extra virgin olive oil

5   whole cloves garlic, peeled and destemmed

2   slices day-old bread, such as ciabatta, cubed and crusts removed

1   medium yellow onion, peeled, destemmed, and cut into small dice

1   medium red onion, peeled, destemmed, and cut into small dice

Kosher salt, to taste

5   choricero peppers *en adobo*, puréed*

4   tablespoons (50 g) diced *Jamón* (see recipe on p. 143)

1   tablespoon (12 g) granulated sugar

¼   cup (50 mL) sherry vinegar, plus more to taste

* You can find choricero peppers *en adobo* at tienda.com or the food store Despaña in New York City. If you can't get them, consider substituting ancho or guajillo chile peppers.

CONTINUED ON NEXT PAGE

BACALAO EN SALSA VIZCAINA CONTINUED FROM PREVIOUS PAGE

8   Add the peppers, *Jamón,* and sugar to the skillet and cook for 10 minutes, until the peppers soften and change color to a darker red. Add the garlic and bread to the skillet and cook for 10 minutes. (See Notes.)

9   Add the vinegar to the skillet and stir. Remove from the heat. Transfer the sauce to a bowl by passing it through a food mill fitted with a medium blade. (If you do not have a food mill, you can process the sauce in a blender or food processor or use an immersion blender to get the sauce to a smooth consistency.)

10   Taste and reseason the sauce as necessary with more sherry vinegar and salt.

11   Transfer the *Bacalao* to a serving platter. Pour the sauce over the *Bacalao.* Serve warm.

NOTES: If desired, substitute half the quantity of bread called for in the recipe with ground *galletas Marias* or graham crackers. If so, toast only the garlic and add the crackers at the end of step 8, when you put the fried garlic in the skillet to finish the sauce.

The heretics out there can join me in adding ½ cup (120 mL) of *Tomate Frito* (see recipe on p. 407) after step 8, and then simmering for 20 minutes or until the acid is cooked out of the tomatoes. Tradition be damned. It's delicious.

XATÓ

If you follow the *Ruta del Xató* along the coast between Barcelona and Tarragona, you'll have no problem finding the most famous and traditional Catalan fisherman's salad in Spain: *Xató* (pronounced like *chateau).* The *Xató* route was established via a pact between different municipalities and some 85 participating restaurants to compete for tourism dollars and to preserve the salad's culinary heritage.

But that doesn't mean everyone gets along: Just about every city along the coast lays claim to the "original, authentic" form of the salad, and each city's residents firmly believe that anyone from anywhere else doesn't have a clue about how to make it.

As for the salad, the base is pretty easy: escarole and/or frisée; *Bacalao*; anchovies, *Boquerones,* and a pungent *picada* that resembles *romesco,* a traditional sauce from Catalonia that's heavy on garlic, tomato, almonds, and hazelnuts. From that point on, every interpretation is different and the cause of much Catalonian pontificating. I particularly love this modern take, with oil-cured tuna, lemon, and shaved *Mojama.*

1   Preheat the oven to 375°F (190°C).

2   Place the dried chilies in a small mixing bowl. Cover with the boiling water and soak the chilies for 30 minutes, until soft.

3   Drain the chilies. Transfer to a cutting board. Cut them open, scrape out the pulp, and place the pulp in a small bowl. Discard the rind. Set aside.

4   Place the almonds and hazelnuts on a baking sheet. Toast in the oven for 10 minutes, until golden brown and fragrant. Remove from the oven and set aside.

5   Warm 2 tablespoons of the oil in a small skillet over medium–high heat until just rippling. Place the bread in the skillet and toast on each side for 2 minutes, until golden on both sides. Remove from the heat. Set aside.

6   Grate the tomato halves into a bowl using the medium holes of a box grater. Set aside.

7   In a mortar and pestle, crush together the garlic and salt to form an *ajosal.* If desired, you can finish the *ajosal* in a food processor fitted with the "S" blade.

YIELD

4 servings

4	dried ñora peppers, seeded*
3	cups (750 mL) boiling water
¼	cup (35 g) whole almonds, skins removed
¼	cup (35 g) whole hazelnuts, skins removed
¼	cup (60 mL) extra virgin olive oil, divided
1	slice country-style bread, such as ciabatta, crusts removed
1	whole Roma tomato, halved
1	clove garlic, peeled and destemmed
	Generous pinch kosher salt, plus more to taste
1	*Anchoa en Salazón* (see recipe on p. 129), rinsed
1	tablespoon (7 g) *pimentón picante*
2	tablespoons (30 g) sherry vinegar, plus more to taste
¼	cup (15 g) chopped flat-leaf parsley
1	head bitter greens (chicory or endive), cored and trimmed
1	head frisée (leaves of endive), destemmed and trimmed
6	ounces (170 g) *Bacalao* (see recipe on p. 124), soaked overnight, skinned, and deboned
3	ounces (85 g) *Atún en Aceite* or *Atún en Escabeche* (see recipes on pp. 205 and 193)
2	tablespoons (25 g) *Atún en Aceite* curing oil
8	*Boquerones* (see recipe on p. 110), halved into chunks
3	ounces (85 g) oil-cured pitted black olives
	Juice and zest of 1 lemon
	Freshly ground black pepper, to taste
3–4	ounces (85–113 g) freshly grated *Mojama* (see recipe on p. 127), for sprinkling

CONTINUED ON NEXT PAGE

XATÓ CONTINUED FROM PREVIOUS PAGE

8   Add the *Anchoa en Salazón* to the *ajosal* and continue crushing or processing until fully incorporated. Add the toasted nuts to the *ajosal* and continue crushing or processing until fully incorporated. Add the toasted bread and pepper pulp to the *ajosal* and continue crushing or processing until the consistency reaches a homogenous, lumpy paste. Transfer to a large measuring cup or other pourable container.

9   To the measuring cup, add the grated tomato, *pimentón,* vinegar, parsley, and the remaining olive oil. Stir to combine, creating a *picada.* Taste and reseason with the salt and sherry vinegar. Set aside.

10 Tear the greens and frisée into bite-sized pieces and place them in a separate large serving bowl.

11 Tear the *Bacalao* into shreds and the *Atún en Aceite* into chunks. Add them to the serving bowl and toss. Add the curing oil, *Boquerones,* and olives to the serving bowl and toss again.

12 Drizzle the *picada* into the bowl, along with the lemon juice, and toss. Season to taste with the salt and black pepper.

13 Divide the salad among 4 plates. Sprinkle each portion with the *Mojama* and lemon zest. Serve.

* You can find ñora peppers at tienda.com or the food store Despaña in New York City. If you can't get them, consider substituting ancho or guajillo chile peppers.

JOSEP MERCADER'S *GARUM* ANCHOVY SPREAD

As a young cook curious about culinary history, I discovered that *garum* is considered by some to be the Western world's first fish sauce. Essentially little more than salted and pressed fresh anchovies, the *garum* recipe was developed by the ancient Romans, who elevated it to become a beloved condiment that is still sold today under the name Colatura di Alici.

Years later, while thumbing through Colman Andrews' *Catalan Cuisine* at my local *chiringuito,* I was surprised to see a recipe for an anchovy spread also called *garum.* Andrews explained that the great Catalan chef Josep Mercader created the spread in his kitchen at the Hotel Ampurdán and named it *garum* as a sort of joke—a wink and nod to the Roman sauce of old, which was often made using anchovies from the Catalan coast.

Anyway, this is a delicious spread. I've adapted and adopted the recipe for different events, though I like to add lemon juice and zest. The acid really helps balance the otherwise rich, fatty spread, and the citrusy flavors help the anchovy-averse to put on their big-boy pants and try something deliciously outside of their comfort zone.

1   In a blender or the bowl of a food processor fitted with the "S" blade, combine the olives, oil, *Anchoas en Salazón, Yema Curada,* mustard, capers, garlic, parsley, chives, tarragon, chervil, and lemon zest. Blend or process on high, scraping down the sides, until the mixture becomes a fluffy purée.

2   Press the purée through a food mill into a bowl (if you don't have a food mill, you can omit this step—you will just have to scrape down your bowl or blender a little more until the mixture becomes homogenous). Return the purée to the blender or food processor and blend or process it until it forms a smooth paste. Transfer the paste into a small serving bowl. Taste and season with the salt and black pepper. Add the lemon juice and stir until combined.

3   Serve with the grilled bread.

YIELD

4 servings

3	ounces (85 g) oil-cured pitted black olives
½	cup (120 mL) extra virgin olive oil, divided
15	*Anchoas en Salazón* (see recipe on p. 129), rinsed
1	*Yema Curada* (see recipe on p. 113), minced
2	tablespoons (30 g) stone-ground or Dijon mustard
¼	cup (50 g) capers
2	cloves garlic, destemmed and grated or pressed
2	tablespoons (5 g) minced fresh curly parsley
2	tablespoons (5 g) minced fresh chives
2	tablespoons (5 g) minced fresh tarragon
2	tablespoons (5 g) minced fresh chervil
	Zest and juice of 1 lemon
	Kosher salt, to taste
	Freshly ground black pepper, to taste
4	slices grilled bread rubbed with garlic, for garnish

ESQUEIXADA

Back in 2005, Colman Andrews wrote *Catalan Cuisine,* one of the best books available on regional Catalan cooking. It's an inspiring guidebook for a lot of American cooks who have worked in Spanish kitchens.

In his book, Andrews makes a great comparison of *Esqueixada,* a Catalan favorite of salt cod paired with a mixture of summer vegetables in a very acidic dressing, to *ceviche,* the acid-cooked seafood preparations of Mexico and Latin America. This comparison is spot on, and it puts a salad with a funny name into familiar territory—like *ceviche,* the acid in the *Esqueixada* dressing "cooks" the *Bacalao* a little, changing the texture of the fish ever-so-slightly as the acid does its magic.

To make *Esqueixada,* you'll need to take the time to season and "cure" the veggies and desalinated salt cod with some salt, sugar, and a sharp "inverse" sherry vinaigrette. The resulting liquid leaches out of the veggies and fish (*ceviche* cooks would call this liquid the *leche de tigre*) and helps finish the dressing.

1. In a large mixing bowl, cover the *Bacalao* with the water. Soak in the refrigerator for 1 day, changing the water at least twice during the soaking period.

2. Remove the *Bacalao* from the water and pat it dry. Shred the *Bacalao* with your fingers into tiny threads, discarding any skin or bones, and place in a small bowl. Set aside.

3. Grate the tomato halves into a large mixing bowl using the medium holes of a box grater. Set aside.

4. To the bowl containing the tomatoes, add the *Bacalao,* onion, bell peppers, chile peppers, garlic, sugar, and 3 tablespoons (45 mL) of the *Vinaigretta de Jerez Invertido.* Taste and season with the salt. Set aside to marinate for at least 30 minutes at room temperature. During this period, you will see that liquid will be drawn out of the veggies.

5. Drain and reserve the liquid from the *Bacalao*–vegetable mixture. To the measuring cup containing the remaining *Vinaigretta de Jerez Invertido,* add the olives and the reserved liquid. Whisk thoroughly, creating a dressing for the salad. Taste and reseason with the salt, black pepper, and sugar if needed.

6. Divide the salad among 4 plates. Drizzle the dressing over the top. Garnish each portion with the parsley, lemon zest, and *Yema Curada.* Serve.

YIELD

4 servings

2.2 pounds (1 kg) *Bacalao* (see recipe on p. 124), skin on and cut into 4 large squares

Water, to cover

2 medium plum tomatoes, halved

1 medium red onion, thinly sliced and separated into rings

1 medium red bell pepper, seeded and thinly sliced

1 medium green bell pepper, seeded and thinly sliced

2 Fresno or other red chile peppers, seeded and thinly sliced

1 clove garlic, destemmed and grated or pressed

2 tablespoons (24 g) granulated sugar, plus more to taste

1/3 cup (80 mL) *Vinaigretta de Jerez Invertido* (see recipe on p. 411)

Kosher salt, to taste

3 ounces (85 g) oil-cured pitted black olives

Freshly ground black pepper, to taste

1/4 cup (15 g) minced fresh flat-leaf parsley, for garnish

Zest of 1 lemon, for garnish

1 *Yema Curada* (see recipe on p. 113), grated, for garnish

CHAPTER

5

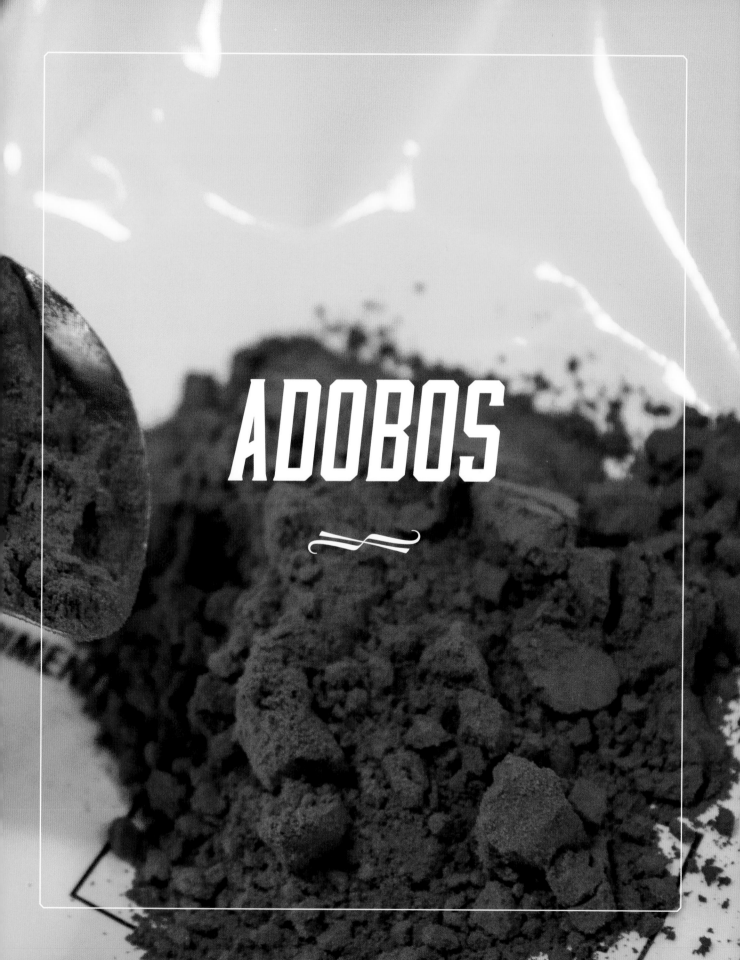

ADOBOS

N *ADOBO* IS a simple marinade that is often used to keep and flavor meats until they can be cooked. In *sabias'* kitchens, *adobo* recipes almost always include garlic, *pimentón,* spices, salt, and olive oil, and sometimes also vinegar. The resulting bright red paste gets slathered on nearly everything to hold it over for further cooking (*pimentón* purportedly has antibacterial properties) and to imbue a redness into the final dish (not to mention the cook's hands, if he or she forgets to wear gloves).

This marinade is particularly famous in the south of Spain, where it's the flavoring of choice for everything from seafood dishes like *cazón en adobo* to the Moorish-inspired skewered meats called *pincho moruno.*

As with the previous chapter, always remember that the recipes that include a column for the *Charcutier's* Percentage are based on 2.2 pounds (1 kg) of the protein involved, and that you should use the *Charcutier's* Percentage as a guide to figure out the proper percentages of each ingredient for the actual size of the protein you'll be preparing.

BASIC *ADOBO*

As you will see in the recipes throughout this section, the basic *adobo* is a pretty standard marinade that can be applied to pork, fish, or any number of game meats.

Fair warning: Oil-soaked *pimentón* is quite the potent, permanent stainmaker, so consider wearing gloves and anything other than the color white.

1   Using a mortar and pestle, crush together the garlic and salt to form an *ajosal*. If desired, you can finish the *ajosal* in a food processor fitted with the "S" blade.

2   One at a time, add the oregano, *pimentón,* oil, sherry vinegar, and water to the *ajosal* and continue crushing or processing, until the mixture is a bright red paste.

3   Transfer the paste into a small covered bowl. Use the *adobo* immediately or store in the refrigerator for up to 7 days.

YIELD

1 cup (200 mL)

INGREDIENTS

per 2.2 pounds (1 kg) of protein

5   cloves garlic, peeled, destemmed, and crushed

1   tablespoon (14 g) kosher salt

1   tablespoon (3 g) dried oregano

2   tablespoons (15 g) *pimentón dulce*

¼   cup (50 mL) extra virgin olive oil

¼   cup (50 mL) sherry vinegar

¼   cup (50 mL) water

LOMO ADOBADO

Spaniards like their pork loin one of two ways: Dry cured and stuffed into a casing, as with *Lomo Embuchado* (see recipe on p. 313), or semicured, marinated, and then seared *a la plancha*-style, as with this *Lomo Adobado*.

During my time in Toledo, I enjoyed many a family meal over slices of *lomo adobado* seared on the *plancha,* stuffed into a sliced baguette, and topped with melted Manchego cheese and caramelized onions. It was like a Brooklyn version of a Philly cheese steak made by Manchego chefs—picture Don Quixote in skinny jeans and a fedora.

1   To a medium foodsafe container, add the Basic *Salmuera*. While the *salmuera* is still warm, whisk in the kosher and curing salts until fully combined.

2   Seal the container and refrigerate the brine for at least 2 hours, or until chilled.

3   Transfer the chilled *salmuera* to a foodsafe container large enough to accommodate the *lomo* and a few ceramic plates for weighing down the meat. Add the *lomo* to the chilled *salmuera* and place the plates on top to keep it to keep the meat fully submerged in brine. Seal the container. Refrigerate the container for 2 days.

4   Using a mortar and pestle, crush together the oregano, garlic, *pimentón,* and choricero peppers until a paste forms. Add the vinegar and *adobo* and stir to combine.

5   Remove the *lomo* from the *salmuera* and pat it dry. Place the *lomo* on a baking sheet and smear the *adobo* all over it. At this point, you can either cook the *lomo* immediately or return it to the refrigerator to marinate for up to 3 days.

6   Warm the oil in a large sauté pan over medium–high heat for 4 minutes, until just rippling.

7   Slice the loin into thick *lonchas,* or ¼-inch (6-mm) slices. Season the slices with the salt and black pepper.

8   Add the *lonchas* to the pan. Sear the *lonchas* for 3 to 5 minutes per side, until cooked through. Remove from the heat.

9   I love serving *lomo adobado* stuffed into a sandwich with caramelized onions and melted Manchego cheese, but it also works as an entrée accompanied by something like rice and veggies.

INGREDIENTS	CHARCUTIER'S PERCENTAGE
per 2.2 pounds (1 kg) of *lomo de cerdo* (pork loin), cleaned of silverskin	**100%**
1 recipe Basic *Salmuera* (see recipe on p. 105)	**50%**
1 ounce (25 g) kosher salt	**2.5%**
.14 ounces (4 g) TCM #1 or DQ #1 curing salt mix (see p. 83)	**.4%**
1 tablespoon (3 g) minced fresh oregano	
3 cloves garlic, destemmed and grated using a Microplane	
2 tablespoons (15 g) *pimentón dulce*	
3 choricero peppers in *adobo,* puréed*	
½ cup (120 mL) apple cider vinegar	
1 recipe Basic *Adobo* (see recipe on p. 177)	
2 tablespoons extra virgin olive oil	

* You can find choricero peppers *en adobo* at tienda.com or the food store Despaña in New York City. If you can't get them, consider substituting ancho or guajillo chile peppers.

PIMENTÓN

EVERY TIME I see those jars of fake-me-out "Spanish Smoked Paprika" at the grocery store a little piece of my soul dies. You know the ones: There's not a word of actual Spanish on the label, and if you look on the back, it says something like "Made in Hoboken."

True *pimentón's* origin story starts with Columbus's return from his second voyage to the Americas, whereupon vibrant peppers were presented to King Fernando and Queen Isabel of Spain at a ceremony in the royal monastery of Santa María de Guadalupe in Extremadura.

The monks of the monastery passed the seeds from the peppers onto their brethren across Spain—which is how the vibrant red color of these peppers permeated much of Spanish cuisine.

And as the monks in different regions experimented and discovered whether the peppers would grow in their native soils, two major types of Spanish paprika evolved; these are the ones you will most likely find today and illustrates that—even in Spain—not all *pimentón* is created equally.

The first type is from Murcia, which is essentially the vegetable garden for much of Spain. Here, peppers are left to dry in the natural sun that beats down on the fields—so the *pimentón* that you find from Murcia is likewise fruitier, rather than smokier, since no actual smoke is used in the process of drying this style of paprika.

The second type of *pimentón,* called *pimentón de la Vera,* is more popular since it has a hauntingly smoky flavor that we have come to associate with the paprika of Spain. That smokiness is something that evolved as a necessity of preservation—since the area in and around La Vera tends to be wetter during harvest seasons, the monks there needed to find a way to dry their peppers before they rotted.

And that's what begat the tradition and now-booming modern business of loading ancient sheds with the peppers, literally lining the ceilings with removable wooden slats for lining up the fresh-picked fruit, and placing smoldering holm oak fires below for a period of 12 to 15 days.

Real-deal *pimentón* is a thing of brick-red beauty imported exclusively from Spain and typically sold in decorative tins. The paprika is made from either the sweeter, rounder ñora peppers or the more common, long choricero (*jaranda*) peppers, and it's the proportion of those peppers that yields three standard *pimentón* flavors: *dulce* (sweet), *agridulce* (bittersweet), or *picante* (spicy—in kind of a wussy Spanish sense).

Once the mix is made, a slow and methodical grinding process begins and results in a powder that—at least for me—winds up as something completely worth the extra buck or three over the price of that flaccid red powder otherwise known as "paprika."

PAPADA EN ADOBO

Jowl meat is one of the most amazing and coveted pork products on earth. The Italians cure it to make *guanciale,* Asian cultures use it for everything from sausages to roasting, and here in the United States, you'll often find it cured and smoked as a welcome addition to Southern collard greens and black-eyed pea dishes.

In Spain, I was surprised to find that cured *papada* isn't all that popular in *charcutería* shops. Instead, it's mainly incorporated in sausage grinds, since it is such a flavorful meat, but I finally saw the *sabias* cure some at a few *matanzas* in Extremadura. In this recipe, the *papada* gets coated in a basic *adobo* that also gets a splash of sherry.

At the *matanzas,* we smoked the *papada* by hanging it near a wood-burning fire and letting nature take its course. Feel free to take the more controlled smoking approach used in the recipe, however—we don't all have the luxury of an Extremeñan farmhouse to hang meat in.

1   Weigh the pork jowl and record its green weight for later.

2   Place the Basic *Salazón* in a measuring cup. If using, whisk in the curing salt.

3   Transfer the salazón to a ziptop bag. Add the sherry and Basic *Adobo.* Place the pork jowl in the ziptop bag. Close the bag, removing as much air as possible. Toss the bag to distribute the cure evenly while pressing the cure into the meat.

4   Place the ziptop bag into a medium baking dish and weigh it down with a few cans. Refrigerate the dish for 1 day per 2.2 pounds (1 kg) of green weight, turning the meat over each day to redistribute the cure.

5   Stock a large cold-smoking chamber with oak wood and bring the temperature to 90°F (30°C) and the relative humidity to 80% to 85%. (See Note.)

6   Remove the *papada* from the cure and rinse it lightly under cold water. Pat the *papada* dry and set it aside on a clean plate.

INGREDIENTS | CHARCUTIER'S PERCENTAGE

per 2.2 pounds (1 kg) of *papada* (pork jowl), cleaned of all lymph nodes and glands **100%**

2½ ounces (70 g) Basic *Salazón* (see recipe on p. 111) **7%**

¼   cup (50 mL) fino or oloroso sherry

1   recipe Basic *Adobo* (see recipe on p. 177)

OPTIONAL

⅛   ounce (3 g) DQ #2 or Instacure #2 curing salt mix .... **.3%**

A *matanza* in full swing—note the hanging *papadas* and other meats in the chimney area.

7 Tie one end of the *papada* securely with butcher's twine and secure a loop in order to hang it. Hang the *papada* in the smoking chamber and smoke for 1 day, or break it up into 4-hour-long sessions for 3 days, placing the *papada* in a drying chamber set at 54°F to 60°F (12°C to 16°C) and 80% to 85% relative humidity in between sessions.

8 Place the *papada* in a drying chamber set at 54°F to 60°F (12°C to 16°C) and 80% to 85% relative humidity for 1 to 2 months, until it has lost about 35% of its green weight.

NOTE: These instructions are for larger-format smoking chambers; smaller chambers may require less smoking time. See Chapter 3 for more information about smoking meat.

CAZÓN EN ADOBO

Some of my Madrileño chef friends have a saying that compares their cuisine with the cooking found in the rest of Spain. It's sort of a backhanded compliment, but it also succinctly describes each region's historical penchant for certain cooking methods, while (not surprisingly) pointing out that Madrileños consider their own food to be far superior. The saying goes a little something like this: "In the north, they grill; in the south, they fry; and here, we cook."

Of course, to any Spaniard residing outside of Madrid, these words mean war—even though they are historically accurate in terms of the regionally favored cooking methodologies preferred in the north and south. In the grill-happy north, for example, you'll find one of the finest grilling restaurants and tastiest *chuletones* anywhere on the planet at Asador Etxebarri in Axpe. And in the south, you'll find some of the best fried fish anywhere in the humble *chiringuitos*—with this *Cazón en Adobo* recipe being perhaps one of the best-known preparations of all.

Cazón en Adobo, another Andalucían dish owing its roots to the Moorish conquest, uses a highly seasoned *pimentón* and vinegar-spiked *adobo* very similar to the one used for *Pincho Moruno* (see recipe on p. 187). The fish pieces are then breaded and fried, making for an absolutely delicious and popular *tapa* that you'll find everywhere in Andalucían coastal towns where *cazón* is fished. Since *cazón* is a type of shark, you may not be able to find it where you live, but any firm-fleshed white fish will do. Try anything from monkfish to swordfish with this assertive marinade.

INGREDIENTS

per 2.2 pounds (1 kg) of *cazón*

FOR THE *ADOBO:*

1   tablespoon (5 g) whole coriander seed

Peel of 1 orange, removed in 1 piece

Peel of 1 lemon, removed in 1 piece

1   recipe Basic *Adobo* (see recipe on p. 177)

¼   cup (50 mL) sherry vinegar

¼   cup (50 mL) extra virgin olive oil

FOR DREDGING THE FISH:

Cazón or other firm, meaty white fish

½   cup (50 g) chickpea flour (see Notes)

2½ tablespoons (20 g) kosher salt

Olive oil, for frying

FOR THE FINISHED DISH:

Kosher salt, to taste

2   whole lemons, for garnish

¼   cup (50 mL) *Alioli* (see recipe on p. 413), for serving

1   In a small, dry skillet over medium heat, toast the coriander for 2 minutes, until fragrant. Using a mortar and pestle or spice mill, grind the toasted coriander into a fine powder.

2   In a medium baking dish, combine the ground coriander, citrus peels, Basic *Adobo,* sherry vinegar, and oil. Stir and set aside.

3   Clean the fish thoroughly, removing all bloodlines, pinbones, and skin. Cut the fish into 1-inch (2.5-cm) cubes.

4   Toss the fish in the *adobo* and chill in the refrigerator for at least 4 hours or overnight.

5   In a deep saucepan over medium–high heat, warm the frying oil to 375°F (190°C). Line a baking sheet with paper towels. Remove the fish from the baking dish and discard the *adobo.*

6   In a medium shallow bowl, combine the flour and the salt. Toss the fish cubes in the flour. Shake off any excess flour from the fish pieces and set them on a plate.

7   In the saucepan, fry the fish in batches for 2 to 3 minutes, until golden brown and cooked through. Place the fish on the lined baking sheet and cover with foil. Repeat the process with the remaining fish. Season the fish with the salt to taste.

8   Transfer the *Cazón en Adobo* cubes to a serving platter. Garnish with the lemon and serve with the *Alioli.*

NOTES: Andalucían fried fish is traditionally coated in chickpea flour (you can find it at Whole Foods or at an Indian grocery under the name *besan*). If you can't find any, substitute a half-and-half blend of all-purpose flour and durum semolina flour, the type used to make fresh pasta.

To spice up your *cazón,* add 1 tablespoon *pimentón picante* to the *adobo.*

PINCHO MORUNO

The city of Logroño is—to food nerds in the know—one of the best-kept secrets of the Spanish culinary world thanks to a winding little street that's a foodie haven. Calle de Laurel is a street in the *casco viejo* that is wholly devoted to bars and restaurants. Even more interesting, however, is the fact that each restaurant on the street does one thing exceptionally well and does only that one thing all day, every day, forever-and-ever, amen. There's Bar Soriano, famous for its garlic-slathered mushroom *pintxos,* and a constant line around the corner; Bar Blanco y Negro, famous for its anchovy and green pepper *pintxos;* and then there's a secret, hard-to-find little slice of meat-eaters' heaven called Bar Paganos.

You don't really need to ask directions to Bar Paganos. Just follow your nose to the unofficial home of all things charcoal-grilled in this part of the world. The meat here—typically pork *secreto* or *pluma,* or lamb shoulder—gets a bath in a Moorish-inspired marinade called *moruno* before being skewered and grilled over hardwood. The relationship to Moorish kebabs is pretty obvious, as is the *moruno* marinade's roots in the *ras el hanout* spice mixtures so common to Moroccan cuisine.

The marinade itself is killer on any traditional grilling meats, but I'm a huge fan of using it with *foie* and offal like beef hearts or chicken livers. Just make sure to plan ahead if using bamboo or wooden skewers: They need to soak in water for 20 minutes before using, or they'll catch fire.

1   In a large nonreactive bowl, combine all the ingredients except the meat. Stir.

2   Add the meat to the bowl and coat everything in the marinade. (If you choose to use a selection of different meats, divide the marinade into separate bowls for each type of meat.) Cover the bowl and refrigerate for a minimum of 4 hours or overnight.

3   Light a charcoal grill and let the coals burn until they become white, or heat a broiler or grill pan on high heat (and prepare your smoke detectors for a workout).

4   Skewer the meats on metal or presoaked bamboo or wooden skewers (if the meat is cut into strips, thread the pieces on the skewers). Season the meats with the kosher salt and grill or broil them to your desired doneness.

5   Serve hot. (Bar Paganos serves these skewers simply on a plate, with a warning that they are hot and will burn your mouth...so eat with caution.)

INGREDIENTS

per 2.2 pounds (1 kg) of meat, such as *panceta* (pork belly); pork *pluma;* pork *secreto; solomillo* (pork tenderloin); *paleta* (lamb shoulder); boneless skinless chicken thighs; chicken hearts, beef hearts, sweetbreads, or other offal; and/or *Foie Salada en Torchon* (see recipe on p. 375), cut into strips or bite-sized pieces

1   recipe Basic *Adobo* (see recipe on p. 177)

1   ounce (25 g) *Moruno* Spice (recipe follows)

1   medium yellow onion, julienned

¼   cup (50 mL) extra virgin olive oil

2   tablespoons (25 mL) sherry vinegar

Kosher salt, to taste

MORUNO SPICE

YIELD

1¼ cups (100 g) spice mix

1   In a small, dry skillet, toast the cumin, coriander, and fennel seeds; peppercorns; and saffron threads for 2 to 3 minutes, until fragrant and toasted.

2   In a mortar and pestle or spice mill, grind the toasted spices into a fine powder. Transfer the ground toasted spices to a small mixing bowl.

3   Add the remaining spices to the bowl and whisk to combine (see Note).

4   Place the spice mixture in an airtight container and store in a cool, dark place for up to 6 months.

4	tablespoons (24 g) cumin seeds
2	tablespoons (12 g) coriander seeds
2	tablespoons (12 g) fennel seeds
2	tablespoons (10 g) black peppercorns
	Pinch (1 g) saffron threads
2	tablespoons (6 g) dried oregano
2	tablespoons (6 g) dried thyme
2	tablespoons (12 g) *pimentón dulce* (see Note)
2	tablespoons (12 g) freshly ground black pepper
1	tablespoon (6 g) ground turmeric

NOTE: To spice up your *moruno,* add 1 tablespoon *pimentón picante* to the mix.

CHAPTER

6

ESCABECHE

⌇

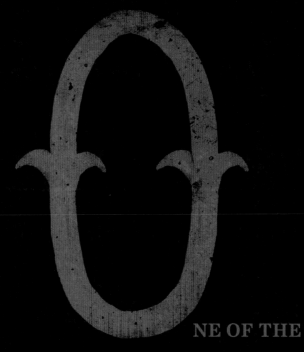

NE OF THE

Moors' primary contributions to Mediterranean cuisine was the introduction of *Escabeche*. Derived from an Arabic sweet-sour dish called *al-sikbaj*—a meat-based dish that has a sweet component from a syrup, like date molasses or honey, and a sour component from vinegar—*Escabeche* is an old way of preserving meats in a vinegary marinade that is made in kitchens across the Mediterranean, Latin America, the Caribbean, and even the Pacific islands.

BASIC *ESCABECHE*

To make an *escabeche,* you start with the star of the show: A main ingredient that's typically meat or poultry, as in *Perdiz en Escabeche* (see recipe on p. 197). But that said, *escabeches* can also be made from fish (any oily, stinky seafood like tuna or mackerel is awesome *en escabeche),* vegetables (heartier veggies like eggplant are a good choice), or fruits. One of my favorite dishes from my time cooking in Adolfo Muñoz's kitchen is *ijada* of tuna. It's a super fatty cut from the belly of the fish that we served with a sauce made from mango *escabeche.*

The next step is to cook the main ingredient in some fashion. In Spain, that typically means giving proteins a quick sear in lots of olive oil to get some color. In that same pan, you then caramelize some aromatic veggies before adding the *escabeche* trifecta of stock, wine, and vinegar, along with some spices and herbs. Simmer it all until done, and then either eat immediately or chill and store for a rainy day.

Since the sauce is somewhat acidic and the olive oil floats to the top, providing a nice sealant against oxygen intrusion once chilled, *escabeche* was one of the preservation methods of choice for proteins in the age before modern refrigeration.

1   In a large sauté pan over medium–high heat, warm ¼ cup of the oil until just smoking.

2   If searing a protein or vegetable to serve with the *escabeche,* season it and place it in the hot oil—skin-side down, if applicable—for 4 to 5 minutes, until golden brown but not even close to cooked through. Remove from the pan and set aside.

3   Add the remaining olive oil to the pan. When the oil is almost smoking, add the carrots, leeks, and garlic to the pan. Sear the vegetables for 10 minutes, until they start to turn golden brown. (The veggies are not salted at this point, since salt would draw out their water and steam them instead of having them take on a nice brown color.)

4   Deglaze the pan with the vinegar, wine, and stock. Add some of the salt, herbs, bay leaves, and peppercorns to the pan. Bring the *escabeche* to a boil. Reduce the heat to medium and allow it to reach a simmer.

5   If using a protein or vegetable to serve with the *escabeche,* place it in the pan and poach until it reaches the desired level of doneness in the simmering liquid (see recipes on pp. 195, 197, and 200 to get an idea of cooking times for different types of protein). Remove from the heat.

6   Remove the protein or vegetable from the pan and set aside, covered with foil, until the *escabeche* reaches room temperature. Taste and season the *escabeche* with the salt.

7   Return the protein or vegetable to the cooled *escabeche* and chill in the refrigerator until ready to use. Whether or not you made the *escabeche* with a protein, it makes a great salad dressing, pickling liquid, or sauce.

YIELD

1 quart (1 L)

INGREDIENTS

per 2.2 pounds (1 kg) of meat, fish, vegetables, or fruit

2   cups (500 mL) extra virgin olive oil, divided

2   medium carrots, peeled and cut into small dice

1   medium leek, cleaned and cut into small dice

10   whole cloves garlic, peeled and destemmed

1   cup (250 mL) sherry vinegar

1   cup (250 mL) white wine

1   cup (250 mL) stock or water (if making fish, use a fish stock; if making poultry, use a poultry stock; and so on)

Kosher salt, to taste

5   sprigs fresh thyme

1   sprig fresh rosemary

2   fresh bay leaves

2   tablespoons (10 g) black peppercorns

SETAS EN ESCABECHE

Jean Luc Figueras is a living chef-legend in Barcelona. He ran his eponymous Michelin-starred restaurant for years and garnered all sorts of praise. Perhaps more importantly, however, he is rightfully known as a generous mentor and a solid guy. I was very fortunate when, during my time at the *matanzas* in Extremadura, he showed up with a cabal of young chefs.

While Jean Luc made a killer impromptu *Pâté de Cerdo* (see recipe on p. 363) for lunch, we struck up a friendship over conversation about some pork intestines and how Jean Luc likes to call them *"un condom prehistorico"*—sausages and dick jokes being an international language of culinary kinship.

Years later, Jean Luc asked me to be his sous chef at a dinner he cooked at the Sun Valley Wine Auction—a yearly summer event that draws some great chefs to cook wine-inspired dinners in the beautiful Sun Valley area of Idaho. We didn't know it until we arrived but we showed up in the middle of morel mushroom season—so, like any great chef, Jean Luc saw nature's mushroom-bounty and made a decision on the spot: We would make a *tapa* of *Setas en Escabeche*—a very delicious mixture of pickled wild mushrooms.

You will notice some similarities between this *escabeche* recipe and the basic one used for proteins—the only difference is that stock isn't required here. If you have some porcini stock kicking around, feel free to add it, but otherwise the olive oil, wine, and vinegar will do their magic.

You will also notice that I give some pretty specific directions on how to sear the 'shrooms—this technique comes from the *maestros* of Bar Soriano, a famous mushroom *tapas* bar in the town of Logroño. These *maestros* cook mushrooms every day of their lives, so their technique results in a perfect mushroom every time.

The trick involves first searing the 'shrooms unsalted and stem-side down/cap-side up. Then, once you get the first side nice and brown, you flip the 'shroom and sear stem-side up/cap-side down. Only then do you add salt—this allows the mushroom to steam itself lightly, and fills the little caps with delicious mushroom liquid for the *escabeche*.

Of course, you can just as easily make this recipe with any other kind of edible, non-cap mushroom—I particularly like using shiitakes and oyster mushrooms, or morels if they are as abundant as in Sun Valley. Whatever you use, just make sure you get a good caramelization when you sear the 'shrooms.

YIELD

4–6 servings

INGREDIENTS

per 2.2 pounds (1 kg) of cap-style mushrooms such as crimini or button, cleaned, trimmed, and stemmed

2	cups (500 mL) extra virgin olive oil, divided
	Kosher salt, to taste
2	medium carrots, peeled and thin cut on the bias
1	medium leek, tender white parts only, cleaned and thin cut on the bias
10	whole cloves garlic, peeled and destemmed
1	cup (250 mL) sherry vinegar
1	cup (250 mL) white wine, such as a Verdejo
1	cup (250 mL) water
5	sprigs fresh thyme
1	sprig fresh rosemary
2	fresh bay leaves
2	tablespoons (10 g) black peppercorns

CONTINUED ON NEXT PAGE

SETAS EN ESCABECHE CONTINUED FROM PREVIOUS PAGE

1   In a large sauté pan over medium–high heat, warm ¼ cup (60 mL) of the oil until just smoking.

2   Sear the mushrooms in one layer, cap-side up, in the pan, without crowding them (repeat the process if you have more than one batch). Once the first side is browned, flip the 'shrooms and continue searing. Season with the salt and continue cooking until the mushrooms have released some of their water into their stem holes. Remove from the pan, place in a large, shallow, nonreactive serving bowl and set the mushrooms aside.

3   Add the remaining olive oil to the pan. When the oil is almost smoking, add the carrots, leek, and garlic to the pan. Sear the vegetables for 10 minutes, until golden brown. (The veggies are not salted at this point, since salt would draw out their water and steam them instead of having them take on a nice brown color.)

4   Deglaze the pan with the sherry vinegar, wine, and water. Then add some of the salt, herbs, bay leaves, and peppercorns to the pan. Bring the *escabeche* to a boil. Reduce the heat to medium and allow it to reach a simmer. Simmer the *escabeche* for 10 minutes.

5   Return the mushrooms to the pan and remove the *escabeche* from the heat. Taste and reseason the *escabeche* with the salt.

6   Store the mushrooms in the cooled *escabeche* in the refrigerator for at least 24 hours. Serve as a killer *tapa,* or either warm or cold as a first-course salad. Also, the *escabeche* liquid makes a great dressing.

PERDIZ EN ESCABECHE

A few years back, King Juan Carlos declared Adolfo Muñoz's *perdiz* (partridge) as the best rendition in all of Spain, and ever since, *Perdiz en Escabeche* has been one of Adolfo's signature dishes. I made this *perdiz* more times than I can count while cooking for Adolfo at his restaurant, Restaurante Adolfo, in Toledo—so yes, I am admittedly and whole-heartedly biased in favor of Adolfo, his son Javier, and the rest of my *familia manchega.* You simply will not find a more gracious, hospitable family anywhere in Spain, and I count myself truly lucky to know them.

For me, the trick to Adolfo's *perdiz* lies not just in cooking the partridge right and getting the *escabeche* proportions spot on, but rather the depth you can get from a rich and perfectly made partridge stock.

Also, what's cool about this dish is that—once everything is cooked and then chilled down—you can hold it for a while. The *escabeche* technique is a means of preservation, but there's a bonus, too: The flavor of the dish only gets better with a little aging and maturation.

1   On a large cutting board, break each partridge down as you would a chicken, removing the leg and thigh segments first, followed by the breast and wing segments. Reserve the meat for the *escabeche* and the carcasses for the partridge stock.

2   Preheat the oven to 400°F (204°C).

3   Place the carcasses, onions, carrots, celery, and garlic in a roasting pan and drizzle with the oil. Roast for 40 minutes, until all are deep brown. Remove from the oven.

4   Place the carcasses and vegetables in a large stockpot. Pour the wine into the hot roasting pan and scrape the fond with a wooden spoon. Pour this residue into the stockpot as well.

5   Barely cover the carcasses and vegetables with cold water. Place the stockpot over medium–high heat and bring to a light boil. Skim all foam and impurities from the water's surface. Reduce the heat to medium and simmer the stock for 3 hours.

6   Add the thyme, bay leaves, and peppercorns and simmer for 30 minutes. Remove the stock from the heat.

YIELD

4–6 servings

INGREDIENTS

per 2.2 pounds (1 kg) of *perdiz* (partridges), pin feathers removed, gutted and rinsed

2   medium yellow onions, peeled and cut into medium dice

4   medium carrots, peeled and cut into medium dice

4   medium stalks celery, cut into medium dice

1   whole head garlic, cut in half

½   cup (125 mL) extra virgin olive oil

1   cup (250 mL) white wine or water

Water, to cover

5   sprigs fresh thyme

2   fresh bay leaves

2   tablespoons (10 g) black peppercorns

Kosher salt, to taste

1   quart (1 L) Basic *Escabeche* (see recipe on p. 193), made with partridge stock

CONTINUED ON NEXT PAGE

PERDIZ EN ESCABECHE

CONTINUED FROM PREVIOUS PAGE

7   Strain the stock at least 3 times through a strainer into another pot, finishing with a final strain using a fine *chinois*.* Discard the solids.

8   Place the pot containing the stock over medium–high heat. Return to a boil, reduce the heat to medium, and simmer for 15 to 20 minutes, until the stock's volume is reduced by at least ½ (reduce it further if you want a more concentrated *perdiz* flavor in your *escabeche)*. Remove from the heat. Taste and season the stock with the salt. Transfer to a pourable container and set aside to cool to room temperature.

9   Prepare the Basic *Escabeche* recipe through Step 4, using the partridge segments and partridge stock.

10 To finish cooking the partridges, first add the leg and thigh segments to the simmering liquid. Simmer for 10 minutes. Add the breast and wing segments. Continue simmering for 10 minutes, until the meat is just barely cooked through. Remove the meat from the pan and place in a deep serving dish covered with foil to cool. Allow the *escabeche* to cool to room temperature. Taste and reseason with the salt.

11 Store in the cooled *escabeche* in the refrigerator for at least 1 day. Serve the *perdiz* in either a salad, as a first course, or as a warmed main course.

* If you do not have a *chinois,* consider straining the stock through muslin, cheesecloth, or another tightly woven fabric.

NOTES: Adolfo typically serves the *perdiz* with some aromatic herbs and simple sautéed vegetables.

If your partridges still have their hearts and/or livers, Adolfo would be proud of you. Reserve them for the *escabeche.* Sear them with the partridge segments until just cooked through, then add them back in to marinate overnight with the meats.

MEJILLONES EN ESCABECHE

Canned seafood is a rock star in Spain. I know this seems weirdly incongruous to us Americans, mainly because our idea of canned seafood is pretty much restricted to the tuna aisle, and we damn near flee from those scary cans of mussels, clams, and anchovies exiled in the grocery store's corner nether regions.

But walk into Barcelona's Quimet i Quimet or Taverna Ca L'Espinaler and you will see canned seafood elevated to a culinary art form. The former is a bar and *tapas* joint that only "cooks" with stuff from a can, while the latter serves its own proprietary canned seafood that can cost upward of $100 apiece!

What I'm getting at here is that you should forget everything you know about canned seafood. In Spain, the canned varieties are as coveted and cherished as the fresh stuff. This is one of my favorite recipes for canned mussels, a really popular *tapa* in Barcelona: mussels pickled in a spicy, vinegary liquid full of garlic, citrus, and *pimentón*.

One caveat: In Spain, the mussels go right from the boat to the canning room's boilers, so whether you intend to store these beauties or keep them on hand for snacking, make sure to use the freshest mussels you can find.

1   In a medium stockpot over medium–high heat, bring the water to a boil. Place a colander inside a large bowl. Set aside.

2   Place a sieve containing ⅓ of the mussels in the boiling water for 2 to 5 minutes, until they open. Once they open, transfer them to the colander. Repeat the process for the remaining mussels. Remove from the heat. Allow to cool to room temperature.

3   Remove the mussels from the shells. Discard the shells (but see Note) and place the mussels in a covered bowl. Refrigerate the mussels until ready to use.

4   Using cheesecloth, strain the mussel liquor into a saucepan. Place the saucepan over medium–high heat and bring to a boil. Reduce the heat to medium. Simmer for 4 to 6 minutes, until the liquor reduces in volume to about ¼ cup (60 mL).

YIELD

4–6 servings

INGREDIENTS

per 2.2 pounds (1 kg) of Prince Edward Island mussels, debearded and scrubbed

1½   quarts (1.5 L) water

1   cup (250 mL) extra virgin olive oil

5   cloves garlic, peeled, destemmed, and lightly crushed

Zest and juice of 1 lemon

Zest and juice of 1 orange

Pinch red pepper flakes

2   sprigs fresh thyme

2   sprigs fresh rosemary

2   fresh bay leaves

½   cup (125 mL) sherry vinegar

1   tablespoon (5 g) *pimentón dulce*

1   tablespoon (5 g) *pimentón picante*

½   tablespoon (3 g) black peppercorns

Kosher salt, to taste

5   In a large saucepan over medium heat, warm the oil. Add the garlic, lemon zest, orange zest, red pepper
flakes, thyme, rosemary, and bay leaves and sauté for 2 minutes, until it begins to smell fragrant. Remove
from the heat.

6   Add the orange and lemon juices, vinegar, *pimentones,* and peppercorns to the saucepan. Add the reserved
mussel liquor and mix well. Allow the *escabeche* to cool to room temperature. Taste and season with the
salt.

7   Transfer the *escabeche* to a large, deep serving dish. Add the mussels and refrigerate for at least 12 hours.
Serve in a salad or as a small first-course dish.

NOTE: The shells make a great serving vessel for the mussels. Save them if you are serving a crowd.

CHAPTER

7

CONSERVAS Y CONFITS

BOTH OF THESE
methods cover a broad range of preservation techniques
where food eventually winds up getting canned or jarred.
A *conserva* is a technique used for canning and processing
a lot of different foods—anything from the famous canned
seafood that comes from Spain's coastal regions to preserved
vegetables like white asparagus that come from the vegetable-
rich region around inland Murcia.

The term *confit,* on the other hand, technically refers to proteins that are cured, cooked, and then aged under a layer of their own fat. The fat, which hardens, protects the protein from rancidity, since oxygen can't permeate the solid fat barrier. A great example of this is duck *confit,* a technique that crossed the border from Gascony in southwestern France and is now popular in much of País Vasco. At this point in modern gastronomy, however, the Spaniards (and, to be fair, most of the rest of the culinary world) have stretched the meaning of *confit* to include anything cooked under any layer of fat, so you'll often see olive oil–*confited* vegetables as a somewhat modern interpretation of this ancient preservation technique.

ATÚN EN ACEITE

Fish from the tuna family conserved in olive oil is a major export of Spain—and one of my favorite snacks. Typically, the fish is line-caught around the Bay of Biscay or Cantabria in the summertime and then quickly brought back for processing. Right on the docks, it's filleted, poached in a seawater *cocción* (poaching liquid), and canned in olive oil. Any oily, fishy relative of tuna works well with this technique, but *bonito* (aka white or albacore tuna) is the most commonly found fish. Skipjack or yellowfin (ahi) are also great for this recipe.

If you are making this at home, you should definitely try to make more than you think you need and keep it on hand. Oil-packed tuna is a major ingredient in *Ensaladilla Rusa* (see recipe on p. 216), which happens to be one of the best *tapas* of all time.

1 Sterilize some Mason jars (see p. 130 for details) if you plan to store the *atún* for longer than 1 month. Otherwise, a small to medium-sized storage container will do.

2 Place the *atún* in a container that just holds it and cover it with the water. Reserve the water it took to cover the *atún,* as you will use it to make the *salmuera.*

3 Make the Basic *Salmuera* using the reserved water.

4 Place the fish in the stockpot. Warm it over medium heat until it is just below a simmer. Reduce the heat to low.

5 Continue poaching the fish until it is just cooked through and pulling away from the bones, about 25 minutes. Drain the fish and reserve.

6 Break the fish into large chunks and pack into the container or jar. If you used bone-in *atún,* discard any bones. Add the garlic, bay leaves, and thyme to the container and fill with the oil.

7 Close the container and hold in the refrigerator for up to 1 month to age the *atún* in the oil. (If holding for longer, use the Mason jars and follow the complete canning process.)

NOTE: For more information on canning, see page 130.

YIELD

6–8 *tapa* servings

INGREDIENTS

per 2.2 pounds (1 kg) bone-in tuna or *bonito* steaks or loins, bloodline removed

Water, to cover

Basic *Salmuera* (see recipe on p. 105)

5   cloves garlic, peeled and crushed

5   fresh bay leaves

10   sprigs fresh thyme

High-quality extra virgin olive oil (preferably a Spanish varietal, like Arbequina or Cornicabra), to cover

QUESO EN ACEITE

At most high-end restaurants in Spain, a gigantic cheese cart gets wheeled around at the end of the meal. You point out a few stinky delicacies, someone cuts them up and arranges them on a plate, and voila: You're transported into the middle of a Frenchman's wet dream.

Well, you should know that before these cheeses get served, someone has to baby them a little. The rinds need to be washed, any dried ends need to be cut off, and the cheeses generally need to be tended to and made to look as delicious as possible.

During my time cooking in Andalucía, that role fell to our Cheese Cart Girls. Now, it goes without saying that the women of Andalucía are gorgeous, and these Ladies of Lactose were especially hot, given their exotic looks and Demi Moore-esque Spanish baritones. The bane of the Girls' perfectly manicured existence, however, was prepping their cart of slowly decaying milk products, since hot girls, no matter where they reside in the world, don't really like to tend to stinky things.

So when it came time to clean up their cheese cart, the Girls would bat their eyelashes at us, their knights in blue aprons.

Cooks are naturally inclined to befriend any female in the vicinity of our kitchen, so these damsels in dairy-induced distress really didn't have to work too hard to gain our interest and service. And, in return for our eager assistance, we were rewarded with the gratitude of some cute girls and a few awesome samples from their cheese cart.

I particularly fell in love with a local goat's milk cheese that was cured in olive oil. It's very awesome, very sexy, and very simple—not unlike those Andalucían Cheese Cart Girls.

1   Sterilize some Mason jars (see p. 130 for details) if you plan to store the cheese for longer than 1 month. Otherwise, a small storage container will do.

2   If using a semifirm cheese, like Manchego, cut it into bite-sized pieces. If using a soft cheese, like goat cheese, form it into discs or balls that will fit in the container.

3   Place the cheese pieces, herbs, and spices into the storage container. Cover with the oil.

4   Close the container and hold in the refrigerator for up to 1 month to age the cheese in the oil. (If holding for longer, use the Mason jars and follow the complete canning process described on p. 130.)

YIELD

6–8 *tapa* servings

1.1	pounds (500 g) fresh, semisoft, or semifirm goat's or sheep's milk cheese, such as Manchego or Queso de Cabra
5	sprigs fresh thyme
3	sprigs fresh rosemary
2	sprigs fresh marjoram
2	whole cloves garlic, peeled and destemmed
10	black peppercorns
	High-quality extra virgin olive oil (preferably a Spanish varietal, like Arbequina or Cornicabra), to cover

NOTE: I'm a fan of eating this right out of the jar with some good bread and a glass of wine. Why make something so delicious difficult?

CONSERVA DE CANGREJO

By now, everyone and their grandmother in the food world knows that País Vasco in general and San Sebastián in particular offer some of the best food in the world, most notably in the small *pintxo* bars that line the streets of the *casco viejo*.

One of the staple dishes found in these bars is *txangurro,* the Basque name for a species of spider crab as well as the famous preparation from it. For *txangurro,* the meat of the crab is removed, mixed with a tomato- and onion-rich *sofrito* laced with brandy and local wine, and then stuffed back into the shell and broiled.

You will find *txangurro* in many guises throughout the area, which is how I came across this amazing potted crab dish. The crab spread is great with warm, grilled bread slices—just make sure your clarified butter layer is as thin as possible and seals the entire top of the crab meat. This sealant will keep the *txangurro* fresh for several days.

1 In a stockpot over medium–high heat, bring the water and salt to a boil. Fill a large mixing bowl with ice and water. Set aside.

2 Add the crabs to the boiling water and cover the stockpot. Once the water returns to a boil, start the clock on the cooking time. For 2 pounds (900 g) of crabs, cook for 15 minutes; for 3 pounds (1.4 kg) of crabs, cook for 20 minutes. Add 5 minutes of cooking time per extra pound (450 g) of crabs. Remove from the heat.

3 Remove the crabs from the stockpot and plunge them into the ice bath to stop them from cooking further.

4 When cool enough to handle, remove all the meat from the crabs, working over a bowl to collect any juices that spill out. Reserve both the crab meat and juice (strained, if necessary) in a foodsafe container and set aside.

5 In a medium skillet over medium–high heat, bring the Basic *Sofrito* to a simmer. Add the brandy and white wine (you should get some flambé action) and reduce the heat to medium.

YIELD

6–8 *tapa* servings

INGREDIENTS

per 2.2 pounds (1 kg) fresh crabs, such as Maryland blue or Dungeness

1   gallon (3.8 L) water

2   ounces (56 g) kosher salt, plus more as needed

½   cup (125 g) Basic *Sofrito* (see recipe on p. 418)

¼   cup (50 mL) brandy

¼   cup (50 mL) white wine, preferably an acidic Basque white wine like Txakoli

7   ounces (200 g) unsalted butter, cold and cubed

     Zest and juice of 1 lemon

¼   cup (15 g) minced fresh flat-leaf parsley

2   tablespoons (7 g) minced fresh chervil

2   tablespoons (7 g) minced fresh tarragon

2   tablespoons (7 g) minced fresh chives

     Freshly ground black pepper, to taste

3½ ounces (100 g) clarified unsalted butter, melted and warm

6 Simmer the *sofrito* for 8 to 10 minutes, until most of the liquid has evaporated. Remove from the heat.

7 Fold the butter cubes into the *sofrito* until the sauce is creamy and emulsified. Add the crab meat, lemon zest, parsley, chervil, tarragon, and chives into the *sofrito*. Taste and season the *txangurro* with the salt, freshly ground black pepper, and lemon juice. Set aside and allow to cool completely.

8 Spoon the *txangurro* into the container, smoothing its surface with the back of the spoon. Chill the mixture for 30 minutes, until it has started to set up.

9 Pour the clarified butter over the surface. Close the container and hold in the refrigerator for up to 1 week.

TRUCOS DE LA COCINA: SALTED WATER "JOSÉ'S WAY"

EVERYONE KNOWS YOU should boil veggies, pasta, and potatoes in salted water. It makes the food taste better by preseasoning whatever it is that you are cooking. It also makes greens brighter and keeps nutrients from leaching out of the veggies.

While the ideal ratio of salt to water depends on your taste, most people agree that the water should be "salty like the ocean"—technically around 4 percent salinity, but that's a problematic comparison since ocean water tends to be saltier in some areas than others.

For example: One day I was getting a lesson on cooking *cigalas,* a type of langoustine, from Chef José Andrés. I was instructed to make the water salty "like my local ocean." I thought I did as instructed when I emptied a good sized-box of salt into the water...until Jóse came over to taste.

"¡Amigo! I do not know what ocean *you* swim in, but this does not taste like *my* ocean!" he said in a booming voice and flashed his trademark wry smile.

With that, José dumped easily five times more salt into the water. Apparently, the oceans of Spain are saltier than those of California. It's a fair point and a lesson well learned for me, since Jose's *cigalas* were delicious and perfectly seasoned right out of the pot.

HABAS EN CONSERVA

Habas is a general term that can mean anything from fresh fava to lima to butter beans. On their own, these beans aren't exactly the sexiest foods on the planet, but pair them with some *jamón* and mint—as in *Habas con Jamón* (see recipe on p. 219)—and you've got yourself a dish that distills the very essence of summertime in Navarra.

I'm writing this recipe with fresh favas in mind since that's what we used in Navarra. Try to get them as soon as they become available (usually in late spring or early summer) when the secondary skin on the bean is so thin that it doesn't always need to be peeled. You can substitute butter or lima beans if you want, but just know that the cooking time will vary to get those beans tender enough.

Whatever bean you choose—if you get your hands on more than you can handle, this conservation method is the best way to enjoy a taste of summer throughout the year.

YIELD

1 cup (250 g)

INGREDIENTS

**per ½ pound (250 g)
fresh *habas* (fava beans)**

1   quart (1 L) water

2   ounces (50 g) kosher salt,
    plus more as needed

    Fresh mixed herbs (mint, bay leaf,
    thyme, marjoram), as needed

    High-quality extra virgin olive oil
    (preferably a Spanish varietal, like
    Arbequina or Cornicabra), to cover

1   Sterilize some Mason jars (see p. 130 for details) if you plan to store the *habas* for longer than 1 month. Otherwise, a small- to medium-sized storage container will do.

2   In a stockpot over medium–high heat, bring the water and salt to a boil. Fill a large mixing bowl with ice and water. Set aside.

3   Add the *habas* to the boiling water and cook for 1 to 5 minutes, depending on the size and tenderness of the beans. Be patient and keep tasting the beans. They will be done when they are softly cooked but retain their structure (especially if they will be cooked again later). Remove from the heat.

4   Remove the *habas* from the stockpot and plunge them into the ice bath to stop them from cooking further.

5   If the *habas* are older and need to be skinned, wait until the beans are cool enough to handle, and then slip them from their skins. Discard the skins. (If the beans are young and tender enough, disregard this step.)

6   Spoon the *habas* into the container. Add any herbs you plan to include (I'm a big fan of using lots of mint). Pour the oil on top. Close the container and hold in the refrigerator for up to 1 month to age the *habas* in the oil. (If holding for longer, use the Mason jars and follow the complete canning process.)

NOTES: For more information on canning, see page 130.

I love cooking these up with a little ham and mint, like in the *Habas con Jamón* recipe (see p. 219), but they are a great addition to salads or as a veggie side dish to an entrée as well.

COSTILLAS DE LA MATANZA

While cooking in Spain, I noticed culinary philosophical differences between Americans and Spaniards when it comes to grilling. While Americans slather their smoked ribs in barbecue sauce, the Spanish like to let the flavor of their meat do the talking, especially when *Ibérico* pigs are the meat in question.

I'm guessing the difference is explained by our history of smoked barbecue meats and the world of sauces and condiments that complement them, contrasted with the Spaniards' heritage of acorny *Ibérico* fat, and letting its subtle flavor come through. The Spanish accent this natural flavor only lightly, with ingredients like herbs and spices.

This recipe, from the *sabias* of Extremadura, demonstrates the Spanish penchant for soft flavors like lemon and cinnamon that are found in everything from *flans* and ice creams to meat and other savory recipes. For those who must have sauce, I'm including a killer glaze that we prepared while the ribs were cooking. The acid in the sauce helps cut the richness of the *confited* meat.

1   In a large baking dish, combine the salt, black pepper, sugar, crumbled bay leaves, and ground cinnamon, creating a cure.

2   Place the ribs in the baking dish and toss them with the cure, coating them evenly. Sprinkle with the oloroso sherry. Cover with plastic wrap and refrigerate for 24 to 36 hours.

3   Preheat the oven to 250°F (120°C).

4   Remove the ribs from the cure. Rinse well and pat dry with paper towels. Place the ribs in a large Dutch oven with the lemon rind, cinnamon sticks, garlic, and whole bay leaves. Cover with the rendered pork fat.

5   Place the Dutch oven over medium heat and bring the fat to a bare simmer. Remove from the heat.

6   Cover and place in the oven for 2 hours, until the ribs are fork tender and the bones pull easily away from the meat. Remove from the oven.

7   Allow the ribs to cool to room temperature in the Dutch oven, and then place in the refrigerator to chill in the *confit* overnight.

YIELD

2 racks ribs

INGREDIENTS

per 2 racks ribs

2   ounces (50 g) kosher salt

2   tablespoons (20 g) freshly ground black pepper

1   tablespoon (12 g) granulated sugar

3   dried bay leaves, crumbled

2   teaspoons (4 g) ground cinnamon

½   cup (100 mL) oloroso sherry

Rinds of 4 lemons, each removed in 1 strip

3   cinnamon sticks

5   cloves garlic, smashed

3   whole fresh bay leaves

Rendered pork fat, to cover

½   cup (100 mL) Pedro Ximénez (PX) sherry*

1   cup (200 mL) Pedro Ximénez (PX) sherry vinegar

* This type of sherry (and vinegar) is made from Pedro Ximénez white grapes. The varietal yields an intensely sweet dessert wine.

CONTINUED ON NEXT PAGE

COSTILLAS DE LA MATANZA CONTINUED FROM PREVIOUS PAGE

8   In a small saucepan over medium–high heat, combine the PX sherry and the PX sherry vinegar. Bring to a boil and then reduce to a simmer. Simmer for 8 to 10 minutes, until the sauce's volume is reduced to ⅓ cup (80 mL). Remove from the heat and set aside.

9   Light a charcoal grill or heat a broiler on high heat (fair warning: The ribs will splatter a lot in an oven, so use caution!). Remove the ribs from the fat, wiping off any excess. Place the ribs, meat-side down, and cook them for 10 minutes, until cooked through and starting to brown.

10 Glaze the ribs with the sauce on the bone side. Flip them over and glaze the other side. Continue cooking for 5 minutes, until nicely glazed and just charred a little in spots. Remove from the heat and serve.

NOTE: We ate these ribs right off the grill during the *matanzas*—finger food at its best! For more civilized affairs, these ribs would make a great main course, especially if you don't have a smoking rig handy for doing American-style ribs.

CONSERVAS Y CONFITS

TRADITIONAL AND MODERN RECIPES

ENSALADILLA RUSA

Just about every country in the world has its own version of potato salad. In the United States, picnics would be remiss without some take on our ubiquitous mixture of potato, mayonnaise, and hard-boiled eggs. And in Spain, you'll likely find that most *tapas* bars of merit always offer some form of the Spanish homage to potato salad: *Ensaladilla Rusa. Ensaladilla* is essentially a mayonnaise-based potato salad with the addition of olive oil–packed *bonito del norte,* hardboiled eggs, and some cooked carrots and peas.

As the name denotes, the salad is linked to the cuisine of Russia—specifically, to the Moscow restaurant Hermitage, where Chef Lucien Olivier served a similar dish that he not-so-humbly called "Olivier Salad" back in the 1860s. That salad was one of the more famous, if not inappropriately garish, salads of the day—think of it as the Paris Hilton of nineteenth-century Russian gastronomy. We're talking about a dish containing layers of potato, caviar, capers, game hen, poached veal tongue, crayfish tails, smoked duck, and other expensive goodies layered together with jellied broth. The salad was huge with the 1-percenters of old Moscow culture until Chef Olivier took the recipe to his grave and the restaurant closed in 1905.

Copycats persisted throughout the years, eventually spreading the recipe across the continent and into Spain. But when times of war and famine came, fancier ingredients were deleted and replaced with humbler fare. Peas tagged in to replace the capers and caviar, regular pickles were substituted for fancy gherkins, cheaper meats and hard-boiled eggs took over for the luxury proteins, and eventually supermarket-quality mayonnaise joined the party to replace Olivier's lavish secret dressing.

But fancified or on the cheap, this is one salad where the sum of the whole is much greater than its inexpensive parts— *Ensaladilla Rusa* is rightfully a *tapas* classic.

YIELD

4–6 *tapa* servings

3   medium russet potatoes, peeled and halved

2   medium carrots, peeled

Water, as needed

Kosher salt, to taste

½   cup (72 g) frozen peas, thawed

6   whole hardboiled eggs, peeled, chilled, and coarsely chopped

15   ounces (425 g) *Atún en Aceite* or *Atún en Escabeche* (see recipes on pp. 205 and 193)

1   cup (225 mL) *Alioli* (see recipe on p. 413)

¼   cup (15 g) roughly chopped fresh flat-leaf parsley

¼   cup (15 g) minced fresh chives

Zest of 1 lemon

4   ounces (115 g) green olives, pitted and chopped

Freshly ground black pepper, to taste

1   tablespoon (6 g) freshly grated *Mojama* (see recipe on p. 127), for sprinkling

1   In a medium stockpot over medium–high heat, place the potatoes and carrots. Add just enough water to cover. Remove the carrots to a bowl and set aside. Season the water with the salt (the water should taste like the ocean), and bring to a boil.

2   Once the water reaches a boil, reduce the heat to medium. Simmer the potatoes for 15 minutes. Add the carrots to the stockpot and cook for 15 minutes, until both the potatoes and carrots are soft. With a slotted spoon, remove the veggies from the heat, leaving the water boiling for the moment.

3   Cut both the potatoes and carrots into a medium dice and place in a large mixing bowl. Set aside.

4   Fill a large mixing bowl with enough cold water and ice to completely cover the peas. Set aside.

5   In the stockpot with the boiling water, place the peas. Cook for 1 minute, until just heated through. Remove from the heat.

6   Remove the peas from the boiling water and immediately plunge them into the ice bath. (Discard the water.) Chill for 5 to 10 minutes, until cold. Once chilled, drain the peas, place them in the bowl containing the potatoes and carrots, and set aside.

7   Add the eggs to the bowl containing the potatoes, carrots, and peas. Break the *Atún en Aceite* or *Atún en Escabeche* into large chunks and add it to the bowl.

9   Fold the *Alioli* into the tuna–vegetable mixture, taking care not to break up or mash the ingredients too much. Fold in the parsley, chives, lemon zest, and olives.

10 Taste and season the salad with the salt and black pepper. Sprinkle with the *Mojama* and serve.

NOTES: I like to top my *Ensaladilla Rusa* with shavings of *Mojama,* the cured tuna loin of Andalucía.

For extra deliciousness, use some of the oil the *Atún en Aceite* was preserved in to make the *Alioli.*

HABAS CON JAMÓN

I can think of few better ways to enjoy *habas,* either fresh or preserved, than in this deceptively simple example of *sabía* fast food at its finest. Preserved and fresh foods share the same plate, time and time again, in the classic Spanish kitchen. It's a sort of formulaic harmony in old-school Spanish cooking that puts to use whatever's on hand—generally some kind of cured meat as a flavoring and salting agent paired with whatever is available in the garden—to make delicious and nutritious dishes.

Specifically, these recipes contain one or two preserved items (in this case, the *jamón),* a deep *sofrito* as a sauce, and something überfresh, like mint, parsley, or lemon zest, to elevate the dish. Just a few minutes of cooking, and you've got the sort of simple lunchtime fare that beats anything made by kings, colonels, or clowns.

YIELD

4–6 entrée servings

1	cup (225 g) fresh *habas* (fava beans) or *Habas en Conserva* (see recipe on p. 211)
3	tablespoons (45 mL) extra virgin olive oil
½	cup (125 g) Basic *Sofrito* (see recipe on p. 418)
4	ounces (114 g) diced *Jamón* (see recipe on p. 143)
1	spring onion, scallion, or white part of leek, cleaned of all dirt and sliced thinly
5	whole stems fresh mint, with leaves on
	Kosher salt, to taste
	Freshly ground black pepper, to taste

IF USING FRESH FAVA BEANS:

1   In a stockpot over medium–high heat, bring 1 quart (1 L) water and 2 ounces (50 g) kosher salt to a boil. Fill a large mixing bowl with ice and water. Set aside.

2   Add the *habas* to the boiling water and cook for 1 to 5 minutes, depending on the size and tenderness of the beans. Be patient and keep tasting the beans. They will be done when they are softly cooked but retain their structure (especially if they will be cooked again later). Remove from the heat.

3   Remove the *habas* from the stockpot and plunge them into the ice bath to stop them from cooking further.

4   If the *habas* are older and need to be skinned, wait until the beans are cool enough to handle, and then slip them from their skins. Discard the skins. (If the beans are young and tender enough, disregard this step.)

IF USING *HABAS EN CONSERVA*:

1   Drain the *Habas en Conserva* in a bowl and set aside. Discard the oil.

TO PREPARE THE DISH:

1   In a large skillet over medium heat, warm the oil for 2 to 3 minutes, until it moves easily in the pan and ripples. Add the *sofrito* and *jamón* to the skillet. Cook for 3 to 4 minutes, until warmed through.

2   Add the spring onion, scallion, or leek and continue cooking for 1 minute, until it softens and is fragrant. Add the fava beans or *Habas en Conserva* and whole mint stems. Stir to combine. Cook for 4 to 5 minutes, until the beans are warmed through. Remove from the heat.

3   Allow the mint stems to steep in the skillet for 10 minutes. Remove and discard.

4   Transfer the *Habas con Jamón* to a serving bowl. Season with the salt and black pepper and serve.

CHAPTER FEATURE

PINTXOS

THE BASQUES LOVE their toothpick-skew-ered *pintxos*; the *tapas* of the north, which are often splayed out on bar tops for passersby to nibble on alongside a glass of Basque wine like Txakoli. In towns like San Sebastián, you can enjoy a *pintxo* crawl, walking from bar to bar and grazing on a few nibbles at each place. The number of toothpicks and napkins you find on the floor of bars serves as an informal litmus test of how good the place is. There's a million and one *pintxos* to try in Basque country, but the truth is that anything you can stick on a toothpick qualifies as a proper *pintxo*—and here are just two of the most famous ones.

The origin story of the first *pintxo,* called the Gilda, can be filed under Weird Urban Legend. The story goes that this very famous *pintxo,* created at the Bar Vallés in San Sebastián, is named after Rita Hayworth's 1946 film *Gilda,* since both were "green, salty, and a little spicy." Yeah, I don't get it either, but this *pintxo* is very famous and a tasty way to get everyone at your party drinking.

The second *pintxo* is called the *Banderilla.* Its name comes from the spiky stick that matadors use to piss off bulls during bullfights and turn them into rampaging war machines.

These are only two of the countless *pintxos* you'll encounter in Basque country, so feel free to get creative with almost any of the cured and pickled recipes in this book and stick them on the end of a toothpick. Rest assured: Your creation probably has a close relative lying on a *pintxo* bar somewhere in San Sebastián.

GILDA

1   Using 6 wooden skewers, arrange 1 *Guindilla,* 2 olives, 1 *Pepinillo,* 1 *Cebollita,* and 1 *Boquerone* on each skewer. Serve.

YIELD

6 *tapa* servings

6	medium Cured Guindillas (see recipe on p. 383)
12	large green Spanish olives, cured and marinated
6	Cured *Pepinillos* (see recipe on p. 392)
6	Cured *Cebollitas* (see recipe on p. 391)
6	*Boquerones* or *Anchoas en Salazón* (see recipes on pp. 110 and 129)

BANDERILLAS

1   Using 6 wooden skewers, arrange 1 *Guindilla,* 2 olives, 1 *Pepinillo,* 1 *Cebollita,* 1 *Boquerone,* 1 *Huevo de Cordoniz*, and 1 piece of tuna on each skewer. Serve.

NOTE: *Guindilla* peppers are small, pickled peppers from the Basque region. They are sometimes called *piparras*, and you can find them at most Spanish markets or grocers like La Tienda. Easy substitutions include Italian pepperoncini or any other medium-hot pickled pepper.

YIELD

6 *tapa* servings

6	medium Cured Guindillas (see recipe on p. 383)
12	large green Spanish olives, cured and marinated
6	Cured *Pepinillos* (see recipe on p. 392)
6	Cured *Cebollitas* (see recipe on p. 391)
6	*Boquerones* or *Anchoas en Salazón* (see recipes on pp. 110 and 129)
6	*Huevos de Cordoniz* (see recipe on p. 393), pickled
6	bite-sized pieces olive oil–cured tuna

CHAPTER

8

EMBUTIDOS

**E**MBUTIDOS, **DIFFERENT**

types of sausage that encompass an important part of Spain's culinary lineage, are the meat-stuffed soul of Spanish *charcuterie.*

Every region of Spain has its own unique *embutido* varieties that the locals take great pride in producing. And some of these varieties, like the Catalan ones, are vastly different and wholly unique compared to what you will find anywhere else in the world.

By definition, an *embutido* is pretty much anything that has been forced into a casing. Thus, the term is a catchall for the countless encased meat products found throughout Spain. Few are exported; all told, less than 1 percent of those Spanish *embutidos* ever reach American shores, so our exposure to much of Spain's *charcuterie* traditions is woefully lacking.

The *embutidos* I cover in this chapter are pretty much the most popular varieties I found in my travels and studies of Spain's seventeen autonomous regions. I list the recipes in ascending order of difficulty, so the early group—the fresh *embutidos*—are little more than ground meat with some spices. Also in this group, you'll find a semicured favorite from the Basque region called *Chistorra,* which is the first (of many) fermented sausage recipe you'll come across in this book.

The section that follows includes *embutidos* that are typically cooked in some manner and then chilled for easy slicing or further grilling. There, you'll discover why the Spanish are so addicted to their *morcillas,* delicious blood sausages with a variety of fillings and cooking techniques.

The last style of *embutidos* involves full-on dry curing. Before attempting this section—which includes a bunch of popular *chorizo* formulations among other unique recipes—you should definitely wrap your head around Chapter 3 and have a manageable curing rig ready to go.

FRESH EMBUTIDOS

CHORIZO FRESCO

This is *Chorizo* 101: Class is in session.

You're going to see this *masa* in various guises throughout the *chorizo* recipes for this book, as it's the go-to Spanish sausage base for raw, semicured, and dry-cured *embutidos*.

This incarnation is the most basic form of a Spanish-style *chorizo* sausage, a raw amalgam of its more famous dry-cured cousin with all the smoky-garlicky punch and none of the waiting times or space concerns. It's perfect for those who are averse to dry curing or for beginners who want something tasty for dinner.

1   Place the *aguja, panceta,* and *papada* meats and grinder parts in the freezer for 30 minutes to par-freeze before attempting to grind.

2   Using a mortar and pestle, crush together the garlic and salt to form an *ajosal*. If desired, you can finish the *ajosal* in a food processor fitted with the "S" blade.

3   In a mixing bowl, combine the meats and *ajosal*. Toss together and set aside as you set up the grinder.

4   Fill a large bowl with ice, and place a smaller bowl inside the ice-filled bowl. Grind the meat mixture once through a medium-coarse (3/8 inch [9.5 mm]) die into the smaller bowl. Be careful: The meat mixture is wet, so it may squirt and pop out of the grinder.

5   In a small mixing bowl, combine the wine, water, *pimentones,* and oregano, making a slurry. Keep the bowl containing the slurry chilled until ready to use.

6   Place the ground meats in the bowl of a stand mixer fitted with the paddle attachment (or you can just mix in a mixing bowl with a sturdy spoon). Begin mixing on low speed. As the mixer runs, pour the wine slurry into the bowl in a steady stream.

7   Continue mixing on medium speed for 1 to 2 minutes, until the wine slurry has been fully incorporated into the mixture, a white residue forms on the sides of the bowl, and the mixture firms up. Place the bowl containing the ground meat mixture in the refrigerator to keep it cold until you are ready to stuff the sausage into casings.

YIELD

3–4 loops or 6–8 links of sausage

per 2.2 pounds (1 kg)

INGREDIENTS	CHARCUTIER'S PERCENTAGE
per 2.2 pounds (1 kg) of the following blend of meats, cut into large cubes: 40% *aguja* (pork collar), 40% *panceta* (pork belly), and 20% *papada* (pork jowl)	**100%**
3/4 ounce (20g) whole cloves garlic, peeled and destemmed	**2%**
1 ounce (25 g) kosher salt	**2.5%**
1/4 cup (50 mL) dry white wine, such as a Verdejo, chilled	
1/4 cup (50 mL) water, chilled	
1/3 ounce (10 g) *pimentón dulce*	**1%**
1/3 ounce (10 g) *pimentón picante*	**1%**
1/8 ounce (2 g) dried oregano	**.2%**
3 tablespoons (45 mL) extra virgin olive oil, for frying, divided	

OPTIONAL

2   feet (60 cm) 1 1/4–1 1/2-inch (32–36-mm) hog casings, soaked, or more as needed

Caul fat, as needed

8   To make a *prueba*, in a small skillet over medium–high heat, warm 1 tablespoon of the oil. Place a small piece of the meat mixture in the skillet and fry for 3 to 4 minutes, until cooked through. Remove from the heat. Taste and adjust the seasonings to your liking.

TO FERMENT THE SAUSAGES:

1   *If stuffing:* Stuff the mixture into the casings and tie into 12-inch (30-cm) loops or 6-inch (15-cm) links. Using a sterile pin or sausage pricker, prick each sausage several times. Place in the refrigerator to ferment overnight. (See Notes.)

If not stuffing: Form the mixture into 8-ounce (226-g) patties. Wrap in plastic wrap or caul fat, if using. Place in the refrigerator to ferment overnight. (See Notes.)

TO COOK THE SAUSAGES:

1   *If stuffing:* If you have stuffed the sausages into links or loops, warm the remaining oil in a large skillet over medium–high heat and fry for 8 to 10 minutes, until they register an internal temperature of 150°F (65°C). You can also oven roast or grill the sausages at 350°F (180°C) for 20 to 25 minutes, until they reach the same internal temperature.

If not stuffing: Warm the remaining oil in a large skillet over medium–high heat and fry the sausage patties for 8 to 10 minutes, until they register an internal temperature of 150°F (65°C).

Remove the sausages from the heat and serve.

NOTES: Since this sausage will be cooked, you don't need to be too concerned about the degree of acidulation (that's the pH level—see Chapter 3 for more info). You are just looking to ferment the mixture for a little flavor.

You can ferment the sausages either before or after stuffing. It's really a matter of preference, since the meat firms up during the fermentation process and makes stuffing a little easier. On the other hand, it might be more efficient for you to make and stuff the sausages all in 1 day.

See photo of Chorizo Fresco on p. 225.

TRUCOS DE LA COCINA: PICADILLO

NOTE THAT IN many parts of Spain, *Chorizo Fresco* is eaten without bothering to stuff it into a casing; it's just fried up as loose, ground meat. In this form, the *chorizo* is called *picadillo,* and competitions are held throughout Spain to compare the best *picadillos* of the land. The quality of the pork, flavors from the different muscles, and the texture of the meat grind leads to a competition-worthy dish.

BUTIFARRA FRESCA (CATALAN WHITE SAUSAGE)

A basic white sausage that you will find most everywhere in and around Catalonia, this *embutido* has relatives all over the world, from the sweet and hot Italian sausages we have come to know in America, to the famous Toulouse sausages of France. Why? *Butifarra*—or *botifarra,* in Catalan—is very closely related to the original sausage recipe mentioned in Apicius, the Roman culinary journal written around AD 900.

In places like Barcelona, *botifarra* is a religion. There's an annual fair in the *butifarra's* honor every February at the famous Barcelona *mercado* La Boquería (the fair is called the *butifarrada),* and a ton of restaurants and shops in the market area serve the sausage in raw and cooked forms. (Check out La Botifarreria, near the Santa Caterina market. You'll find a bigger sausage party there than at a Vegas strip club.) Each shop has its own proprietary mix of the sausage's fat and lean content, different ratios of cuts of pork, and grind sizes.

Just know that Catalan purists thumb their noses at seasoning *butifarra* with anything other than salt and pepper. Anything else, in their opinion, interferes with your ability to taste the quality of the pork. That said, we live in a global society, and modern Catalan chefs are creative people. As a result, these days you can find every flavor of the *butifarra* rainbow in Barcelona, from truffle to *foie* to pizza to even chocolate.

This recipe keeps it simple: salt and pepper. If you choose to buck tradition, consider some of the optional seasoning options in the recipe. They're mostly Medici-based spices that work well alone or combined.

1   Place the *aguja, panceta,* and *papada* meats and grinder parts in the freezer for 30 minutes to par-freeze before attempting to grind.

2   Place the meats in a large mixing bowl with the salt and black pepper. Toss together and set aside as you set up the grinder.

3   Fill a large bowl with ice, and place a smaller bowl inside the ice-filled bowl. Grind the meat mixture once through a medium-coarse (3/8 inch [9.5 mm]) die into the smaller bowl. Be careful: The meat mixture is wet, so it may squirt and pop out of the grinder.

YIELD

3–4 loops or 6–8 links of sausage per 2.2 pounds (1 kg)

INGREDIENTS	CHARCUTIER'S PERCENTAGE
per 2.2 pounds (1 kg) of the following blend of meats, cut into large cubes: 60% *aguja* (pork collar), 20% *panceta* (pork belly), and 20% *papada* (pork jowl)	**100%**
2/3 ounce (20 g) kosher salt	**2%**
1/3 ounce (10 g) freshly ground black pepper	**1%**
1/4 cup (50 mL) dry sherry, such as a fino, chilled	
1/4 cup (50 mL) water, chilled	
3 tablespoons (45 mL) extra virgin olive oil, for frying, divided	

OPTIONAL

1 teaspoon (2 g) ground cinnamon	**.2%**
1 teaspoon (2 g) whole allspice, toasted and ground	**.2%**
1 teaspoon (2 g) whole cloves, toasted and ground	**.2%**
1 teaspoon (2 g) ground white pepper	**.2%**
1/2 teaspoon (1 g) freshly grated nutmeg	**.1%**
2 feet (60 cm) 1 1/4–1 1/2-inch (32-mm–36-mm) hog casings, soaked, or more as needed	
Caul fat, as needed	

4   In a small mixing bowl, combine the sherry, water, and any of the optional spices you choose, making a slurry. Keep the bowl containing the slurry chilled until ready to use.

5   Place the ground meats in the bowl of a stand mixer fitted with the paddle attachment (or you can just mix in a mixing bowl with a sturdy spoon). Begin mixing on low speed. As the mixer runs, pour the sherry slurry into the bowl in a steady stream.

6   Continue mixing on medium speed for 1 to 2 minutes, until the sherry slurry has been fully incorporated into the mixture, a white residue forms on the sides of the bowl, and the mixture firms up. Place the bowl containing the ground meat mixture in the refrigerator to keep it cold until you are ready to stuff the sausage into casings.

7   To make a *prueba,* in a small skillet over medium–high heat, warm 1 tablespoon of the oil. Place a small piece of the meat mixture in the skillet and fry for 3 to 4 minutes, until cooked through. Remove from the heat. Taste and adjust the seasonings to your liking.

TO STUFF THE SAUSAGES:

1   *If stuffing:* Stuff the mixture into the casings and tie into 12-inch (30-cm) loops or 6-inch (15-cm) links. Using a sterile pin or sausage pricker, prick each sausage several times. Place in the refrigerator to ferment overnight. (See Notes.)

If not stuffing: Form the mixture into 8-ounce (226-g) patties. Wrap in plastic wrap or caul fat, if using. Place in the refrigerator to ferment overnight. (See Notes.)

TO COOK THE SAUSAGES:

1   *If stuffing:* If you have stuffed the sausages into links or loops, warm the remaining oil in a large skillet over medium–high heat and fry for 8 to 10 minutes, until they register an internal temperature of 150°F (65°C). You can also oven roast or grill the sausages at 350°F (180°C) for 20 to 25 minutes, until they reach the same internal temperature.

If not stuffing: Warm the remaining oil in a large skillet over medium–high heat and fry the sausage patties for 8 to 10 minutes, until they register an internal temperature of 150°F (65°C).

Remove the sausages from the heat and serve.

NOTES: *Butifarra* recipes vary from family to family and shop to shop; here, I provided a base recipe and then typical spice additions to try. Add your choice(s) to the sherry slurry in Step 4.

You can ferment the sausages either before or after stuffing. It's really a matter of preference, since the meat firms up during the fermentation process and makes stuffing a little easier. On the other hand, it might be more efficient for you to make and stuff the sausages all in 1 day.

CHISTORRA

Chistorra is the redheaded stepchild of the Spanish sausage world—underappreciated, misunderstood, and virtually unheard of outside of its native País Vasco and Navarra. Maybe that's because it's one of the only semicured sausages in the Spanish arsenal. But for that same reason, it's the perfect gateway drug to the addictive world of dry-cured meat.

The recipe begins by creating a *masa* similar to the one used in *Chorizo Fresco,* with the addition of a fermentation agent. This natural bacterial culture inoculates the sausage with the sort of friends needed for a quick lactic acid fermentation. Basically, the fermentation culture lowers the pH of the sausage, giving it a nice, acidic tang.

I learned to make this recipe using only 100% pork, but depending on which region of País Vasco a *Chistorra* comes from, it might be made from beef or various blends of pork and beef. Feel free to change it up a little—just know your meat and where it comes from.

1 Place the *cabecero, panceta,* and *papada* meats and grinder parts in the freezer for 30 minutes to par-freeze before attempting to grind.

2 In a small mixing bowl, combine the F-RM-52 culture and the distilled water, making a slurry. Set aside for a minimum of 10 minutes at room temperature to bloom.

3 Using a mortar and pestle, crush together the garlic and salt to form an *ajosal.* If desired, you can finish the *ajosal* in a food processor fitted with the "S" blade.

4 In a mixing bowl, combine the meats with the *ajosal,* dextrose, and curing salt. Toss together and set aside as you set up the grinder.

5 Fill a large bowl with ice, and place a smaller bowl inside the ice-filled bowl. Grind the *cabecero, panceta,* and *papada* meats once through a medium-coarse (3⁄8 inch [9.5 mm]) die into the smaller bowl. Be careful: The meat mixture is wet, so it may squirt and pop out of the grinder. Return the ground meat to the freezer for 15 to 20 minutes, until it is par-frozen.

6 Grind the *cabecero, panceta,* and *papada* meats again through a medium (1⁄4 inch [6 mm]) die into the same bowl over ice.

7 In a small mixing bowl, combine the wine, water, and *pimentones,* making a slurry. Keep the bowl containing the slurry chilled until ready to use.

YIELD

About 8–10 loops or 16–20 links of sausage per 2.2 pounds (1 kg) of meat

INGREDIENTS	CHARCUTIER'S PERCENTAGE
per 2.2 pounds (1 kg) of the following blend of diced meats: 40% *cabecero* (coppa/head of the pork loin), 40% *panceta* (pork belly), and 20% *papada* (pork jowl)	**100%**
1¾ ounces (10 g) F-RM-52 culture (see p. 94)	**1%**
½ cup (100 mL) distilled water	
⅓ ounce (10 g) minced garlic	**1%**
⅔ ounce (20 g) kosher salt	**2%**
⅛ ounce (5 g) dextrose	**.5%**
⅛ ounce (2.4 g) TCM #1 or DQ #1 curing salt mix (see p. 83)	**.24%**
¼ cup (50 mL) dry white wine, such as Verdejo, chilled	
¼ cup (50 mL) water, chilled	
⅓ ounce (10 g) *pimentón dulce*	**1%**
⅓ ounce (10 g) *pimentón picante*	**1%**
3 tablespoons (45 mL) extra virgin olive oil, for frying, divided	
2 feet (60 cm) 1–1¹⁄₁₆-inch (24–26-mm) lamb casings, soaked, or more as needed	

CONTINUED ON NEXT PAGE

CHISTORRA CONTINUED FROM PREVIOUS PAGE

8   Place the ground meats in the bowl of a stand mixer fitted with the paddle attachment (or you can just mix in a mixing bowl with a sturdy spoon). Begin mixing on low speed. As the mixer runs, pour the wine slurry into the bowl in a steady stream.

9   Continue mixing on medium speed for 1 to 2 minutes, until the wine slurry has been fully incorporated into the mixture, a white residue forms on the sides of the bowl, the meat appears sticky, and the mixture firms up.

10   Reduce the mixer speed to low and add the F-RM-52 slurry. Continue mixing for 1 minute, until the F-RM-52 slurry is fully incorporated into the mixture. Place the bowl containing the ground meat mixture in the refrigerator to keep it cold until you are ready to stuff the sausage into casings.

11   To make a *prueba,* in a small skillet over medium–high heat, warm 1 tablespoon of the oil. Place a small piece of the meat mixture in the skillet and fry for 3 to 4 minutes, until cooked through. Remove from the heat. Taste and adjust the seasonings to your liking.

12   Stuff the mixture into the casings and tie into 24-inch (60-cm) loops. Using a sterile pin or sausage pricker, prick each sausage several times. Weigh each sausage loop to obtain a green weight; record the weights and tag each sausage with its green weight.

13   Ferment the sausages in a drying chamber set at 86°F (30°C) and 80% to 85% relative humidity for at least 2 days. During this time, check the pH of the meat (see p. 95) to ensure that the level has dropped below 5.3 by the third day of drying.

14   Hang the sausages in a drying chamber set at 60°F to 70°F (16°C to 22°C) and 65% to 75% relative humidity for 5 to 10 days, until the sausages have lost about 15% of their green weight.

15   In a large skillet over medium–high heat, warm the remaining oil. Fry the sausages for 6 to 10 minutes, until they register an internal temperature of 150°F (65°C). You can also oven roast or grill the sausages at 350°F (180°C) for 10 to 15 minutes, until they reach the same internal temperature. Remove the sausages from the heat and serve.

NOTES: *Chistorra* has a shelf life of around 15 days held in a refrigerator (remember, it's semicured, so it isn't shelf stable). It is delicious on its own or in other recipes (for example, see *Bollo Preñao* on p. 254).

You can ferment the sausages either before or after stuffing. It's really a matter of preference, since the meat firms up during the fermentation process and makes stuffing a little easier. On the other hand, it might be more efficient for you to make and stuff the sausages all in 1 day.

Since you need to cook this *embutido* prior to eating it, the drying time is not essential. It's simply a matter of taste.

TRUCOS DE LA COCINA: CASINGS

I PREFER LAMB casings for my *Chistorra,* but similarly sized collagen casings are often easier to track down. Just make sure to always check the package label, because the collagen type isn't always edible.

BUTIFARRA DULCE

At first glance, a sausage made with honey, lemon, and cinnamon seems a little strange to an American palate.

Okay, really strange.

This delicacy from the Catalan town of Girona is a favorite for special meals, like Christmas dinner, and even for dessert. The Catalans are in good company with their penchant for sweetened meats, as these types of sausage mixtures have been around for centuries in Europe: Consider British mincemeat, the sweet Italian blood sausage *sanguinaccio,* or the Riojana *morcilla dulce.*

Butifarra Dulce is made with the same base *butifarra masa,* but includes the additions of honey, sherry, lemon, and cinnamon to sweeten things up. It's particularly delicious when cooked with apples.

1   Place the *aguja, panceta,* and *papada* meats and grinder parts in the freezer for 30 minutes to par-freeze before attempting to grind.

2   In a large mixing bowl, combine the meats, salt, black pepper, lemon, cinnamon, and honey. Toss together and set aside as you set up the grinder.

3   Fill a large bowl with ice, and place a smaller bowl inside the ice-filled bowl. Grind the meat mixture once through a medium-coarse (3⁄8 inch [9.5 mm]) die into the smaller bowl. Be careful: The meat mixture is wet, so it may squirt and pop out of the grinder.

4   In a small mixing bowl, combine the sherry and cream.

5   Place the ground meats in the bowl of a stand mixer fitted with the paddle attachment (or you can just mix in a mixing bowl with a wooden spoon). Begin mixing on low speed. As the mixer runs, pour the sherry and cream mixture into the bowl in a steady stream.

6   Continue mixing on medium speed for 1 to 2 minutes, until the sherry and cream has been fully incorporated into the mixture, a white residue forms on the sides of the bowl, and the mixture firms up. Place the bowl containing the ground meat mixture in the refrigerator to keep it cold until you are ready to stuff the sausage into casings.

7   To make a *prueba,* in a small skillet over medium–high heat, warm 1 tablespoon of the oil. Place a small piece of the meat mixture in the skillet and fry for 3 to 4 minutes, until cooked through. Remove from the heat. Taste and adjust the seasonings to your liking.

YIELD

3–4 loops or 6–8 links of sausage per 2.2 pounds (1 kg)

INGREDIENTS	*CHARCUTIER'S* PERCENTAGE
per 2.2 pounds (1 kg) of the following blend of diced meats: 60% *aguja* (pork collar), 20% *panceta* (pork belly), and 20% *papada* (pork jowl)	**100%**
2⁄3 ounce (20 g) kosher salt..................	**2%**
2⁄3 ounce (20 g) freshly ground black pepper	**2%**
Zest of 1⁄2 lemon	
1 teaspoon (3 g) ground cinnamon..........................	**.3%**
2⁄3 ounce (20 g) acacia or wildflower honey	**2%**
1⁄4 cup (50 mL) amontillado sherry, chilled	
1⁄4 cup (50 mL) heavy cream, chilled	
3 tablespoons (45 mL) extra virgin olive oil, for frying, divided	

OPTIONAL

2   feet (60 cm) 11⁄4–11⁄2-inch (32–36-mm) hog casings, soaked, or more as needed

Caul fat, as needed

CONTINUED ON NEXT PAGE

BUTIFARRA DULCE CONTINUED FROM PREVIOUS PAGE

TO FERMENT THE SAUSAGES:

1   *If stuffing:* Stuff the mixture into the casings and tie into 12-inch (30-cm) loops or 6-inch (15-cm) links. Using a sterile pin or sausage pricker, prick each sausage several times.

If not stuffing: Form the mixture into 8-ounce (226-g) patties. Wrap in plastic wrap or caul fat, if using.

TO COOK THE SAUSAGES:

1   *If stuffing:* If you have stuffed the sausages into links or loops, warm the remaining oil in a large skillet over medium–high heat and fry for 8 to 10 minutes, until they register an internal temperature of 150°F (65°C). You can also oven roast or grill the sausages at 350°F (180°C) for 20 to 25 minutes, until they reach the same internal temperature.

If not stuffing: Warm the remaining oil in a large skillet over medium–high heat and fry the sausage patties for 8 to 10 minutes, until they register an internal temperature of 150°F (65°C).

Remove the sausages from the heat and serve.

NOTE: If you want to go a different route, you can dry cure this sausage. If you do, bump up your *charcutier* percentages: You'll need at least 2.3% salt, ¼ of a packet of a fermentation starter culture (such as T-SPX), and .24% of either DQ #2 or TCM #2 curing salt mix per 2.2 pounds (1 kilogram) of meat. Follow the same curing schedule shown in the *Fuet* recipe (see p. 311). As always, carefully read Chapter 3 before giving dry curing a go.

FRESH *EMBUTIDOS*
TRADITIONAL AND MODERN RECIPES

BOTIFARRA AMB MONGETES

Just about every region in Spain has a take on pork and beans. This one—a down-home Catalan soul-food specialty—is particularly interesting because it involves two techniques you don't typically find in other bean dishes:

➡ The sausages are cooked separately from the beans, thus the beans' flavor develops from stock and aromatics.

➡ The cooked beans are fried for just a bit in hot, flavored oil, giving them a "crunchy" outside.

Whatever you do, don't cheat and use canned beans in this recipe. The Spanish Bean Gods demand dried beans for a properly crispy finished product, and those canned mushballs just won't get the job done. If you do it right, you will be richly rewarded with one of the most classic of all Catalan dishes.

Serve this with an unapologetically garlicky *Alioli*.

YIELD

4–6 entrée servings

INGREDIENTS

1¼	cups (250 g) dried small white beans, such as cannellini or white kidney
1	gallon (4 L) water, divided
8¾	ounces (250 g) *Panceta Curada* (see recipe on p. 122)
1	ham bone, *Hueso Salado* (preserved pork bone, see recipe on p. 114), or *Codillo en Salmuera* (preserved pork shank, see recipe on p. 109)
10	stems fresh parsley
1	fresh bay leaf
20	whole black peppercorns
½	white onion, studded with a clove
5	whole cloves garlic, peeled
½	cup (100 mL) extra virgin olive oil
6	*Butifarra Fresca* sausages (see recipe on p. 228), pricked with a fork, pin, or small skewer
3	sprigs fresh rosemary, finely minced
½	cup (125 mL) Basic *Sofrito* (see recipe on p. 418)
	Kosher salt, to taste
	Alioli (see recipe on p. 413), for serving

1 In a large stockpot, cover the beans with 2 quarts of the water and soak overnight at room temperature. In a separate large bowl, cover the *panceta* and ham bone with the remaining 2 quarts of water and soak for the same amount of time at room temperature.

2 Place the *panceta* and ham bone in the stockpot with the beans and bean-soaking water. Add enough water to cover.

3 Make a *bouquet garni* by placing the parsley, bay leaf, and peppercorns in a piece of cheesecloth, forming a pouch, and tying a long piece of kitchen twine to close the pouch.

4 Add the studded onion and the garlic to the stockpot. Drop the *bouquet garni* in the stockpot with a tail of twine hanging on the side of the stockpot. Place the stockpot over medium–high heat. Bring to a boil and skim off the foam. (From this point forward, do not stir the stockpot, since doing so will break up the beans. Instead, just give the stockpot a shake from time to time.)

5 Reduce the heat to medium. Simmer for 60 to 90 minutes (depending on the type of beans), gently shaking the stockpot on occasion, until the beans are tender and cooked through. Remove from the heat and discard the ham bone, other solids, and *bouquet garni*. Drain the beans and transfer to a serving bowl to cool.

6 In a large sauté pan over medium–high heat, warm the oil until just smoking. Add the sausages and fry for 8 to 10 minutes, until golden brown on all sides and cooked through. Remove the sausages to a plate.

7 Return the heat to medium–high. Add the beans and the rosemary to the sauté pan. Fry for 5 to 10 minutes, until the beans get a little crispy. Add the Basic *Sofrito* to the pan and cook for 1 minute, stirring up any browned bits. Taste and season with the salt. Remove from the heat.

8 Place the beans on a deep serving platter or plate. Slice the sausages and top the beans with them. Serve warm, with a dollop of the *Alioli*.

CANELONE CREPES

Canelones, or *canalons,* are the Catalans' favorite adopted child. A dish brought over by Italian immigrants, it has become an emblematic dish of the Barcelona food scene. *Canelones* are rolled pasta sheets or crepes—either works—that are then filled, covered with a sauce, and baked.

You can find the pasta sheets in most Spanish, Italian, or Latin markets, and all you need to do is drop them in salted, boiling water and cook until *al dente* before chilling them down in ice water. Once they've cooled, you roll them into a cylinder, fill them, and place them, seam-side down, in a baking dish. Top 'em with a sauce and bake until bubbly.

The below crepe recipe might seem more time consuming than cooking premade pasta sheets, but I prefer them to the pasta because I can flavor the crepe batter however I want. Also, you can prepare a bunch of them ahead of time and freeze them easily with sheets of parchment paper between each crepe. The crepes freeze very well, keep for a long time, and defrost quickly. After the crepe recipe, I provide a number of different types of *canelone* filling recipes common to Catalan kitchens.

Béchamel is the most common choice of sauce, so I've included a basic recipe. Depending on the *canelones* you make, however, you can play around with that base sauce. My favorite variant is to add some grated Manchego cheese to the béchamel, thus making what the Frenchies call a Mornay sauce. It's a great addition to Crab and Scallop *Canelones* (see recipe on p. 247).

TO MAKE THE CREPES:

1   In a blender, combine the flour, eggs, oil, milk, and coriander seeds. Add a bit of the salt and white pepper for seasoning. Pulse the blender until the mixture is completely combined (it should resemble a thin pancake batter). Pour the batter into a bowl and refrigerate for 30 minutes.

2   Warm a medium nonstick skillet over medium heat. Set up a work station by the cooktop with a plate and plastic wrap or parchment paper for the finished crepes.

YIELD

25–30 crepes

INGREDIENTS

FOR THE CREPES:

8¾   ounces (250 g) unbleached all-purpose flour

4   whole large eggs

2   tablespoons (30 mL) extra virgin olive oil, plus more for greasing

2   cups (500 mL) whole milk

1   teaspoon (2 g) coriander seeds, ground and toasted

Kosher salt, as needed

White pepper, as needed

FOR THE BÉCHAMEL SAUCE:

4¼   cups (1 L) whole milk

1   medium white onion, halved and studded with 1 clove on each half

2   fresh bay leaves

1¾   ounces (50 g) *manteca* (pork fat) or butter, plus more for greasing

1¾   ounces (50 g) unbleached all-purpose flour

1   teaspoon (2 g) freshly grated nutmeg

1   recipe *Canelone* filling of your choice (see recipes on pp. 241, 243, 245, and 247)

1   cup (113 g) freshly grated Manchego cheese, for sprinkling

CONTINUED ON NEXT PAGE

CANELONE CREPES CONTINUED FROM PREVIOUS PAGE

3   Once the pan is hot, dip a paper towel in some oil and grease the pan well with it. Using a ¼ cup (60 mL) measure, pour some of the batter in the skillet. Tip the skillet in a circular motion, coating the entire bottom of the skillet with a thin layer of batter. After 30 seconds, small bubbles should begin to form. When you see them, gently lift the edges of the crepe with a large spatula to loosen it from the pan. Flip the crepe and cook for 30 seconds. (Don't worry if the first one is more crap than crepe; this always happens.) Remove from the pan and place on the prepared plate. Place a sheet of plastic wrap or parchment paper over the crepe.

4   Repeat with the rest of the batter until all of it has been used. From here, you can either chill the crepes and freeze them for later or continue the recipe.

TO MAKE THE BÉCHAMEL SAUCE:

1   In a medium saucepan over medium–high heat, place the milk, onion halves, and the bay leaves. Bring to a boil and reduce the heat to medium. Simmer for 15 minutes. Remove from the heat, remove and discard the onion and bay leaves, and transfer to a large measuring cup. Set aside and wipe out the pan.

2   Return the same saucepan to medium heat and add the *manteca* or butter. Once the fat is hot, whisk in the flour to make a roux. Continue cooking, whisking constantly, for 6 to 8 minutes, until the roux is a light, sandy color.

3   Slowly pour the reserved milk, 1 cup (240 mL) at a time, into the roux, whisking constantly until fully incorporated and smooth. Raise the heat to medium–high and bring to a boil. Reduce the heat to medium and cook for 10 minutes, creating a béchamel sauce. Remove from the heat and season to taste with the salt. Add the nutmeg and stir. Set aside to cool.

4   Preheat the oven to 400°F (200°C). Grease a baking dish large enough to hold all of the crepes in one layer with some *manteca.* Pour a thin layer of béchamel sauce into the dish and spread it across the entire bottom.

5   On a clean work surface, lay out 1 crepe and spoon 3 tablespoons (45 mL) of the filling of your choice onto the crepe. Spread the filling out until it extends about ½ inch (13 mm) from the edges of the crepe. Roll the crepe into a cylinder, and then trim the excess. Cut each crepe in half, giving you 2 filled *canelones* per crepe. Place the *canelones* seam-side down in the baking dish in a single layer. Repeat until you have used all the crepes and all of the filling.

6   Spoon the remaining béchamel sauce over the top of the *canelones.* Sprinkle with the cheese, and bake for 15 minutes, until warmed through.

7   Raise the oven temperature to the broil setting. Gratinée the tops of the *canelones* for 1 to 2 minutes, until golden brown. Serve immediately.

NOTE: If you want to make a Mornay sauce from the béchamel, just stir in 1 cup (100 g) shredded Manchego cheese at the end of making the béchamel. Stir until melted and cheesy delicious.

CANELONE FILLING: *BARCELONESA*

This is the most traditional *canelone* preparation in Barcelona: It's the alpha and the omega, based somewhat loosely on the fancy Rossini preparation for steak made famous by the French master chef Marie-Antoine Carême. This dish features richly flavored meats, *foie gras,* and traditionally a truffle-laden béchamel sauce, so consider this recipe for a special occasion or for impressing a hot date.

Think about the calories tomorrow. You can diet then.

1   In a medium mixing bowl, cover the brains with water. Soak in the refrigerator for 8 hours or overnight to purge them of any blood. Drain.

2   In a large saucepan over medium–high heat, bring the 2 cups water, white wine vinegar, thyme, bay leaf, rosemary, the 1 ounce salt, and the peppercorns to a simmer.

3   Fill a large bowl with ice water. Set aside.

4   Add the brains to the saucepan and cook for 10 minutes, until tender. Remove the brains from the saucepan and plunge them in the ice water to cool. Discard the simmering liquid and solids. Once cool, chop the brains into small pieces and place in a bowl. Set aside.

5   Return the same saucepan to medium–high heat. Prepare the Basic *Sofrito,* using *manteca* as the fat and including the garlic and bay leaf options from the recipe. Remove from the heat, transfer the *sofrito* to a mixing bowl, remove and discard the bay leaf, and set aside.

6   Return the same saucepan to medium–high heat, add the *manteca* and heat it for 3 to 5 minutes, until the fat renders and is nearly smoking. Add the sausages, livers, ground chicken, and ground veal to the saucepan. Season with the salt and Four-Spice Blend. Sear the meat for 4 minutes on one side without moving it, until golden.

7   Deglaze with the wine, stirring up any browned bits and reducing the wine by half. Then, add the reserved *sofrito* to the saucepan and cook for 8 to 10 minutes, until the mixture reaches a high simmer.

YIELD

4 cups (950 mL) filling

INGREDIENTS

2	pork or lamb brains (see Notes)
2	cups (500 mL) water
½	cup (120 mL) white wine vinegar
5	sprigs fresh thyme
1	fresh bay leaf
1	sprig fresh rosemary
1	ounce (25 g) kosher salt, plus more to taste
10	black peppercorns
1	cup (250 g) Basic *Sofrito* (see recipe on p. 418)
1¾	ounces (50 g) *manteca* (pork fat)
4	*Butifarra Fresca* sausages (see recipe on p. 228), removed from the casings and crumbled
2	chicken livers, deveined and very finely chopped
6	ounces (170 g) ground chicken breast
6	ounces (170 g) ground veal
	Four-Spice Blend (see recipe on p. 409), to taste
¼	cup (50 mL) dry white wine, such as Albariño
	Freshly ground black pepper, to taste
9	ounces (250 g) *Foie Salada* (see recipe on p. 270), cubed
¼	cup (27 g) bread crumbs
2	whole large eggs, beaten
¼	cup (15 g) finely chopped fresh flat-leaf parsley
1¾	ounces (50 g) black truffle, halved

CONTINUED ON NEXT PAGE

CANELONE FILLING: *BARCELONESA*

CONTINUED FROM PREVIOUS PAGE

8 Add the brains to the saucepan and cook for 10 minutes, until much of the liquid has evaporated. Remove from the heat. Reseason to taste with the salt, black pepper, and Four-Spice Blend. Set aside for 10 to 15 minutes, until the filling cools slightly.

9 Fold the *Foie Salada,* bread crumbs, eggs, and parsley into the mixture. Grate ½ of the truffle into the mixture and fold again.

10 Chill the filling in the refrigerator until you are ready to make the *Canelones*. Grate the remaining ½ truffle as garnish when serving the *Canelones*.

NOTES: As for the brains, I know that they are both hard to find and may gross some people out—if you wind up using them, 10 points to Gryffindor because they are both traditional and delicious. If not, you can substitute some sweetbreads, liver, or even just chicken breast meat.

Yes, truffles are expensive and definitely not mandatory for this dish. But if you're trying to impress someone or get laid, go for it. Just please don't use truffle oil...that shit needs to remain shamefully buried in our culinary past.

CANELONE FILLING: *BUTIFARRA NEGRA,* TROTTERS, SPINACH, AND PINE NUT

Dollars to donuts: If you walk into your local *tapas* joint right now, there will be some variant of the oft-repeated, ad nauseam classic known as *a la catalana*—a sauté of greens with garlic, pine nuts, and dried fruit, such as raisins or currants—on the menu. It's a standard *gringo tapas* bar go-to combo, but in Barcelona, you aren't likely to find it as a standalone dish anymore. The Catalans are *so* over it.

Instead, you might find these classic flavors used as a component of another dish, like these *canelones* my cook friends in Barcelona love so much. Let's call it *A La Catalana* Version 2.0...trust me, this is a hell of a lot more exciting than veggies with fruits and nuts.

1   In a large stockpot over medium–high heat, cover the *manitas* with cold water. Bring the water to a boil. Drain, reserving the trotters in the stockpot.

2   Return the stockpot to the heat. Add the onions, carrots, garlic, and bay leaves. Cover with the *Caldo Blanco* and bring to a boil. Reduce the heat to medium. Simmer for at least 1½ hours, until the meat can be easily pulled from the bones. Remove from the heat. Drain, reserving ¼ cup (60 mL) of the cooking liquid. Place the *manitas* in a bowl and refrigerate for 8 to 10 minutes, until they are cool enough to separate all the meat from the bones.

3   Remove the meat from the bones. Discard the bones, cut the meat into bite-sized pieces, and return the meat to the bowl. Set aside.

4   In a small mixing bowl, soak the peppers in the boiling water for 30 minutes, until softened. Drain. Cut open the peppers, scrape out their pulp, and return the pulp to the bowl. Discard the rinds.

5   In a large sauté pan over medium–high heat, warm the oil until just rippling. Add the bread and toast for 2 minutes on each side, until golden. Remove from the heat. Place the toasted bread on a plate and set aside.

6   Using a mortar and pestle, crush the pine nuts with a pinch of the salt until coarsely ground. (If desired, you can use a food processor fitted with the "S" blade.) Add the toasted bread and the peppers. Continue grinding until the mixture forms a coarse paste, making a *picada.*

CONTINUED ON NEXT PAGE

YIELD

4 cups (950 mL) filling

INGREDIENTS

4	*manitas* (pork trotters), rinsed well
	Cold water, to cover
2	medium yellow onions, halved and studded with 1 clove on each half
4	medium carrots, peeled and cut into large chunks
1	head garlic, split in half
2	fresh bay leaves
	Caldo Blanco (see recipe on p. 159), chicken stock, or water, to cover
4	dried ñora peppers
	Boiling water, as needed
¼	cup (60 mL) extra virgin olive oil
1	slice ciabatta or other country bread, crusts removed
¼	cup (35 g) toasted pine nuts
	Kosher salt, to taste
¼	cup (40 g) golden raisins
¼	cup (60 mL) Pedro Ximénez (PX) sherry
1	cup (250 g) Basic *Sofrito* (see recipe on p. 418)
1¾	ounces (50 g) *manteca* (pork fat)
4	*Butifarra Negra* sausages (see recipe on p. 267), removed from the casings and crumbled
	Four-Spice Blend (see recipe on p. 409), to taste
1	pound (450 g) frozen spinach, thawed, drained, squeezed, and finely chopped
	Freshly ground black pepper, to taste

CANELONE FILLING: *BUTIFARRA NEGRA,* TROTTERS, SPINACH, AND PINE NUT

CONTINUED FROM PREVIOUS PAGE

7   Moisten the *picada* with some of the reserved cooking liquid and blend until the paste is smooth. (Taste it to ensure it isn't gritty. If it is, moisten it a little more with more of the reserved cooking liquid and blend further.) Set aside.

8   In a small mixing bowl, soak the raisins in the sherry for 30 minutes, until plumped. Set them aside (and try not to snack on all of them).

9   In a large saucepan over medium–high heat, prepare the Basic *Sofrito,* using *manteca* as the fat and including the garlic and bay leaf options from the recipe. Remove from the heat, transfer the *sofrito* to a mixing bowl, and set aside. Wipe out the saucepan.

10   Return the same saucepan to medium–high heat. Add the *manteca* and heat it for 3 to 5 minutes, until the fat renders and is nearly smoking. Add the sausages to the saucepan. Season with the salt and Four-Spice Blend. Sear the meat for 4 minutes on one side without moving it, until golden.

11   Add the reserved *sofrito* and *manitas* to the saucepan and cook for 8 to 10 minutes, until the mixture reaches a high simmer. Add the raisins and sherry, and simmer for 3 minutes more, until slightly reduced.

12   Reduce the heat to medium. Add the reserved *picada* to the saucepan and stir to combine. Remove from the heat.

13   Add the spinach to the saucepan and fold it into the mixture until it wilts, about 1 minute. Reseason to taste with the salt, black pepper, and Four-Spice Blend.

14   Chill the filling in the refrigerator until you are ready to make the *Canelones.* Make sure the filling is completely drained of any liquid before preparing the *Canelones,* as moisture can make them soggy.

CANELONE FILLING: *FOIE* AND MUSHROOM

One of the many lessons I learned while cooking for Daní Garcia in Andalucía was how to make an awesome mushroom *duxelles*. It's really pretty simple and smart. You'd normally make *duxelles* by grinding the mushrooms and then sautéing them until they're quasi-dry, but Daní's first step is to slowly dry the whole 'shrooms at a low temperature in the oven. (It takes a while to get all the moisture out.)

Once the 'shrooms are so dry that no water comes out when you squeeze them, you're ready to put them in a pan with caramelized shallots and some cream. Cook with some herbage until reduced and fragrant, and blitz! You've got amazing mushroom *duxelles* with brilliant Spanish-influenced technique, which we used for stuffing all sorts of pastas.

1   Preheat the oven to 225°F (110°C). Line a baking sheet large enough to contain all the 'shrooms in 1 layer (if you have to, use 2 sheets).

2   Place the mushrooms on the baking sheet, cap-side down. Place in the oven for 2 hours, until all their moisture is gone. Remove from the oven and set aside.

3   In a large saucepan over medium–high heat, warm half the butter for 3 to 5 minutes, until foamy. Reduce the heat to medium and add the shallots and garlic. Season the shallot–garlic mix with the salt. Sweat for 10 minutes, until the shallots and garlic are translucent and without color.

4   Add the dried mushrooms, thyme, bay leaves, and rosemary to the saucepan. Just cover with the cream. Bring to a boil. Reduce the heat to medium and cook for 20 minutes, until the sauce volume is reduced to $1/2$ cup (120 mL) and the mushrooms are softened. Remove from the heat and discard the herbs.

5   Transfer the mushrooms to the bowl of a food processor fitted with the "S" blade, saving the cream separately. Pulse the *duxelles* with just enough cream to help them form a coarsely textured purée. Season, and set the *duxelles* aside to cool.

6   In a large saucepan over medium–high heat, prepare the Basic *Sofrito,* using butter as the fat and including the garlic, bay leaf, and cinnamon stick options from the recipe. Remove from the heat, transfer the *sofrito* to a mixing bowl, and set aside. Wipe out the saucepan.

YIELD

4 cups (950 mL) filling

INGREDIENTS

8¾   ounces (250 g) crimini mushrooms, destemmed and rinsed

8¾   ounces (250 g) button mushrooms, destemmed and rinsed

4   ounces (115 g) unsalted butter, divided

5   medium shallots, peeled and minced

5   cloves garlic, peeled, destemmed, and minced

    Kosher salt, as needed

5   sprigs fresh thyme

2   fresh bay leaves

1   sprig fresh rosemary

1   cup (250 mL) heavy cream

1   cup (250 g) Basic *Sofrito* (see recipe on p. 418)

1.1   pounds (500 g) ground duck or chicken

    Four-Spice Blend (see recipe on p. 409), to taste

¼   cup (60 mL) oloroso sherry or cognac

1.1   pounds (500 g) *Foie Salada* (see recipe on p. 270), cubed

    Freshly ground black pepper, to taste

CONTINUED ON NEXT PAGE

CANELONE FILLING: *FOIE* AND MUSHROOM

CONTINUED FROM PREVIOUS PAGE

7   In the same large saucepan over medium–high heat, warm the remaining butter for 3 to 5 minutes, until foamy. Add the ground duck and season with the salt and Four-Spice Blend. Sear the meat for 4 minutes on one side without moving it, until golden.

8   Add the reserved *sofrito* to the saucepan and cook for 8 to 10 minutes, until the mixture reaches a high simmer. Add the sherry.

9   Reduce the heat to medium. Cook for 10 minutes, until much of the liquid has evaporated. Remove from the heat and set the mixture aside to cool for 10 minutes.

10 Fold the *Foie Salada* and reserved *duxelles* into the duck mixture. Remove from the heat. Reseason to taste with the salt, black pepper, and Four-Spice Blend.

11 Chill the filling in the refrigerator until you are ready to make the *Canelones*. Make sure the filling is completely drained of any liquid before preparing the *Canelones,* as moisture can make them soggy.

CANELONE FILLING: CRAB AND SCALLOPS

Shellfish are a big deal in the coastal areas of Catalonia, which is why one of the most popular *canelone* variants takes advantage of the local *marisco* population. *Canelones de mariscos* are typically a mixture of shelled seafood goodies—scallops, crab, shrimp, and mussels—smothered in a seafoody béchamel, Mornay, or tomato sauce.

Here is a great place to employ that potted crab that you tucked away for just such an occasion (see recipe on p. 208). You can either make the potted crab *Txangurro* mix and omit the storage steps, use some fresh crab meat, or just use whatever shellfish you have on hand.

1   In a large saucepan over medium–high heat, prepare the Basic *Sofrito,* using olive oil as the fat and including the garlic, leeks, red bell peppers, chile pepper, and tomato paste options from the recipe. Remove from the heat, transfer the *sofrito* to a mixing bowl, and set aside.

2   Warm a large sauté pan over medium–high heat. Dry the scallops very well and then season them with the salt and black pepper. Set aside. Line a plate with paper towels for the finished scallops.

3   Add the oil to the sauté pan and warm it until it ripples and is almost smoking. Carefully add the scallops to the sauté pan. Sear for 2 to 3 minutes, until golden brown. Flip them and sear for 2 to 3 minutes more. Remove the finished scallops to the prepared plate. Remove from the heat and discard the oil the in pan.

4   Blot the scallops dry and dice them into large chunks. Set aside.

5   Return the same sauté pan to medium–high heat. Deglaze with the brandy (you should get a little flambé action). Add the reserved *sofrito* to the sauté pan and cook for 3 to 5 minutes, until the mixture reaches a high simmer.

6   Reduce the heat to medium. Add the crab meat and lemon zest to the saucepan and cook for 2 to 3 minutes, until warmed through. Fold in the reserved scallops. Remove from the heat and allow to cool to room temperature. Reseason to taste with the salt, black pepper, and lemon juice.

7   Chill the filling in the refrigerator until you are ready to make the *Canelones.* Make sure the filling is completely drained of any liquid before preparing the *Canelones,* as moisture can make them soggy.

YIELD

4 cups (950 mL) filling

INGREDIENTS

1   cup (250 g) Basic *Sofrito* (see recipe on p. 418)

1.1   pounds (500 g) scallops, adductor muscle removed and patted dry

       Kosher salt, to taste

       Freshly ground black pepper, to taste

¼   cup (50 mL) extra virgin olive oil

¼   cup (60 mL) brandy

2.2   pounds (1 kg) potted crab or fresh crab meat

       Juice and zest of 1 lemon

TRINXAT

If you are visiting the Catalan mountain town of Puigcerdà on a cold winter day, your lunch will need to be warm and comforting. If you're lucky, it'll be *Trinxat,* a stinky local cabbage specialty with a festival of its very own every February. *Trinxat* is part of a long line of European cabbage and potato mash-ups, including comfort-food favorites like the English Bubble and Squeak, Swiss *rösti,* Irish colcannon... the list goes on.

These recipes have common ground: They're an easy way to use up leftovers, particularly back in the days when refrigeration was scarce and food was never wasted. Today, of course, these sorts of dishes are sought out as opportunities to try old-school cooking at its finest. It's a throwback to the days of hearty, unfussy, fat-soaked dishes, from when body fat percentage wasn't quite as important as it is now. This *Trinxat* recipe is especially porky and delicious with the inclusion of *butifarra.*

Also, I say, go big or go home—take the time to fry the cabbage cake in foaming butter, like a real fine-dining cook. Worry about the calories another day—this is cold weather mountain fare, after all.

1   In a large saucepan, prepare the Basic *Sofrito,* using *manteca* as the fat and including the garlic, bay leaf, and *panceta* options from the recipe. Remove from the heat, transfer the *sofrito* to a mixing bowl, and set aside. Wipe out the pan.

2   In the same saucepan, cover the cabbage and potatoes with cold water. Season with the salt until the water tastes like the ocean, and bring the water to a rolling boil. Reduce the heat to medium and simmer for 20 minutes, until the potatoes are just tender. Drain but reserve the veggies in the pot.

3   Return the saucepan to medium heat and stir the veggies until you see that all of the residual water has evaporated. Once the mixture is dry, remove from the heat. Transfer the vegetables to a large mixing bowl.

4   Using a potato masher, mash the cabbage and potatoes and set aside to cool to room temperature (the more steam that is released, the less moisture will remain).

YIELD

4 entrée servings

INGREDIENTS

½   cup (125 g) *Basic Sofrito* (see recipe on p. 418)

1   head napa or savoy cabbage, cored and cut into medium dice

1.1   pounds (500 g) medium russet potatoes, peeled and cut into medium dice

Water, to cover

Kosher salt, to taste

½   cup (125 mL) + 3 tablespoons (45 mL) extra virgin olive oil, as needed

3   *Butifarra Blanca* or *Negra* sausages (see recipes on pp. 269 or 267), removed from the casing and crumbled

White pepper, to taste

Freshly grated nutmeg, to taste

All-purpose flour, as needed

Unsalted butter, as needed

5   In a sauté pan over medium–high heat, warm the 1/2 cup (125 mL) of oil until rippling but not smoking. Add the sausages and sear, breaking them up with a spoon, for 8 to 10 minutes, until thoroughly cooked. Stir and add the reserved *sofrito*. Sauté for 8 to 10 minutes, or until simmering.

6   Add the contents of the sauté pan to the bowl containing the veggies. Mix well. Taste and season the mixture with the salt, white pepper, and freshly grated nutmeg as needed. Cover the bowl with plastic wrap and chill in the refrigerator for 20 to 30 minutes, or until cool.

7   Set a baking sheet on the counter. Using a 1/2 cup (120 mL) measure, scoop out some of the mixture and form it into a ball. Place the ball on a clean work surface. Smash the ball down to form a cake-like shape (like a crab cake). Rest the cake on the baking sheet. Repeat until all of the mixture has been used. Lightly dust the *Trinxat* cakes with the flour and set aside.

8   In a large sauté pan over medium–high heat, warm the remaining 3 tablespoons (45 mL) of oil until rippling. Fry the *Trinxat* cakes in the sauté pan for 8 to 10 minutes, until seared on one side. Flip the cakes, add 2 tablespoons of the butter to the pan, and baste with a spoon, "fine-dining style." Cook for 6 to 8 minutes on the second side, until warmed through. Repeat as needed for the remaining cakes. Serve hot.

NOTE: This dish is pretty heavy, so it goes well with an acidic salad to cut its richness. Otherwise, definitely serve it as they do in the mountains, with a garlicky *alioli* for dipping.

CHORIZO AL INFIERNO

Meat + stick + fire. This, possibly the easiest recipe ever, is a favorite in Andalucían *tapas* bars that (presumably) have solid fire insurance policies. All you need to do is skewer a *chorizo,* place it over a terra cotta vessel, add warmed *aguardiente* or another high-proof liquor, and then **CAREFULLY** set the liquor alight with a long match. Cook and spin, cook and spin, cook and spin...and now eat. Repeat as needed until full, or the beer runs out.

1   Warm a terra cotta or other flameproof dish over high heat for 10 to 15 minutes, until it is very hot. Using a metal skewer long enough to suspend the sausage over the terra cotta dish, impale the sausage. Remove the dish from the heat and place on a heatproof trivet.

2   Place the skewered sausage over the dish and carefully pour the liquor into the dish. With a long match, set the liquor alight. Cook the sausage over the strong flame (watch your fingers, eyebrows, and other body parts) for 6 to 8 minutes, turning as needed, until the sausage is charred and cooked through. Serve hot.

YIELD

1 serving

INGREDIENTS

1   *Chorizo Fresco* or *Chistorra* sausage (see recipes on pp. 226 and 231)

¼   cup (50 mL) *orujo, aguardiente,* or other high-proof neutral liquor

BUTIFARRAS DULCE CON MANZANA

You won't find a better way to enjoy the sweeter style of *embutidos* in the Catalan arsenal than this Christmastime ode to pork, apples, cinnamon, and caramel.

The recipe, which hails from the great restaurant town of Girona, is traditionally served with *Butifarra Dulce* (see recipe on p. 233), but you can also use a *morcilla dulce* with great results. If you do, I would recommend substituting a fruity, sweet, red wine like a Garnacha for the Muscatel.

1   Place a large sauté pan over medium–high heat. Poke several holes in the sausages with a cake tester or small paring knife and get a paper towel-lined plate handy for draining and reserving the sausages.

2   Add the butter to the pan and heat it for 3 to 5 minutes, until it gets foamy. Add the sausages and apples to the pan and brown them on one side for 2 minutes. Flip and cook them on the other side for 2 minutes, but take care not to burn any fond that you build, as you're going to need it for the sauce. Remove from the heat. Place the sausages on the prepared plate and the apples in a small mixing bowl. Set aside.

3   Return the same sauté pan to medium–high heat. Add the sugar, lemon peel, and cinnamon stick. Stirring as little as possible, heat the sugar for 6 to 8 minutes, until it takes on a light amber color.

4   Carefully add the wine and water to the sauté pan—it will sputter violently at first. Return the pan to medium–high heat and bring to a simmer.

5   Add the sausages to the sauté pan and continue simmering for 15 to 20 minutes, until the sauce is reduced by ½ (to a syrup consistency) and the sausages are cooked through. Remove from the heat. Remove and discard the lemon peel and cinnamon stick.

6   Cut the sausages into large chunks and transfer to a serving platter. Pour the sauce over the chunks. Add the apples to the plates. Season with the salt flakes to taste and serve with the crusty bread or brioche for dunking in the sauce.

NOTE: If pears are in season, a firm type like a Bosc makes a great substitute for the apples in this recipe.

YIELD

4 entrée servings

INGREDIENTS

4   *Butifarra Dulce* sausages (see recipe on p. 233)

4   tablespoons (60 g) unsalted butter

4   tart apples, such as Granny Smith, cored and cut into large wedges

½   cup (100 g) granulated sugar

Peel of 1 lemon, in a single strip without any white pith

1   cinnamon stick

1   cup (250 mL) sweet wine, such as Muscatel

½   cup (125 mL) water

Sea salt flakes (such as Maldon), to taste

Crusty bread or brioche

BOLLO PREÑAO

If a Spaniard was going to make a bagel dog, this would be it. Essentially a baguette "impregnated" with a *chistorra* or small *chorizo* sausage (*pan preñao* is also called *pan embarazado*— literally translated as "pregnant bread"), these little beauties are a common and welcome sight after long Madrid evenings of imbibing that bleed into early Madrid mornings of impending hangover.

You can also swing a cheat on these if you are not the bread-baking sort: In a pinch, make *bollos* with everything from pizza dough to that unmentionable popping "dough in a can."

1   In the bowl of a stand mixer fitted with the whisk attachment, mix together the water, yeast, honey, and sugar on low speed for 2 to 3 minutes. Set aside at room temperature for 10 minutes, until the yeast blooms and gets foamy.

2   Replace the whisk attachment with the dough hook attachment. Add the flour to the bowl and knead on medium speed for 1 minute, until the dough forms a shaggy mass. If the dough is too sticky, add a little flour; if it is not sticky enough, add a little more water. Let the dough rest in the bowl to hydrate for 30 minutes at room temperature.

3   Add the salt and continue kneading on medium speed for 5 to 10 minutes, until the *masa* is smooth, elastic, and bounces back immediately when pressed with a finger.

4   Using the olive oil, lightly grease a large mixing bowl. Place the dough ball in the bowl and cover with a damp cloth. Let it rise in a warm place for 30 minutes, until it roughly doubles in size.

5   Remove the dough from the bowl. Punch it down and divide it into 7 equal-sized sections (each roughly 4¼ ounces [120 g]). Form each section into a *barra* (baguette) shape slightly longer than the sausages, and place the *barras* onto a baking sheet or wood board. Re-cover the *barras* with the damp cloth, and set aside for 20 to 30 minutes until the dough is relaxed and proofed by double.

YIELD

7 servings

INGREDIENTS	BAKER'S PERCENTAGE
per 17 ounces (500 g) bread flour, plus more as needed	**100%**
11½ ounces (325 g) warm water	**65%**
¾ teaspoon (3 g) instant or active dry yeast	**.6%**
¾ teaspoon (5 g) honey	**1%**
1½ teaspoons (6 g) granulated sugar	**1.2%**
½ ounce (15 g) kosher salt, plus more to taste	**3%**
Olive oil, for greasing	
7 *Chistorra* or Cantimpalos-Style *Chorizo* (see recipes on pp. 231 and 299)	
1 cup (240 mL) water	

6 Take a *barra*, make a deep slit in the middle of it, and lay 1 of the sausages inside. Pinch the dough around the sausage, crimping together the sides of dough and completely sealing it. Repeat for all of the remaining *barras*.

7 Place the *bollos* on a floured cloth or piece of baker's canvas. Cover with a clean cloth, and set aside to ferment the *bollos* for 20 minutes, or until they have risen slightly.

8 Preheat the oven to 450°F (230°C) for at least 30 minutes, making sure 2 racks are in the oven. Place a ceramic baking stone on the top shelf and place a cast iron pan, baking pan, or second ceramic baking stone on the bottom rack.

9 Remove the cloth from the *bollos*, then place them onto the top ceramic baking stone in the oven (you can also roll the *bollos* onto a floured board or pizza peel to transfer them to the oven). Immediately pour the water into the pan or stone on the bottom rack and quickly shut the door. (The water will create steam to help the bread rise.)

10 Bake the *bollos* for 15 minutes, until they are golden brown. Remove from the oven and cool for 10 to 20 minutes on a rack. Serve warm.

COOKED
EMBUTIDOS

MORCILLA

IF YOU PARTICIPATE in a *matanza* in Spain, one fact is simply unavoidable: There will be blood...lots of blood.

And any *sabía or matancero* worthy of his or her good name will tell you that blood, a historic and beloved source of nutritional value and sustenance, will always, always be collected in a *lebrillo* so one of Spain's great contributions to both ancient and modern gastronomy can be made: *morcilla,* the Spanish take on blood sausage.

Virtually every culture that reveres the pig has found a way to turn the perishable blood of porcine sacrifices into an edible delicacy—be they French *(boudin),* British (black pudding), Polish *(kiszka),* German *(blutwurst),* Italian *(biroldo),* Portuguese *(morcela),* or any other. Most of the time, that means finding a way to lace the blood with fragrant fillings of onions, seasonings, and a binding agent of some kind (potatoes, rice, flours, and cereals are the most common) before cooking or drying the mixture and hanging the sausage up for safekeeping in the winter larder.

Spanish *morcillas* are almost always made in one of two ways: either (1) with rice and/or other starchy binding agents or (2) with a ton of onions as a sort of loose binder. From there, flavorings really depend on whatever region you're in, since more variations on this bloody theme exist than anyone could ever hope to compile and justly represent in this, or any other, book.

The most popular of these recipes is the famous rice-based *morcilla* from Burgos, and you'll find the recipe for it on the very next page. It's amazingly good at a summertime BBQ sliced into hunks and grilled until the rice gets a little crispy.

Replace the rice in the Burgos-style sausage with a whole mess of chopped onions, and you'll have *Morcilla de* Cebolla, the next most commonly found *morcilla* in Spain. It's a favorite to add to most soups and stews, such as *Cocido Madrileño.*

Everything else, in one form or another, is a variation on these two themes. If you add organ meats and other offal to a *Morcilla de Cebolla,* you've made the Catalan variant called *Botifarra Negra* (see recipe on p. 267). If you smoke a *Morcilla de Cebolla* for a while, you'll have a traditional *Morcilla Asturiana* (see recipe on p. 263), one of the secret ingredients in Michelin-starred chef Nacho Manzano's righteously famous *Fabada* (see recipe on p. 287). By lacing a *morcilla* with *chorizo masa,* cooked potatoes, onions, and rice, you'll make my favorite *embutido,* the incomparable *Morcilla Achorizada* (see recipe on p. 322, and when you make it, give me a shout. I'll bring the beer, *amigo!).*

And that's just a small sampling of what the Spanish have to offer on the order of bloody eats and treats, since "nose to tail" in the Spanish countryside means so much more than cooking with just meat and fat.

MORCILLA DE ARROZ

In the northern city of Burgos, you will encounter the greatest *morcilla* known to sausage-kind: *morcilla de Burgos,* a rice-based blood sausage protected under an IGP label. So I won't be calling this recipe *morcilla de Burgos* for fear of landing in sausage jail, but rest assured that this is as close as anything you can find to the real deal here in the United States.

Morcilla de Burgos is a Spanish favorite and one of the few *embutidos* you will find all over the country. Some examples: Simple *tapas* bars in La Rioja throw chunks over a grapevine-clipping fire and cook until crispy. Madrileños serve it for breakfast alongside fried potatoes and eggs. Pamplonans remove the casing and stuff the *masa* into peppers or piquillos before baking. And one of the most creative *pintxo* bars of San Sebastián, Bar Zeruko, coats a "sandwich" of *foie* and *morcilla de Burgos* in pistachios.

The sausage follows a pretty basic *sabía* recipe: A 1:1:1:1 ratio of fat, onions, rice, and blood cooked in the typical murder scene–like blood sausage fashion (see p. 257 for details). Once they've chilled for a bit, you'll have *embutidos* that will last a pretty long time since these sausages chill and freeze very well. And that's a good thing, since you'll find yourself cooking up this *morcilla* with almost everything once you taste its awesomeness.

1   Make a *sofrito:* In a large saucepan, warm the *manteca* over medium heat until rippling but not smoking. Add the minced onions, 1 ounce (25 g) of the salt, and the bay leaf. Reduce the heat to low and allow the onions to sweat for 40 to 60 minutes, until the onions are completely cooked through but not browned (if you bite into a piece, there should be no crunch). Discard the bay leaf and remove from the heat.

2   In a large mixing bowl, combine the *sofrito* and the rice. Cover with plastic wrap and place in the refrigerator to chill for 6 hours to overnight, until the rice has absorbed the flavors and some of the fat. Note that if you don't let the rice absorb all the liquid in this step, you may need to cook the sausages longer later on. If you are rushed for time, you can speed things up by par-cooking the rice before mixing with the sofrito in this step.

3   Place the *sangre,* the *sofrito*–rice mixture, the remaining 2½ ounces (75 g) of salt, the black pepper, the cloves, the anise seed, the nutmeg, and the *pimentones* in the bowl of

YIELD

2–4 loops of sausage per
4¼ cups (1 L) of *masa*

INGREDIENTS	*CHARCUTIER'S* PERCENTAGE
per 4¼ cups (1 L) of *sangre* (pork blood)	**100%**
10½ ounces (300 g) *manteca* (pork lard), divided	**30%**
2.2 pounds (1 kg) minced white onions	**100%**
3½ ounces (100 g) kosher salt, divided, plus more to taste	**10%**
1 fresh bay leaf	
2.2 pounds (1 kg) uncooked bomba* rice	**100%**
2 teaspoons (4 g) freshly ground black pepper, plus more to taste	**4%**
1 teaspoon (2 g) whole cloves, toasted and ground	**.2%**
1 teaspoon (2 g) anise seed, toasted and ground	**.2%**
½ teaspoon (1 g) freshly grated nutmeg	**.1%**
1½ tablespoons (10 g) *pimentón dulce*	**1%**
1½ tablespoons (10 g) *pimentón picante*	**1%**
Water, to cover	
1 tablespoon (15 mL) extra virgin olive oil, for frying	
2 feet (60 cm) 1¼–1½-inch (32–36-mm) hog casings, soaked, or more as needed	

*Bomba rice is a type of rice from Spain typically used for making paella. It is unique because it absorbs 3 times its volume in liquid.

a stand mixer fitted with the paddle attachment (or you can just mix in a mixing bowl with a sturdy spoon). Begin mixing on low speed. Mix for 3 to 5 minutes, until the mixture is fully combined.

4   To make a *prueba,* in a small skillet over medium–high heat, warm the oil. Place a small piece of the *sangre* mixture in the skillet and fry for 3 to 4 minutes, until cooked through. Remove from the heat. Taste and adjust the seasonings to your liking—just bear in mind that the rice won't be fully cooked yet.

5   Fit a casing over the opening of a funnel. Ladle in enough of the *sangre* mixture to loosely fill the casing. Tie the end of the casing, and then tie each sausage into a 12-inch (30-cm) loop. Repeat with the remaining casings. Using a sterile pin or sausage pricker, prick each sausage several times.

6   Place the sausages in a large stockpot. Cover the sausages with the water and place the stockpot over medium–low heat. Once the water temperature reaches 154°F to 158°F (68°C to 70°C), poach the sausages for 45 to 60 minutes, until the internal temperature of the sausages reaches 158°F (70°C) and they are firm to the touch. (At this point, you can gently squeeze the sausage and feel if the rice is cooked.) Remove from the heat.

7   Fill a bowl with ice water. Remove the sausages from the stockpot and immediately plunge them in the water to stop the cooking process. Remove from the water and transfer the sausages to a bowl. Chill overnight in the refrigerator (allowing the sausages to set up) before serving.

TRUCOS DE LA COCINA: CACHONDO

THE COOKING LIQUID left over from poaching this *morcilla* is called *cachondo,* which translates literally as "horny." While I won't guarantee to you the broth's aphrodisiac powers, you can rest assured that it will make a killer broth as a base for *cocido* or other dishes.

MORCILLA DE CEBOLLA

Morcilla de Cebolla is a blood sausage that is bound only by onions, blood, and the angelic soul of the faithful pig. It's a textural experience that is completely different from that of a *morcilla* with a binding agent like rice, so, in my experience, people either absolutely love the texture or are repulsed by it—there really is no middle ground.

Most Spaniards, however, love it, which is why *Morcilla de Cebolla* is the go-to blood sausage in Spain. It's deployed in various guises all over the country for everything from *cocidos* and other stews; paired alongside all sorts of legumes; or used as part of the *sofrito* for sautéed/stewed/simmered/sexified vegetables.

Just know that, yes, this seems like an absolutely ridiculous amount of onions. But remember that onions and blood are the only binding agents for this sausage, and once you cook the onions down, more than half their weight and volume will evaporate. If you're concerned about having a cooking vessel large enough for such a large quantity of onions, home cooks could always opt to split the onions into multiple pots to speed things up. Professionals, meanwhile, have a number of volume cookery tools at their disposal, including jacket kettles and tilt skillets.

1   Make a *sofrito:* In a large stockpot, warm the *manteca* over medium heat until rippling but not smoking. Add the minced onions, 1⅔ ounces (50 g) of the salt, the garlic, and the bay leaves. Reduce the heat to low and allow the onions to sweat for 2 to 3 hours, until the onions are completely cooked through but not browned (if you bite into a piece, there should be no crunch). Discard the bay leaves and remove from the heat.

2   Place the *sangre,* the *sofrito,* the remaining 2⅔ ounces (75 g) of salt, the *tocino,* the black pepper, the cloves, the anise, the nutmeg, the cinnamon, and the *pimentones* in the bowl of a stand mixer fitted with the paddle attachment (or you can just mix in a mixing bowl with a sturdy spoon). Begin mixing on low speed. Mix for 3 to 5 minutes, until the mixture is fully combined.

YIELD

2–4 loops of sausage per 4¼ cups (1 L) of *masa*

INGREDIENTS	*CHARCUTIER'S* PERCENTAGE
per 4¼ cups (1 L) of *sangre* **(pork blood)**	**100%**
4.4 pounds (2 kg) *manteca* (pork lard)	**200%**
33 pounds (15 kg) minced white onions (your grinder's small die would be a great idea for this job!)	**1500%**
4⅓ ounces (125 g) kosher salt, divided, plus more to taste	**12.5%**
1¾ ounces (50 g) garlic, peeled and destemmed	**5%**
2 fresh bay leaves	
6.6 pounds (3 kg) *tocino* (pork fat), chilled and cut into large dice	**300%**
2 teaspoons (4 g) freshly ground black pepper, plus more to taste	**.4%**
¾ teaspoon (1 g) whole cloves, toasted and ground	**.1%**
1 teaspoon (2 g) anise seed, toasted and ground	**.2%**
1 teaspoon (2 g) freshly grated nutmeg	**.2%**
1 teaspoon (2 g) ground cinnamon	**.2%**
1½ tablespoons (10 g) *pimentón dulce*	**1%**
1½ tablespoons (10 g) *pimentón picante*	**1%**
Water, to cover	
1 tablespoon (15 mL) extra virgin olive oil, for frying	
2 feet (60 cm) 1¼–1½-inch (32–36-mm) hog casings, soaked, or more as needed	

3   To make a *prueba,* in a small skillet over medium–high heat, warm the oil. Place a small piece of the *sangre* mixture in the skillet and fry for 3 to 4 minutes, until cooked through. Remove from the heat. Taste and adjust the seasonings to your liking.

4   Fit a casing over the opening of a funnel. Ladle in enough of the *sangre* mixture to loosely fill the casing. Tie into a 12-inch (30-cm) loop. Repeat with the remaining casings. Using a sterile pin or sausage pricker, prick each sausage several times.

5   Place the sausages in a large stockpot. Cover the sausages with the water and place the stockpot over medium–low heat. Once the water temperature reaches 154°F to 158°F (68°C to 70°C), reduce the heat to low and poach for 45 to 60 minutes, until the internal temperature of the sausages reaches 158°F (70°C) and they are firm to the touch. Remove from the heat.

6   Fill a bowl with ice water. Remove the sausages from the stockpot and immediately plunge them in the water to stop the cooking process. Remove from the water and transfer the sausages to a bowl. Chill overnight in the refrigerator (allowing the sausages to set up) before serving.

MORCILLA ASTURIANA

I love Asturias for so many reasons: for its lush green hills; for its earthy, hearty foods served in bucolic *sidra* houses; for its dialect of Spanish so hard to understand that it's basically the Louisiana Swamp People–speak of Spain; and especially for its smoky contribution to the world of *morcillas.*

Morcilla Asturiana is an essential component for *Fabada Asturiana* (see recipe on p. 263), as well as a whole host of other stews and *potajes* famous in the region. And, at its core, this *morcilla* is really just the previous recipe—*Morcilla de Cebolla*—held in a smoker and then aged.

The big question with this recipe, then, is how long should you smoke your *morcilla?* Alas, this is a question simpler than its answer: Smoking meat has been the subject of Asturian essays and arguments across generations, as many Northerners feel that the length of time that an *embutido* sits in the smoker directly impacts its digestibility.

And there's some truth to the argument; it explains why, after eating a *fabada,* you will experience what we Americans colloquially call "The Itis"—a debilitating state of post-meal drowsiness typified by loosened belt buckles and heavy snoozing. In fact, this state of fullness is so infamous with *fabada* that famous food journalist Julio Camba—upon being invited to a *fabada* fete and consuming the stew copiously—joked six months later that he could still feel it in his belly.

That bears out an essential truth and Spanish inside joke about Asturias: Wherever you go, you will never leave hungry—these are a people with gigantic levels of hospitality that is dwarfed only by their idea of portion sizes.

YIELD

2–4 loops of sausage per 4¼ cups (1 L) of *masa*

INGREDIENTS	CHARCUTIER'S PERCENTAGE
per 4¼ cups (1 L) of *sangre* (pork blood) **100%**	
4.4 pounds (2 kg) *manteca* (pork lard) **200%**	
33 pounds (15 kg) minced white onions (your grinder's small die would be a great idea for this job!) **1500%**	
4⅓ ounces (125 g) kosher salt, divided, plus more to taste **12.5%**	
1¾ ounces (50 g) garlic, peeled and destemmed **5%**	
2 fresh bay leaves	
6.6 pounds (3 kg) *tocino* (pork fat), chilled and cut into large dice **300%**	
1⅓ cups (150 g) *pimentón dulce* **15%**	
⅔ cup (75 g) *pimentón picante* **7.5%**	
⅛ ounce (2 g) TCM #1 or DQ #1 curing salt mix (see p. 83) .... **.2%**	
1 teaspoon (2 g) freshly ground black pepper, plus more to taste...... **.2%**	
1 tablespoon (15 mL) extra virgin olive oil, for frying	
2 feet (60 cm) 1¼–1½-inch (32–36-mm) hog casings, soaked, or more as needed	

CONTINUED ON NEXT PAGE

MORCILLA ASTURIANA CONTINUED FROM PREVIOUS PAGE

1 Make a *sofrito:* In a large stockpot, warm the *manteca* over medium heat until rippling but not smoking. Add the minced onions, 1⅔ ounces (50 g) of the salt, the garlic, and the bay leaves. Reduce the heat to low and allow the onions to sweat for 2 to 3 hours, until the onions are completely cooked through but not browned (if you bite into a piece, there should be no crunch). Discard the bay leaves and remove from the heat.

2 Place the *sangre,* the *sofrito,* the remaining 2⅔ ounces (75 g) of salt, the *tocino,* the *pimentones,* the curing salt, and the black pepper in the bowl of a stand mixer fitted with the paddle attachment (or you can just mix in a mixing bowl with a sturdy spoon). Begin mixing on low speed. Mix for 3 to 5 minutes, until the mixture is fully combined.

3 To make a *prueba,* in a small skillet over medium–high heat, warm the oil. Place a small piece of the *sangre* mixture in the skillet and fry for 3 to 4 minutes, until cooked through. Remove from the heat. Taste and adjust the seasonings to your liking.

4 Fit a casing over the opening of a funnel. Ladle in enough of the *sangre* mixture to loosely fill the casing. Tie into a 12-inch (30-cm) loop. Repeat with the remaining casings. Using a sterile pin or sausage pricker, prick each sausage several times. Weigh each sausage to obtain a green weight; record the weights and tag each sausage with its green weight.

TO SMOKE THE SAUSAGES:

1 *To cold smoke, which produces a more subtle sausage with a softer texture:* Stock a cold smoker with oak wood and bring the temperature to 90°F (30°C) and the relative humidity to 80% to 85%. Hang the *morcilla* in the smoking chamber and smoke for 7 to 10 days, or break it up into 4-hour-long sessions for 12 to 14 days, returning the *morcilla* to the drying chamber set at 54°F to 60°F (12°C to 16°C) and 80% to 85% relative humidity in between sessions.

To hot smoke, which produces a smokier, firmer sausage that's a little less refined: Stock a hot smoker with oak wood and bring the temperature to 110°F (43°C) and the relative humidity to 70%. Meanwhile, dry the *morcillas* for at least 2 hours in the refrigerator, until a pellicle forms on the surface (see Chapter 3 for details). Apply smoke to the *morcillas* for 6 hours, gradually increasing the temperature by 10°F (5.6°C) every hour, until the internal temperature of the *morcillas* reaches 158°F (70°C) and is firm to the touch. Place the *morcillas* in ice water to cool them to an internal temperature of at least 110°F (43°C). Chill the sausages overnight in the refrigerator.

TO DRY AND SERVE THE SAUSAGES:

1 *If cold smoking:* Place the *morcillas* in a drying chamber set at 54°F to 60°F (12°C to 16°C) and 80% to 85% relative humidity for 9 to 14 days (depending on their green weight), until they shrivel and have lost about 35% of its green weight. At that point, the sausages will be ready to consume.

If hot smoking: Hot smoking allows the sausages to be ready for consumption immediately, but you can always place them in the drying chamber for awhile once they've cooled if you wish to make them a little firmer and easier to slice. To dry hot-smoked sausages, follow the same directions for drying the cold-smoked sausages.

MORCILLA DULCE

Much like the *butifarra dulce* of Girona, *Morcilla Dulce* hails from that bizarro world to American palates of sweeter sausages containing ingredients like nuts and dried fruits as garnish.

In actuality, *Morcilla Dulce* is a commonplace sausage in La Rioja—one with a long history connected with the *matanzas* performed up there. You will find it almost always served hot in some form—like grilled or pan roasted, or crumbled up as a component in veggie dishes and *tapas*.

1   In a large saucepan over medium–high heat, combine the rice, water, and 1⅓ ounces (40 g) of the salt. Bring to a boil and reduce the heat to medium. Simmer for 25 minutes, until the rice is cooked through. Remove from the heat and set aside to cool.

2   In a small mixing bowl, combine the raisins, figs, and sherry. Set aside and allow the fruits to macerate and plump for 30 minutes to overnight.

3   Place the *sangre,* the reserved rice, the remaining ⅓ ounce (10 g) of salt, the macerated fruits and its wine, the *manteca,* the pine nuts, the sugar, the anise seed, the cinnamon, the cloves, and the white pepper in the bowl of a stand mixer fitted with the paddle attachment (or you can just mix in a mixing bowl with a sturdy spoon). Begin mixing on low speed. After it has mixed for 1 minute, add the bread crumbs and almond flour. Mix for 3 to 5 minutes, until the mixture is fully combined.

4   To make a *prueba,* in a small skillet over medium–high heat, warm the oil. Place a small piece of the *sangre* mixture in the skillet and fry for 3 to 4 minutes, until cooked through. Remove from the heat. Taste and adjust the seasonings to your liking.

5   Fit a casing over the opening of a funnel. Ladle in enough of the *sangre* mixture to loosely fill the casing. Tie into a 12-inch (30-cm) loop. Repeat with the remaining casings. Using a sterile pin or sausage pricker, prick each sausage several times.

6   Place the sausages in a large stockpot. Cover the sausages with the water and place the stockpot over medium–low heat. Once the water temperature reaches 154°F to 158°F (68°C to 70°C), reduce the heat to low and poach for 45 to 60 minutes, until the internal temperature of the sausages reaches 158°F (70°C) and they are firm to the touch. (At this point, you can gently squeeze the sausage and feel if the rice is cooked.) Remove from the heat.

7   Fill a bowl with ice water. Remove the sausages from the stockpot and immediately plunge them in the water to stop the cooking process. Remove from the water and transfer the sausages to a bowl. Chill overnight in the refrigerator (allowing the sausages to set up) before serving.

YIELD

2–4 loops of sausage per 4¼ cups (1 L) of *masa*

INGREDIENTS	CHARCUTIER'S PERCENTAGE
per 4¼ cups (1 L) of *sangre* (pork blood)	**100%**
18 ounces (500 g) uncooked bomba* rice	**50%**
1⅔ quarts (1½ L) water	**150%**
1⅔ ounces (50 g) kosher salt, divided, plus more to taste	**5%**
1 ounce (25 g) golden raisins	**2.5%**
1 ounce (25 g) dried figs, cut into small dice	**2.5%**
½ cup (120 mL) Pedro Ximénez (PX) sherry	
2⅛ cups (½ L) melted *manteca* (pork lard)	**50%**
¼ cup (35 g) toasted pine nuts	**5%**
18 ounces (500 g) granulated sugar	**50%**
1 teaspoon (2 g) anise seed, toasted and ground	**2%**
1 teaspoon (2 g) ground cinnamon	**2%**
1 teaspoon (2 g) whole cloves, toasted and ground	**2%**
1 teaspoon (2 g) freshly ground white pepper	**2%**
2.2 pounds (1 kg) bread crumbs	**100%**
3½ ounces (100 g) almond flour	**10%**
1 tablespoon (15 mL) extra virgin olive oil, for frying	
Water, to cover	
2 feet (60 cm) 1¼–1½-inch (32–36-mm) hog casings, soaked, or more as needed	

* Bomba rice is a type of rice from Spain typically used for making paella. It is unique because it absorbs 3 times its volume in liquid.

From left: *Butifarra Negra*, *Butifarra D'Ou*, and *Butifarra Blanca*

BUTIFARRA NEGRA

This is the quintessential Catalan blood sausage. It's the cousin that most Spaniards don't talk about unless they're from Catalonia, in which case it's the kid they boast to the world about on a bumper sticker.

As with most of Spain's *charcutería* recipes, *Butifarra Negra* has an almost infinite number of variations. I came across this one in Barcelona—a finely diced meat and blood variant laced with delicious little bits of hearts and livers—and it was love at first bloody sight.

Given the proximity of Catalonia to France, is should come as no surprise that this *butifarra* closely resembles some types of French *boudin noir,* the *butifarra's* doppelgänger from across the Franco-Catalan border.

1 Make a *sofrito:* In a medium saucepan, warm the *manteca* over medium heat until rippling but not smoking. Add the minced onions, 1 ounce (28 g) of the salt, and the bay leaf. Reduce the heat to low and allow the onions to sweat for 30 to 40 minutes, until the onions are completely cooked through but not browned (if you bite into a piece, there should be no crunch). Remove from the heat and discard the bay leaf. Allow the *sofrito* to cool to room temperature.

2 Place the *aguja, panceta,* and *papada* meats and grinder parts in the freezer for 30 minutes to par-freeze before attempting to grind.

3 In a large bowl, combine the *aguja, panceta,* and *papada* meats and the *corazon* or *hígado.* Season with the remaining 2⅔ ounces (72 g) of salt and the black pepper.

4 Fill a large bowl with ice. Place a smaller bowl inside the ice-filled bowl. Grind the meat and organ mixture once through a medium-coarse (⅜ inch [9.5 mm]) die into the smaller bowl. Be careful: The meat mixture is wet, so it may squirt and pop out of the grinder.

5 Grind the *corteza* through a fine (⅛ inch [3 mm]) die into the same bowl over ice. Add the *sangre* and *sofrito* to the bowl and stir well, until the meats and *sofrito* are completely integrated. Set aside.

6 In a small mixing bowl, combine the sherry, cinnamon, cloves, white pepper, and nutmeg, making a slurry. Keep the bowl containing the slurry chilled until ready to use.

CONTINUED ON NEXT PAGE

YIELD

2–4 loops of sausage per 4¼ cups (1 L) of *masa*

INGREDIENTS	CHARCUTIER'S PERCENTAGE
per 4¼ cups (1 L) of *sangre* (pork blood)	**100%**
10½ ounces (300 g) *manteca* (pork lard)	**30%**
8¾ ounces (250 g) minced white onions	**25%**
3⅔ ounces (100 g) kosher salt, divided, plus more to taste	**10%**
1 fresh bay leaf	
6.6 pounds (3 kg) diced *aguja* (pork collar)	**300%**
2.2 pounds (1 kg) diced *panceta* (side-cut pork belly)	**100%**
2.2 pounds (1 kg) diced *papada* (pork jowl)	**100%**
1.1 pounds (500 g) diced pork *corazon* (heart) or *hígado* (liver)	**50%**
1 tablespoon (6 g) freshly ground black pepper	**6%**
1.1 pounds (500 g) diced pork *corteza* (skin)	**50%**
¼ cup (50 mL) dry sherry, chilled	
1 teaspoon (2 g) ground cinnamon	**2%**
1 teaspoon (2 g) whole cloves, toasted and ground	**2%**
1 teaspoon (2 g) ground white pepper	**2%**
½ teaspoon (1 g) freshly grated nutmeg	**1%**
1 tablespoon (15 mL) extra virgin olive oil, for frying	
2 feet (60 cm) 1¼–1½-inch (32–36-mm) hog casings, soaked, or more as needed	
Cold water, to cover	

BUTIFARRA NEGRA CONTINUED FROM PREVIOUS PAGE

7   Place the ground meat and organ mixture in the bowl of a stand mixer fitted with the paddle attachment (or you can just mix in a mixing bowl with a sturdy spoon). Begin mixing on low speed. As the mixer runs, pour the sherry slurry into the bowl in a steady stream.

8   Continue mixing on medium speed for 1 to 2 minutes, until the sherry slurry has been fully incorporated into the mixture, a white residue forms on the sides of the bowl, and the mixture firms up. Place the bowl containing the ground meat and organ mixture in the refrigerator to keep it cold until you are ready to stuff the sausage into casings.

9   To make a *prueba,* in a small skillet over medium–high heat, warm the oil. Place a small piece of the *sangre* mixture in the skillet and fry for 3 to 4 minutes, until cooked through. Remove from the heat. Taste and adjust the seasonings to your liking.

10 Fit a casing over the opening of a funnel. Ladle in enough of the *sangre* mixture to loosely fill the casing. Tie into a 12-inch (30-cm) loop. Repeat with the remaining casings. Using a sterile pin or sausage pricker, prick each sausage several times.

11 Place the sausages in a large stockpot. Cover the sausages with the water and place the stockpot over medium–low heat. Once the water temperature reaches 154°F to 158°F (68°C to 70°C), reduce the heat to low and poach for 45 to 60 minutes, until the internal temperature of the sausages reaches 158°F (70°C) and they are firm to the touch. Remove from the heat.

12 Fill a bowl with ice water. Remove the sausages from the stockpot and immediately plunge them in the water to stop the cooking process. Remove from the water and transfer the sausages to a bowl. Chill overnight in the refrigerator (allowing the sausages to set up) before serving.

BUTIFARRA BLANCA

Here again, we see the influence of French *charcuterie* on the Catalans. A Frenchman would look at *Butifarra Blanca* and immediately recognize the DNA of his native *boudin blanc*.

Butifarra Blanca is a pretty basic boiled sausage in Spanish *charcuterie* circles. Like the *Butifarra Negra* (see recipe on p. 267), this sausage is used all over Catalonia as a component in other dishes or as a standalone meal. I'm a big fan of throwing it on the grill and serving with a garlicky *alioli*.

1   Place the *aguja, panceta,* and *papada* meats and grinder parts in the freezer for 30 minutes to par-freeze before attempting to grind.

2   In a large bowl, combine the *aguja, panceta,* and *papada* meats. Season with the salt, black pepper, and curing salt.

3   Fill a large bowl with ice. Place a smaller bowl inside the ice-filled bowl. Grind the meat mixture once through a medium-coarse (3⁄8 inch [9.5 mm]) die into the smaller bowl. Be careful: The meat mixture is wet, so it may squirt and pop out of the grinder.

4   In a small mixing bowl, combine the sherry, water, cinnamon, cloves, white pepper, and nutmeg, making a slurry. Keep the bowl containing the slurry chilled until ready to use.

5   Place the ground meat mixture in the bowl of a stand mixer fitted with the paddle attachment (or you can just mix in a mixing bowl with a sturdy spoon). Begin mixing on low speed. As the mixer runs, pour the sherry slurry into the bowl in a steady stream.

6   Continue mixing on medium speed for 1 to 2 minutes, until the sherry slurry has been fully incorporated into the mixture, a white residue forms on the sides of the bowl, and the mixture firms up. Place the bowl containing the ground meat mixture in the refrigerator to keep it cold until you are ready to stuff the sausage into casings.

7   To make a *prueba,* in a small skillet over medium–high heat, warm the oil. Place a small piece of the meat mixture in the skillet and fry for 3 to 4 minutes, until cooked through. Remove from the heat. Taste and adjust the seasonings to your liking.

YIELD

3–4 loops or 6–8 links of sausage per 2.2 pounds (1 kg)

INGREDIENTS	CHARCUTIER'S PERCENTAGE
per 2.2 pounds (1 kg) of the following blend of diced meats: **60% *aguja* (pork collar), 20% *panceta* (pork belly), and 20% *papada* (pork jowl)**	**100%**
3⁄4 ounce (22 g) kosher salt	**2.2%**
1⁄3 ounce (10 g) freshly ground black pepper	**1%**
1⁄8 ounce (2 g) TCM #1 or DQ #1 curing salt mix (see p. 83)	**.2%**
1⁄2 cup (100 mL) dry sherry, chilled	
1⁄2 cup (100 mL) water, chilled	
1 teaspoon (2 g) ground cinnamon	**.2%**
1 teaspoon (2 g) whole cloves, toasted and ground	**.2%**
1 teaspoon (2 g) ground white pepper	**.2%**
1⁄2 teaspoon (1 g) freshly grated nutmeg	**.1%**
1 tablespoon (15 mL) extra virgin olive oil, for frying	
2 feet (60 cm) 1¼–1½-inch (32–36-mm) hog casings, soaked, or more as needed	
Cold water, to cover	

CONTINUED ON NEXT PAGE

BUTIFARRA BLANCA CONTINUED FROM PREVIOUS PAGE

8   Stuff the mixture into the casings and tie into 12-inch (30-cm) loops or 6-inch (15-cm) links. Using a sterile pin or sausage pricker, prick each sausage several times.

9   Place the sausages in a large stockpot. Cover the sausages with the water and place the stockpot over medium–low heat. Once the water temperature reaches 154°F to 158°F (68°C to 70°C), reduce the heat to low and poach for 45 to 60 minutes, until the internal temperature of the sausages reaches 158°F (70°C) and they are firm to the touch. Remove from the heat.

10 Fill a bowl with ice water. Remove the sausages from the stockpot and immediately plunge them in the water to stop the cooking process. Remove from the water and transfer the sausages to a bowl. Chill overnight in the refrigerator (allowing the sausages to set up) before serving.

NOTE: If you are feeling fancy and have some *Foie Salada en Torchon* (see recipe on p. 375) kicking around, freeze chunks of the *foie* and fold it into the sausage *masa* just before stuffing the *butifarra*. The trick here is twofold: You will need to stuff the *butifarra* with the biggest stuffing tube you have so the *foie* doesn't clog the stuffer. Also, freezing the *foie* allows you to cook the sausage with the *foie* chunks staying somewhat whole as opposed to rendering out of the sausage.

BUTIFARRA D'OU

I remember walking along an outdoor food fair in Barcelona and seeing a commotion at a little *charcuterie* booth. As a properly raised American rubber-necker, I was all over it...

There were cooked and cured sausages in all manner of colors: whites (*butifarra blancas),* blacks and browns (variations on *butifarras negras),* long, skinny reds (*fuet),* and a huge line for the last of some strange, bright yellow, egg-based sausages called *butifarra d'ou.*

The commotion was this: Two little older Catalan ladies, each grabbing onto the end of the horseshoe-shaped yellow sausage, were claiming ownership and intoning—in what I could more-or-less translate as a colorful conversation regarding one another's sexually liberal proclivities with a donkey—for the other to let go. And that was all it took for a crowd to form and watch the show—it turns out Spaniards love a spectacle as much as we do.

In the end, karma prevailed and the sausage gods were merciful: The sausage split in twain, each lady paid for her half, and life moved on.

But any sausage worth fighting over deserves its place of recognition in this book...so here's the recipe.

Note that the yellow color of the sausage is entirely dependent upon the yolks—if you have deep yellow/red yolks like the ones in Europe, your *butifarra* will reflect that color.

1   Place the *aguja, panceta,* and *papada* meats and grinder parts in the freezer for 30 minutes to par-freeze before attempting to grind.

2   In a large bowl, combine the *aguja, panceta,* and *papada* meats. Season with the salt, black pepper, and curing salt.

3   Fill a large bowl with ice. Place a smaller bowl inside the ice-filled bowl. Grind the meat mixture once through a medium-coarse (⅜ inch [9.5 mm]) die into the smaller bowl. Be careful: The meat mixture is wet, so it may squirt and pop out of the grinder.

4   In a small mixing bowl, combine the egg yolks, sherry, water, cinnamon, cloves, white pepper, and nutmeg, making a slurry. Keep the bowl containing the slurry chilled until ready to use.

YIELD

3–4 loops or 6–8 links of sausage per 2.2 pounds (1 kg)

INGREDIENTS	CHARCUTIER'S PERCENTAGE
per 2.2 pounds (1 kg) of the following blend of diced meats: 60% *aguja* (pork collar), 20% *panceta* (pork belly), and 20% *papada* (pork jowl)	**100%**
¾ ounce (22 g) kosher salt	**2.2%**
⅓ ounce (10 g) freshly ground black pepper	**1%**
⅛ ounce (2 g) TCM #1 or DQ #1 curing salt mix (see p. 83)	**.2%**
7 large egg yolks, beaten	
½ cup (100 mL) dry sherry, chilled	
½ cup (100 mL) water, chilled	
1 teaspoon (2 g) ground cinnamon	**.2%**
1 teaspoon (2 g) whole cloves, toasted and ground	**.2%**
1 teaspoon (2 g) ground white pepper	**.2%**
½ teaspoon (1 g) freshly grated nutmeg	**.1%**
1 tablespoon (15 mL) extra virgin olive oil, for frying	
2 feet (60 cm) 1¼–1½-inch (32–36-mm) hog casings, soaked, or more as needed	
Cold water, to cover	

CONTINUED ON NEXT PAGE

BUTIFARRA D'OU CONTINUED FROM PREVIOUS PAGE

5 Place the ground meat mixture in the bowl of a stand mixer fitted with the paddle attachment (or you can just mix in a mixing bowl with a sturdy spoon). Begin mixing on low speed. As the mixer runs, pour the egg yolk slurry into the bowl in a steady stream.

6 Continue mixing on medium speed for 1 to 2 minutes, until the egg yolk slurry has been fully incorporated into the mixture, a white residue forms on the sides of the bowl, the meat appears sticky, and the mixture firms up. Place the bowl containing the ground meat mixture in the refrigerator to keep it cold until you are ready to stuff the sausage into casings.

7 To make a *prueba,* in a small skillet over medium–high heat, warm the oil. Place a small piece of the meat mixture in the skillet and fry for 3 to 4 minutes, until cooked through. Remove from the heat. Taste and adjust the seasonings to your liking.

8 Stuff the mixture into the casings and tie into 12-inch (30-cm) loops or 6-inch (15-cm) links. Using a sterile pin or sausage pricker, prick each sausage several times.

9 Place the sausages in a large stockpot. Cover the sausages with the water and place the stockpot over medium–low heat. Once the water temperature reaches 154°F to 158°F (68°C to 70°C), reduce the heat to low and poach for 45 to 60 minutes, until the internal temperature of the sausages reaches 158°F (70°C) and they are firm to the touch. Remove from the heat.

10 Fill a bowl with ice water. Remove the sausages from the stockpot and immediately plunge them in the water to stop the cooking process. Remove from the water and transfer the sausages to a bowl. Chill overnight in the refrigerator (allowing the sausages to set up) before serving.

MORCILLA BLANCA

Truth be told, I still don't know how I feel about a blood sausage made with no blood. It's kinda like Christian rock for me: I find the whole thing a little confusing and can't help but feel like I'm being cheated out of a more enjoyably depraved experience. But there's no arguing that this *morcilla* tastes pretty good and is popular in a lot of bars in southern Spain, so let's roll with it.

Morcilla Blanca is similar to its blood-soaked cousin *morcilla de burgos,* but in this case, we substitute out the blood for something else that will help bind it together—here, a mixture of pork meat. The British make something similar, white pudding, and it's quite popular in the UK.

I think you'll find this recipe to be pretty versatile, especially if you want to make *morcilla* but can't track down some blood or aren't into the bloodbath it tends to make.

1   Place the *aguja, panceta,* and *papada* meats and grinder parts in the freezer for 30 minutes to par-freeze before attempting to grind.

2   Make a *sofrito*: In a medium saucepan, heat the lard over medium heat. Add the minced onions, ⅓ ounce (10g) of the salt, and the bay leaf. Reduce the heat to low and allow the onions to sweat for 30 minutes, until the onions are completely cooked through but not browned (if you bite into a piece, there should be no crunch).

3   Add the bomba rice to the onions. Cook for 10 minutes, until the grains have absorbed some of the onions' liquid and have plumped.

4   Raise the heat to medium–high. Add 1¼ cups (300 mL) of the water and bring to a boil. Reduce the heat to medium and simmer for 20 to 25 minutes, until the rice is cooked. Remove from the heat and discard the bay leaf. Allow to cool to at least room temperature.

5   In a large bowl, combine the *aguja, panceta,* and *papada* meats. Season with the remaining 1 ounce (25 g) of salt, curing salt, and black pepper.

6   Fill a large bowl with ice. Place a smaller bowl inside the ice-filled bowl. Grind the meat mixture once through a medium-coarse (⅜ inch [9.5 mm]) die into the smaller bowl. Be careful: The meat mixture is wet, so it may squirt and pop out of the grinder. Add the *sofrito* and rice mixture to the bowl and stir well, until the meats and *sofrito* are completely integrated. Set aside.

CONTINUED ON NEXT PAGE

YIELD

3–4 loops or 6–8 links of sausage per 2.2 pounds (1 kg)

INGREDIENTS	CHARCUTIER'S PERCENTAGE
per 2.2 pounds (1 kg) of the following blend of diced meats: 60% *aguja* (pork collar), 20% *panceta* (pork belly), and 20% *papada* (pork jowl)	**100%**
3½ ounces (100 g) *manteca* (pork lard)	**10%**
18 ounces (500 g) minced white onions	**50%**
1⅓ ounces (35 g) kosher salt, divided, plus more to taste	**3.5%**
1 fresh bay leaf	
3½ ounces (100 g) uncooked bomba* rice	**10%**
1¾ cups (420 mL) water, divided	**30%**
1 teaspoon (2 g) freshly ground black pepper	**.2%**
⅛ ounce (2 g) TCM #1 or DQ #1 curing salt mix (see p. 83)	**.2%**
¼ cup (50 mL) dry sherry, chilled	
1 teaspoon (2 g) ground cinnamon	**.2%**
1 teaspoon (2 g) whole cloves, toasted and ground	**.2%**
1 teaspoon (2 g) ground white pepper	**.2%**
½ teaspoon (1 g) freshly grated nutmeg	**.1%**
1 tablespoon (15 mL) extra virgin olive oil, for frying	
2 feet (60 cm) 1¼–1½-inch (32–36-mm) hog casings, soaked, or more as needed	
Cold water, to cover	

MORCILLA BLANCA CONTINUED FROM PREVIOUS PAGE

7 In a small mixing bowl, combine the sherry, the remaining ½ cup (120 mL) water, the cinnamon, the cloves, the white pepper, and the nutmeg, making a slurry. Keep the bowl containing the slurry chilled until ready to use.

8 Place the ground meat mixture in the bowl of a stand mixer fitted with the paddle attachment (or you can just mix in a mixing bowl with a sturdy spoon). Begin mixing on low speed. As the mixer runs, pour the sherry slurry into the bowl in a steady stream.

9 Continue mixing on medium speed for 1 to 2 minutes, until the sherry slurry has been fully incorporated into the mixture, a white residue forms on the sides of the bowl, and the mixture firms up. Place the bowl containing the ground meat mixture in the refrigerator to keep it cold until you are ready to stuff the sausage into casings.

10 To make a *prueba,* in a small skillet over medium–high heat, warm the oil. Place a small piece of the meat mixture in the skillet and fry for 3 to 4 minutes, until cooked through. Remove from the heat. Taste and adjust the seasonings to your liking.

11 Stuff the mixture into the casings and tie into 12-inch (30-cm) loops or 6-inch (15-cm) links. Using a sterile pin or sausage pricker, prick each sausage several times.

12 Place the sausages in a large stockpot. Cover the sausages with the water and place the stockpot over medium–low heat. Once the water temperature reaches 154°F to 158°F (68°C to 70°C), reduce the heat to low and poach for 45 to 60 minutes, until the internal temperature of the sausages reaches 158°F (70°C) and they are firm to the touch. Remove from the heat.

13 Fill a bowl with ice water. Remove the sausages from the stockpot and immediately plunge them in the water to stop the cooking process. Remove from the water and transfer the sausages to a bowl. Chill overnight in the refrigerator (allowing the sausages to set up) before serving.

CHOSCO

Chosco is a hyper-regional, IGP-protected product hand-crafted by a scant few producers in and around the municipality of Tineo, a small town in western Asturias with a population equal to about a square city block in Queens. Compared to more popular *embutidos* like *chorizo* or *morcilla, Chosco* is produced on a miniscule scale, so finding it can be tough.

Chosco is also unique since it is a product made from multiple curing techniques: It's the *cabecero* muscle of the pig combined with the *lengua* (tongue), both of which are first rubbed down with salt and the Spanish *adobo* Superfriends garlic and *pimentón*. It all gets stuffed into a large *ciego* casing, smoked, and sometimes hung up in a drying chamber to cure further.

1   Using a mortar and pestle, crush together the garlic and salt to form an *ajosal*. If desired, you can finish the *ajosal* in a food processor fitted with the "S" blade.

2   In a mixing bowl, combine the meats with the *ajosal,* curing salt (if using), dextrose, and sugar. Toss together and chill in the refrigerator while you prepare the rest of the ingredients.

3   In a small mixing bowl, combine the *pimentón* and wine, making an *adobo*.

4   Wearing disposable gloves (unless you enjoy having red-tinged fingers for the rest of the day), pour the *adobo* over the meats. Mix well with your hands until all of the meat is coated with the *adobo*. Place the meat in a foodsafe ziptop bag or container and refrigerate for 3 days.

5   Tightly stuff the meat pieces into the *ciego* casing. Weigh the *chosco* to obtain a green weight; record and tag each *chosco* with the weight.

6   Using a sterile pin or sausage pricker, prick the *chosco* several times. Tie the *chosco* as you would a *morcón* and secure a loop at one end in order to hang it (see p. 92 in Chapter 3 for an illustration and directions). Smear any excess marinade over the casing to keep it moist and stain it red.

YIELD

1 large *chosco*

INGREDIENTS	*CHARCUTIER'S* PERCENTAGE
per 2.2 pounds (1 kg) of the following blend of meats: 80% *cabecero* (*coppa*/head of pork loin), cut in half lengthwise; and 20% *lengua* (pork tongue), blanched and peeled	**100%**
⅔ ounce (20 g) garlic, peeled and destemmed	**2.0%**
1 ounce (28 g) kosher salt	**2.8%**
⅛ ounce (5 g) dextrose	**.5%**
⅛ ounce (5 g) granulated sugar	**.5%**
1 ounce (25 g) *pimentón dulce*	**2.5%**
¼ cup (50 mL) dry white wine, such as Verdejo, chilled	
1 *ciego de cerdo* (pork bung), soaked	
Cold water, to cover	

OPTIONAL

⅛ ounce (3 g) TCM #2 or DQ #2 curing salt mix	**.3%**

CONTINUED ON NEXT PAGE

CHOSCO CONTINUED FROM PREVIOUS PAGE

7   Stock a cold smoker with oak wood and bring the temperature to 90°F (30°C) and the relative humidity to 70% to 85%. Hang the *Chosco* in the smoking chamber and smoke for 8 to 14 days, or break it up into 6-hour-long sessions for 14 to 21 days, returning the *Chosco* to the drying chamber set at 54°F to 60°F (12°C to 16°C) and 80% to 85% relative humidity in between sessions.

Once your *chosco* is smoked, you have two options for finishing it:

➡ *Chosco* is sometimes cooked, chilled, and then typically boiled in a stew with potatoes.

➡ Otherwise, it is dry cured for a period of time and thinly sliced, like a *chorizo.*

TO COOK THE *CHOSCO:*

1   Place the *Chosco* in a large stockpot. Cover it with the water and place the stockpot over medium–high heat. Once the water temperature reaches 176°F to 212°F (80°C to 100°C), reduce the heat to medium and poach for 45 to 60 minutes, until the internal temperature of the *chosco* reaches 176°F (80°C) and it is firm to the touch. Remove from the heat and chill the *chosco* in the refrigerator overnight.

TO CURE THE *CHOSCO:*

1   Place the smoked *chosco* in a drying chamber set at 54°F to 60°F (12°C to 16°C) and 80% to 85% relative humidity for 7 to 14 days (depending on its green weight), until it has lost about 35% of its green weight. At that point, it will be ready to consume.

NOTE: For spicier *Chosco,* instead of 1 ounce (25 g) *pimentón dulce,* use roughly ½ ounce (12.5 g) of *pimentón dulce* and roughly ½ ounce (12.5 g) of *pimentón picante.*

BOTILLO

Botillo is a product hailing from the area of Bierzo, a small *comarca* in the province of Castilla y León. Much like the *chosco de tineo,* which hails from just north of Bierzo, *Botillo* has no equal—that I am aware of—in the world of stuffed-meat artistry.

This recipe involves a bunch of random meat cuts left over after the *matanza:* tails, ribs, and any other bony trim (yes, bones are stuffed in there too!), all seasoned with Spanish-y flavors heavy on the garlic and *pimentón,* and then stuffed, smoked, and dried.

This *embutido,* much like the *chosco,* often winds up in the winter stew pot alongside tubers, cabbage, and other veggies. Its smoky porkiness works much like Southern cuisine's ham hocks to flavor broth and everything else with essence of piggie.

1   Using a mortar and pestle, crush together the garlic and salt to form an *ajosal.* If desired, you can finish the *ajosal* in a food processor fitted with the "S" blade.

2   In a mixing bowl, combine the meats with the *ajosal,* oregano, dextrose, sugar, and curing salt (if using). Toss together and chill in the refrigerator while you prepare the rest of the ingredients.

3   In a small mixing bowl, combine the *pimentón* and wine, making an *adobo.*

4   Wearing disposable gloves (unless you enjoy having red-tinged fingers for the rest of the day), pour the *adobo* over the meats. Mix well with your hands until all of the meat is coated with the *adobo.* Place the meat in a foodsafe ziptop bag or container and refrigerate for 3 days.

5   Tightly stuff the meat pieces into the *ciego* casing. Weigh the *Botillo* to obtain a green weight; record and tag each *Botillo* with the weight.

6   Using a sterile pin or sausage pricker, prick the *Botillo* several times. Tie the *Botillo* as you would a *morcón* and secure a loop at one end in order to hang it (see Chapter 3 for an illustration and directions). Smear any excess marinade over the casing to keep it moist and stain it red.

7   Stock a cold smoker with oak wood and bring the temperature to 90°F (30°C) and the relative humidity to 70% to 85%. Hang the *Botillo* in the smoking chamber and smoke for 4 to 7 days, or break it up into 6-hour-long sessions for 7 to 10 days, returning the *Botillo* to the drying chamber set at 54°F to 60°F (12°C to 16°C) and 70% to 80% relative humidity in between sessions.

8   Place the smoked *Botillo* in a drying chamber set at 54°F to 60°F (12°C to 16°C) and 80% to 85% relative humidity for 9 to 14 days (depending on its green weight), until it has lost about 35% of its green weight. At that point, it will be ready to consume.

YIELD

1 *Botillo*

INGREDIENTS	CHARCUTIER'S PERCENTAGE
per 2.2 pounds (1 kg) of the following blend of diced meats: 65% *costilla* (pork rib meat); 15% *lengua* (pork tongue), blanched and peeled; 10% *rabo* (bone-in pork tail); and 10% piggy mix (shoulder, cheeks, etc.)	**100%**
⅔ ounce (20 g) garlic cloves, peeled, destemmed, and smashed into a paste	**2%**
1 ounce (25 g) kosher salt	**2.5%**
½ tablespoon (2 g) dried oregano	**.2%**
⅛ ounce (5 g) dextrose	**.5%**
⅛ ounce (3 g) granulated sugar	**.3%**
1 ounce (25 g) *pimentón dulce*	**2.5%**
¼ cup (50 mL) dry white wine, such as Verdejo, chilled	
1 *ciego de cerdo* (pork bung), soaked Cold water, to cover	

OPTIONAL

⅛ ounce (3 g) TCM #2 or DQ #2 curing salt mix	**.3%**

NOTE: For spicier *Botillo,* instead of 1 ounce (25 g) *pimentón dulce,* use roughly ½ ounce (12.5 g) of *pimentón dulce* and roughly ½ ounce (12.5 g) of *pimentón picante.*

BULL CATALAN

Bull Catalan—sometimes called *el bisbe* (the bishop)—is one of the pride-and-joy *embutidos* of Catalan *charcuterie*, so that's why this recipe is pretty involved. All sorts of different techniques and tricks are employed to imbue this enormous sausage with flavor.

Every shop in Barcelona tries to show off by adding different and varied goodies inside of their *Bull*. I've seen it laced with everything from pork blood (called *bull negre)*, to liver and head pieces, to even a whole *foie gras* garnish in the center. This recipe, as well, is pretty labor intensive, but it will yield a showpiece sausage if you decide to take the *Bull* by its horns.

If you are feeling especially brave and want to try something different, make these sausages in the style of classic *andouillettes*, which means adding 5% soaked tripe pieces (aka chitterlings) to the interior garnish. But don't say I didn't warn you: Even the flavorful and fragrant broth won't stop the sausage (and everything else in the area) from taking on the stinky funk that only boiling tripe can deliver.

1   Place the piggy mix (head, trotters, kidneys, liver, or chitterlings) and *lengua,* which make up the interior garnish meats, in a foodsafe container that is just large enough to contain them. Cover with the water. Remove the meats and pour the water into a measuring cup. Make note of the amount of water, as this is the amount of Basic *Salmuera* you will need to have. Discard the water.

2   To the container you measured the water in, add the Basic *Salmuera*. Add to the *salmuera* 1 each of the onions, carrots, and leeks; half of the head of garlic; 1 teaspoon of the black peppercorns; and the toasted coriander seeds, cloves, and allspice berries. Add the garnish meats to the container, seal it, and place it in the refrigerator to brine. Consult the table on p. 107 to determine the right brine time for the mixture of meats you are preparing.

3   Remove the garnish meats from the brine, reserving the brine. Rinse the garnish meats well in cold water.

YIELD

8–10 links or 3–4 large sausages per 2.2 pounds (1 kg)

INGREDIENTS	*CHARCUTIER'S* PERCENTAGE
per 2.2 pounds (1 kg) of the following blend of meats and parts: 60% *panceta* (pork belly), 20% piggy mix (head, trotters, kidneys, liver, or chitterlings), 10% *tocino* (pork fat), and 10% *lengua* (pork tongue) **100%**	

FOR THE *SALMUERA:*

	Basic *Salmuera* (see recipe on p. 105), as needed
2	medium yellow onions, peeled and cut into large dice, divided
2	medium carrots, peeled and cut into large dice, divided
2	medium leeks, peeled and cut into large dice, divided
1	head garlic, divided
2	teaspoons (4 g) toasted black peppercorns, divided
1	teaspoon (2 g) toasted coriander seeds
1	teaspoon (2 g) toasted whole cloves
1	teaspoon (2 g) toasted allspice berries

FOR THE *COCCIÓN* (POACHING LIQUID):

	Water, to cover
1	toasted star anise
10	stems fresh parsley
1	fresh bay leaf
5	sprigs fresh thyme
1	sprig fresh rosemary

continued

INGREDIENTS	CHARCUTIER'S PERCENTAGE

FOR FINISHING THE SAUSAGE:

²⁄₃ ounce (20 g) kosher salt, plus more to taste............................**2%**

1½ tablespoons (10 g) freshly ground black pepper........................ **1%**

⅛ ounce (2 g) TCM #1 or DQ #1 curing salt mix (see p. 83) **2%**

½ cup (100 mL) dry sherry, chilled

½ cup (100 mL) water, chilled

1 teaspoon (2 g) ground cinnamon...........................**2%**

1 teaspoon (2 g) whole cloves, toasted and ground **2%**

1 teaspoon (2 g) ground white pepper...................................**2%**

½ teaspoon (1 g) freshly grated nutmeg................................**1%**

1 tablespoon (15 mL) extra virgin olive oil, for frying

4 feet (1.2 m) 1⅓–1¾-inch (35-mm–45-mm) hog casings or beef middles, soaked, or more as needed

4   Pour the brine (including the vegetables and seasonings) into a measuring cup. Make note of the amount of brine, as this is the amount of *cocción* you will need to have. Discard the brine.

5   In a large stockpot over medium–high heat, combine the *cocción* ingredients—the remaining 1 each of the onions, carrots, and leeks; the remaining half of the head of garlic; and the remaining 1 teaspoon of the black peppercorns; plus the star anise, parsley, bay leaf, thyme, and rosemary—with the interior garnish meats. Pour in water equal to the amount of brine you measured, and bring the meats and poaching liquid to a boil.

6   Reduce the heat to medium and simmer, repeatedly skimming off any foam or impurities that rise to the surface, for 10 to 60 minutes, until the meats are just cooked through to the desired doneness for each one. Bear in mind, however, that cooking times will vary depending on types of meats used.

7   Pull out each piece of the meat as you determine that it is cooked through and set aside in a bowl to cool. (If you are using tongue, head, or trotters, peel them as soon as they are cool enough to handle, as peeling is easiest then.)

8   After all the interior garnish meats have been cooked, strain the *cocción* into a mixing bowl and reserve it. Discard the solids. Set aside.

9   Dice the interior garnish meats by hand into irregular small chunks. If using organ meats, cut them into a smaller dice and set aside in a separate bowl.

CONTINUED ON NEXT PAGE

BULL CATALAN CONTINUED FROM PREVIOUS PAGE

10 Place the *panceta* and *tocino* and the grinder parts in the freezer for 30 minutes to par-freeze before attempting to grind.

11 In a large mixing bowl, combine the meats with the salt, black pepper, and curing salt. Toss together and chill in the refrigerator while you prepare the rest of the ingredients.

12 Fill a large bowl with ice. Place a smaller bowl inside the ice-filled bowl. Grind the meat mixture once through a medium-coarse (⅜ inch [9.5 mm]) die into the smaller bowl. Be careful: The meat mixture is wet, so it may squirt and pop out of the grinder.

13 In a small mixing bowl, combine the sherry, water, cinnamon, cloves, white pepper, and nutmeg and stir, making a slurry. Keep the bowl containing the slurry chilled until ready to use.

14 Place the ground meat mixture in the bowl of a stand mixer fitted with the paddle attachment (or you can just mix in a mixing bowl with a sturdy spoon). Begin mixing on low speed. As the mixer runs, pour the sherry slurry into the bowl in a steady stream.

15 Continue mixing on medium speed for 1 to 2 minutes, until the sherry slurry has been fully incorporated into the mixture, a white residue forms on the sides of the bowl, and the mixture firms up. Add the reserved garnish meats to the bowl and stir gently to combine. Place the bowl containing the ground meat mixture in the refrigerator to keep it cold until you are ready to stuff the sausage into casings.

16 To make a *prueba,* in a small skillet over medium–high heat, warm the oil. Place a small piece of the meat mixture in the skillet and fry for 3 to 4 minutes, until cooked through. Remove from the heat. Taste and adjust the seasonings to your liking.

17 You can stuff this sausage into regular-sized hog casings, like any other *butifarra,* or you can stuff it into larger beef middles to make a true *Bull Catalan* like those found in many Barcelona *charcutería* shops.

If stuffing into regular casings: Stuff the mixture into the casings and tie into 6-inch (15-cm) links. Using a sterile pin or sausage pricker, prick each sausage several times.

If stuffing into jumbo casings: Stuff the mixture into the beef middles and tie each sausage as you would a roast. Tie a number of knots down the length of the sausage, as you would with a *ciego.* This will ensure that the meat stays compact in the casing during the whole cooking period. Using a sterile pin or sausage pricker, prick each sausage several times.

18 Place the sausages in a large stockpot. Cover the sausages with the reserved *cocción* and place the stockpot over medium–low heat. Once the water temperature reaches 154°F to 158°F (68°C to 70°C), reduce the heat to low and poach for 45 to 60 minutes, until the internal temperature of the sausages reaches 150°F (65°C) and they are firm to the touch. Remove from the heat.

19 Fill a bowl with ice water. Remove the sausages from the stockpot and immediately plunge them in the water to stop the cooking process. Remove from the water and transfer the sausages to a bowl. Chill overnight in the refrigerator (allowing the sausages to set up) before serving.

COOKED

EMBUTIDOS

TRADITIONAL AND MODERN RECIPES

GARBANZOS CON BUTIFARRA NEGRA

Hang out in Barcelona's La Boquería market long enough, and you'll wind up on one of the stools at Bar Pinotxo, one of the best places to eat at the market. If you've heard most anything about Barcelona, you know Bar Pinotxo's story: Great food, an unpretentious joint, and Juanito Bayen's vest/bow tie/welcoming smile all make for a fun time.

Plus, the cooks use whatever they find in the market stalls around them, so everything is fresh, in season, and cooked simply.

There's almost always a garbanzo dish like this one on the wintertime menu and it represents everything I love about what they do there; this is basically a super-charged, updated version of the *a la catalana* style of veggie dishes.

1   In a medium mixing bowl, combine the chickpeas and baking soda. Cover with the water and soak overnight at room temperature.

2   Drain the chickpeas. Place them in a large stockpot over medium heat, cover with fresh water (or substitute any stock you have on hand, to add more flavor), and simmer for 1 hour, until cooked through. Remove from the heat.

3   Strain the chickpeas into a mixing bowl. Set aside to cool. (I recommend reserving the broth and reusing it as a terrific vegetarian soup base.)

4   Place the raisins in a small mixing bowl. Cover with the sherry and set aside for 30 minutes, until plumped.

5   In a sauté pan over medium–high heat, warm ¼ cup (60 mL) of the oil until rippling but not smoking. Add the onions, season with the salt, and cook, stirring frequently, for 20 minutes, until they begin to turn brown. Add the brown sugar and continue cooking for 15 more minutes, until the onions are a deep golden brown. Remove from the heat. Transfer the onions to a mixing bowl, set aside, and wipe out the sauté pan.

6   In a blender, purée together the 1 bunch parsley, the ¼ ounce (6 g) mint leaves, the garlic, and another ¼ cup (60 mL) of the oil until the oil is bright green. Strain the oil through a sieve and set aside in a small pitcher. Discard the solids.

YIELD

4–6 entrée servings

INGREDIENTS

5¼   ounces (150 g) dried chickpeas

1   tablespoon (20 g) baking soda

   Cold water, to cover, divided

2¾   ounces (80 g) golden raisins

½   cup (120 mL) Pedro Ximénez (PX) sherry

1½   cups (370 mL) extra virgin olive oil, divided

3   medium yellow onions, cut into medium dice

   Kosher salt, as needed

2   ounces (55 g) dark brown sugar

1   bunch flat-leaf parsley, chopped, plus more as needed for garnish

¼   ounce (6 g) chopped mint leaves, plus more as needed for garnish

2   cloves garlic, peeled and lightly crushed

4   *Butifarra Negra* sausages (see recipe on p. 267), removed from the casings and crumbled

1¾   ounces (50 g) toasted pine nuts

1   ounce (30 g) loosely packed spinach leaves, rinsed

   Freshly ground black pepper, to taste

CONTINUED ON NEXT PAGE

GARBANZOS CON BUTIFARRA NEGRA

CONTINUED FROM PREVIOUS PAGE

7   In the sauté pan over medium–high heat, warm the remaining 1 cup (250 mL) of oil until rippling but not smoking. Add the sausages and sear, without disturbing them, for 1 minute. Stir and add the reserved chickpeas, the onions, the raisins and soaking liquid, and the pine nuts. Cook for 10 minutes, until warmed through and simmering.

8   Add the spinach and stir to combine. Sauté the mixture for 1 minute, until the spinach is just wilting. Taste for seasoning and remove from the heat.

9   Mound the *Garbanzos con Butifarra Negra* onto plates. Drizzle with the reserved parsley–mint oil and season with the salt and black pepper to taste. Garnish with the chopped parsley and mint.

NOTE: In a pinch, you can use canned chickpeas if you rinse them very, very well. They won't have the same texture as soaked ones, but they work well if you are just looking for a quick meal.

TRUCOS DE LA COCINA: COOKING CHICKPEAS

ANY TIME I soak dried chickpeas for cooking, I follow a *sabía* trick of adding a little baking soda into the water. This tradition, which goes back generations, yields chickpeas with a slightly creamier consistency than those soaked in plain water. As explained by food expert Harold McGee in *On Food and Cooking,* the alkalinity of the baking soda helps dissolve the chickpea's cell walls, allowing them to break down quicker. Hence, the chickpeas take on a creamier quality.

NACHO MANZANO'S *FABADA*

If Asturias is the kingdom of *fabada,* then Nacho Manzano is its king. From his mountaintop, Michelin-starred restaurant Casa Marcial, Nacho presides over all sorts of modern and classic comfort dishes of the Asturian repertoire, but his *fabada* is second to none.

Laced with deliciously smoky *Chorizo Asturiano* and *Morcilla Asturiana, fabada* is basically the Asturian version of pork and beans. The pork comes in every conceivable format (the meats are called the *compango)* and the *fabes* are the gigantic, creamy Asturian white beans native to these lands.

This recipe adds a little bit of herbage at the end, in the form of parsley and bay leaf. Purists might scoff at the herbs I add at the end, but I'm a fan of the bit of freshness they add to what is otherwise one of the heaviest dishes in the Spanish culinary vernacular. Even with the herbs, this will put you on the couch—for at least a week—in the happiest food coma you'll ever know.

YIELD

4–6 entrée servings

INGREDIENTS

18	ounces (500 g) dried *fabes* (Asturian white beans)
1	gallon (4 L) room-temperature water, divided, plus more to cover
9	ounces (250 g) *panceta* (cured pork belly)
1	ham bone
9	ounces (250 g) *Jamón* or *Lacón* (see recipes on pp. 143 and 151)
3	*Chorizo Asturiano* sausages (see recipe on p. 308)
2	*Morcilla Asturiana* sausages (see recipe on p. 263)
½	white onion, studded with a clove
5	whole cloves garlic
2	cups (500 mL) ice water, divided
10	stems fresh parsley
1	fresh bay leaf
20	whole black peppercorns
½	cup (100 mL) extra virgin olive oil
1¾	ounces (50 g) *pimentón dulce*
	Pinch saffron
	Kosher salt, to taste

CONTINUED ON NEXT PAGE

NACHO MANZANO'S *FABADA*

CONTINUED FROM PREVIOUS PAGE

1   In a large stockpot, cover the *fabes* with 2 quarts (2 L) of the
    water. Soak overnight at room temperature. In a separate large
    bowl, cover the *panceta* and ham bone with the remaining 2
    quarts (2 L) of water and soak at room temperature for the same
    amount of time.

2   Place the *Jamón,* sausages, onion, and garlic in the stockpot with
    the beans and bean-soaking water. Drain the *panceta* and ham
    bone and add them to the pot as well. Add enough water to cover.

3   Place the stockpot over medium–high heat. Bring to a boil and
    skim off the foam. Add 1 cup (250 mL) of the ice water to "shock"
    the beans. Return to a boil, and then add the remaining 1 cup
    (250 mL) of ice water. Bring to a boil and skim off the foam.

4   Reduce the heat to medium. Make a *bouquet garni* by placing
    the parsley, bay leaf, and peppercorns in a piece of cheesecloth,
    forming a pouch, and tying a long piece of kitchen twine to close
    the pouch. Drop the *bouquet garni* in the stockpot with a tail of
    twine hanging on the side of the stockpot. Add the oil, *pimentón,* and saffron. Gently shake the stockpot to
    incorporate the added ingredients. (From this point forward, do not stir the stockpot. Doing so will break
    up the beans. Instead, just give the stockpot a shake from time to time.)

5   Cook for 40 to 60 minutes, gently shaking the stockpot on occasion, until the beans are tender and cooked
    through. Remove from the heat. Remove and discard the *bouquet garni* and the ham bone. Season with the
    salt, shaking the pot gently to disperse the salt.

6   Remove the *compango* (meats) from the pot and place them on a platter cut into bite-sized pieces. Transfer
    the *fabes* to a serving bowl, and serve the *compango* and *fabada* warm.

NOTE: *Fabes* are perhaps the most famous bean in Spain. It's a large, flat, white bean with an exceptionally creamy and
buttery texture that really has no equal anywhere in the world. You can find *fabes* at most Spanish grocers like tienda.com or
New York City's Despaña.

TRUCOS DE LA COCINA: SHOCKING BEANS

MANY SPANISH BEAN recipes call for "shocking" larger varieties of beans, like the *fabes* used in this
recipe. *Sabias* shock the beans so they cook uniformly. Adding ice water to a simmering pot of beans slows
the cooking process for the beans' exterior, thus allowing everything to cook at an even pace.

CALDO GALLEGO

Whereas *fabada* is the pork and beans specialty of Asturias, *Caldo Gallego* is the Galician version. This wintery, stick-to-your-ribs Gallego comfort food contains a very unique regional ingredient: *grelos,* known in America as the leafy greens of turnips. Simmering bitter, peppery greens in a porky stock is common all over the world, but in Galicia *grelos* are also used in everything from soups and stews to sautéed dishes—it's the go-to veggie. You can substitute broccoli rabe leaves or dandelion leaves if you can't find turnip greens, but it's worth it to go turnip if you can!

If you are cooking this dish for a native Gallego either track down some *Unto Gallego* and *Lacón Cocido* (see recipes on pp. 119 and 151) or risk a little criticism. If that's too difficult, you can cheat with some nice, streaky bacon or cured *panceta* for the *unto* and some country ham for the *Lacón.*

1   In a medium stockpot, cover the beans with 2 quarts (2 L) of the water and soak overnight at room temperature. In a separate bowl, cover the *panceta* and ham bone with the remaining 2 quarts (2 L) of water and soak at room temperature for the same amount of time.

2   Place the *Lacón* and *Unto Gallego* in the stockpot with the beans and bean-soaking water. Drain the *panceta* and ham bone and add them to the pot as well. Add enough water to cover.

3   Place the stockpot over medium–high heat. Bring to a boil and skim off the foam. Add the ice water to "shock" the beans. Return to a boil, and then reduce the heat to a low simmer.

4   Add the sausages, gently shaking the stockpot to incorporate them. (From this point forward, do not stir the stockpot. Doing so will break up the beans. Instead, just give the stockpot a shake from time to time.)

5   Cook for 30 to 40 minutes, gently shaking the stockpot on occasion, until the beans are starting to soften. Add the potatoes, *grelos,* and turnips. Continue cooking for 15 minutes, until the beans are cooked through. Remove from the heat. Remove and discard the ham bone. Season with the salt, shaking the pot gently to disperse the salt.

6   Remove the meats from the pot and cut them into bite-sized pieces. Transfer the *Caldo Gallego* to a serving bowl and add the meats to the bowl. Serve warm.

NOTE: This stew is delicious served with a sharp Albariño wine and hearty, rustic bread for dunking.

YIELD

4–6 entrée servings

INGREDIENTS

18   ounces (500 g) dried small white beans, such as cannellini or white kidney

1   gallon (4 L) room-temperature water, divided, plus more to cover

9   ounces (250 g) *panceta* (cured pork belly)

1   ham bone, *Huesos Salados* (preserved pork bone, see recipe on p. 114), or *Codillo en Salmuera* (preserved pork shank, see recipe on p. 109)

9   ounces (250 g) *Lacón Cocido* (see recipe on p. 151)

3½   ounces (100 g) *Unto Gallego* (see recipe on p. 119)

1   cup (240 mL) ice water

3   *Chorizo Asturiano* sausages (see recipe on p. 308)

5   medium russet potatoes, peeled and quartered

1   bunch *grelos* (turnip greens), stemmed and chopped

10   small turnips, peeled and quartered

Kosher salt, to taste

CARCAMUSA

While I was cooking in Toledo with Adolfo's crew, our local bar of choice was Bar Ludeña. Like many other bars in the area, Ludeña doesn't look like much: The signage is old, the building is older, and the beer selection is generally awful, particularly given the Spanish standard for shitty beer. But unlike many bars in the area, Bar Ludeña is well known for a Manchego specialty that's great bar food and even more fun to say: It's called *Carcamusa*.

Think of *Carcamusa* as the best chili you will ever have: ground pork stewed for a long time with a thick, tomato-heavy *sofrito,* whatever cured meat you have on hand, and finished with peas. This dish is my go-to chili for football season. Serve it with crusty bread and some appropriately terrible Spanish lager.

1   In a medium saucepan, warm the oil over medium–high heat until rippling but not smoking.

2   Season the *lomo* and *aguja* meats with the salt and add it to the saucepan. Sear for 4 to 6 minutes, until the meat is browned. Transfer the meat to a bowl and set aside.

3   Add the sausages and *jamón* to the saucepan. Sauté for 10 minutes, until the sausage's fat has rendered. Transfer the sausages and *jamón* to the bowl containing the meats.

4   Add the onion, garlic, and bay leaves to the saucepan and season with the salt. Sauté for 15 to 20 minutes, until the onions are very soft and starting to brown. Add the Piquillo *Confit* and sauté for 10 minutes, until their liquid has evaporated.

5   Add the *Tomate Frito* and the reserved meats. Bring the *Carcamusa* to a boil and then reduce the heat to medium. Simmer for 15 minutes more, until the meat is soft and cooked through. Add the peas and warm through, then remove the stew from the heat and serve warm.

NOTE: *Carcamusa* is typically served as a communal dish, so pour it into a large, warm terra cotta bowl and place in the center of the table, alongside some bread for dunking.

YIELD

4–6 entrée servings

INGREDIENTS

¼   cup (60 mL) extra virgin olive oil

18   ounces (500 g) *lomo* (pork loin), cut into large dice

18   ounces (500 g) *aguja* (pork collar), cut into large dice

     Kosher salt, as needed

4   Cantimpalos-Style or Riojano-Style *Chorizo* sausages (see recipes on pp. 299 and 302), cut into small dice

5   ounces (150 g) diced *Jamón* or *Lacón Cocido* (see recipes on pp. 143 and 151)

1   medium yellow onion, peeled and cut into small dice

5   cloves garlic, minced

2   fresh bay leaves

4   Piquillo *Confit* peppers (see recipe on p. 408), minced

1   quart (950 mL) *Tomate Frito* (see recipe on p. 407)

11   ounces (300 g) frozen peas

ALUBIAS NEGRAS DE TOLOSA

To do this one right, you'll need to get your hands on some proper black beans from the little town of Tolosa in País Vasco (don't worry, they are available here in the United States).

Regular dried black beans are also doable, but be aware that the starch content is different and less creamy. In fact, Tolosa beans are so special and revered that they have their own *denominación de origin* protection and, at a festival each November, the local "Bean Brotherhood" makes vats of the stuff to pass out on the street.

According to some purists, you should cook the beans with nothing more than olive oil and maybe an onion. That's it—no pork, no tomato, no spices. I say that's no fun, and fortunately, some Spaniards agree with me.

This properly porky and spicy recipe is best consumed alongside a jar of pickled spicy peppers, just like they do in País Vasco. You eat a little stew and then bite into a pepper, allowing the juice of the pepper to drip into the beans and cut their richness. Repeat until full, then repeat some more.

Most importantly, this dried bean is so small that it doesn't need to be presoaked before cooking. This is a big deal: Presoaking requires forethought and planning, constraints that could be prohibitive to a quick and nutritious meal.

1   In a large bowl, cover the *panceta* and ham bone with the 2 quarts (2 L) of water and soak at room temperature for at least 2 hours. Drain.

2   Place the *panceta,* ham bone, and beans in a large stockpot. Add enough water to cover. Place the stockpot over medium–high heat. Bring to a boil and skim off the foam.

3   Add the 3 tablespoons of oil and gently shake the pot until the oil is incorporated. (From this point forward, do not stir the stockpot. Doing so will break up the beans. Instead, just give the stockpot a shake from time to time.)

4   Reduce the heat to medium–low. Make a *bouquet garni* by placing the parsley, bay leaf, and peppercorns in a piece of cheesecloth, forming a pouch, and tying a long piece of kitchen twine to close the pouch. Drop the *bouquet garni* in the stockpot with a tail of twine hanging on the side of the stockpot. Gently shake the stockpot to incorporate the added ingredients.

YIELD

4–6 entrée servings

INGREDIENTS

9	ounces (250 g) *panceta* (cured pork belly)
1	ham bone
2	quarts (2 L) room-temperature water, plus more to cover
18	ounces (500 g) dried *alubias negras de Tolosa* (black beans)
3	tablespoons (45 mL) + ½ cup (100 mL) extra virgin olive oil, divided
10	stems fresh parsley
1	fresh bay leaf
20	whole black peppercorns
1	medium yellow onion, cut into small dice
5	cloves garlic, peeled and minced
	Kosher salt, to taste
5	Piquillo *Confit* peppers (see recipe on p. 408), minced
¾	ounce (20 g) *pimentón picante*
⅓	ounce (10 g) *pimentón dulce*
3	Cantimpalos-Style *Chorizo* sausages (see recipe on p. 299), pricked with a fork
2	*Morcilla de Cebolla* sausages (see recipe on p. 260), pricked with a fork
	Sherry vinegar, as needed

CONTINUED ON NEXT PAGE

ALUBIAS NEGRAS DE TOLOSA CONTINUED FROM PREVIOUS PAGE

5   Meanwhile, make a *sofrito:* To a medium sauté pan over medium–high heat, add the remaining ½ cup (60 mL) of oil, the onion, and the garlic. Season with the salt and sauté for 15 minutes, until the onion is very soft and starting to brown. Add the Piquillo *Confit* to the pan and sauté for 10 minutes, until their liquid has evaporated. Add the *pimentones* and sauté for 1 minute (be careful, as *pimentón* burns easily).

6   Add the *sofrito* and the sausages to the stockpot containing the beans. Cook for 20 minutes, gently shaking the stockpot on occasion, until the beans are tender and cooked through. Remove from the heat. Remove and discard the *bouquet garni* and the ham bone. Taste and season with the salt and sherry vinegar, shaking the pot gently to disperse the ingredients.

7   Transfer the stew to a serving bowl, removing the meats and cutting them into bite-sized pieces. Return the pieces to the stew and serve warm.

NOTES: This recipe is delicious served with some pickled peppers, such as Guindillas (see recipe on p. 383).

Spanish black beans are available at most Spanish grocers, such as tienda.com or New York City's Despaña.

CHOSCO OR *BOTILLO* *CON PATATAS*

Chosco or *Botillo* both work well in this traditional boiled dinner recipe. Since they hail from relatively close locales, you'll find either *embutido* to be interchangeable as the featured meat of this dish.

This stew is commonly found all over the north, where a simple wintertime boiled dinner of cured meats, potatoes, and cabbage are the means to counter the constant winter chill. It's not all that different from the Irish stew of corned beef and cabbage.

Have some bread on hand for dunking and a beer nearby while you eat this elbows-on-the-table kind of meal.

1   In a large saucepan over medium–high heat, cover the *Chosco* or *Botillo* with the water. Bring to a boil.

2   Reduce the heat to medium and simmer for 30 minutes. Add the *Chorizo Asturiano* to the pot and continue simmering for 30 minutes, until the meats are just warmed through.

3   Preheat the oven to 200°F (93°C). Add the cabbage and potatoes to the saucepan. Continue cooking for 20 minutes, until the cabbage has softened. Remove the cabbage to a large baking dish and set aside, but continue cooking the meat and potatoes for 10 minutes, until soft. Remove the potatoes and *chorizo* and place them in the dish containing the cabbage. Splash the veggies and *chorizo* with a little broth, cover them with foil, and place the baking dish in the oven to keep warm.

4   Raise the heat to medium–high and return to a boil. Cook for 8 to 10 minutes, until the liquid volume has been reduced by 1/3. Remove from the heat and remove the baking dish from the oven.

5   Using a slotted spoon, remove the *Chosco* or *Botillo* from the saucepan. Cut the *chorizos* into bite-sized pieces and arrange them in the dish. If using the *Chosco,* slice into pieces and arrange in the dish; if using the *Botillo,* split the *Botillo* open and place on a platter. Moisten the baking dish and the platter (if using) with the cooking liquid and season to taste with the salt and olive oil. Serve warm.

NOTE: This dish goes great with a bottle of Sidra and some crusty bread to mop all of the juice up.

YIELD

6–8 entrée servings

INGREDIENTS

2.2   pounds (1 kg) *Chosco* or *Botillo* (see recipes on pp. 275 or 279), pricked with a fork

Water, to cover

4   *Chorizo Asturiano* sausages (see recipe on p. 308), pricked with a fork

1   head green cabbage, cut into wedges

4   large russet or Yukon Gold potatoes, peeled and cut into large chunks

Kosher salt, as needed

Extra virgin olive oil, as needed

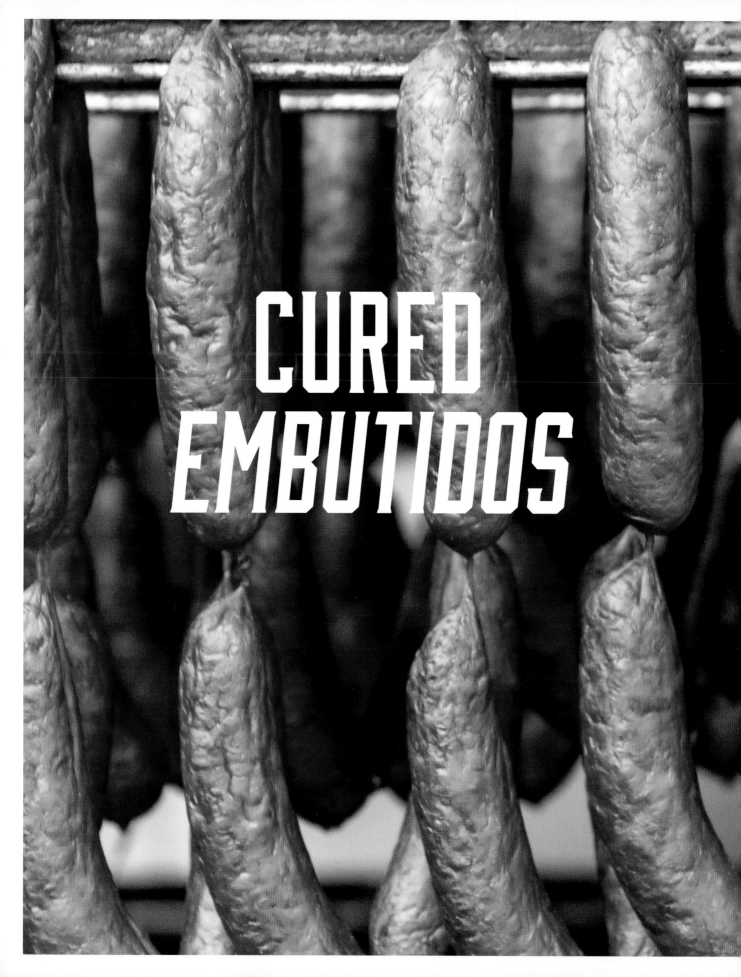

CURED
EMBUTIDOS

CHORIZO

CHORIZO IS THE first station of the holy *charcutería* cross.

Is there any other tubular meat product that has, over the course of history, had greater influence on so much of our world's cuisines? Nearly every Latino culture—and every other nation throughout the New World visited and pillaged by *conquistadors*—has some version of this very basic combination of ground meat, fat, salt, and spices as a part of their modern day gastronomy.

That's not to mention the popularity of *chorizo* within Spain, where every region and dialect has a word for everyone's favorite working-class meal: *chorizo* (Castellano); *chorizu* (Asturias), *txorizo* (Basque), *chourizo* (Gallego), and *xoriço* (Catalan) are just a few of the myriad ways the Spanish spell "sausage deliciousness."

The only problem, however, is that the word *chorizo* itself is rather ambiguous. The word basically translates to "sausage," which leads to confusion. So here's the deal...you'll most likely find Spanish *chorizo* in four different formats: (1) *picadillo,* (2) *fresco,* (3) *semicurado,* and (4) *curado.*

Picadillo is the most basic version of *chorizo*: loose, nonencased meat that's seasoned before being fried up. Take uncooked *picadillo* and stuff it into a sausage casing and you've got *Chorizo Fresco*—essentially *chorizo* in a stuffed, but raw, format. It's perfect for the grill or sliced up and seared with some fried eggs and *pan tostada* in the morning.

If the *fresco* sausage is slightly fermented—meaning held in a warmish, damp, and controlled environment for a short period of time to lower the pH of the meat—then it is a *semicurado* (semicured) *chorizo.* These are typically found in the northern reaches of Spain, in places like Asturias and Galicia where most *embutidos* are cold smoked as an extra layer of protection against the colder, wetter climate.

Last is the dry-cured *chorizo* we all know and really love. It's dry cured by taking the fermentation stage of the *semi-curado chorizo* a little further and drying it slowly in a controlled environment over a much longer period of time.

CANTIMPALOS-STYLE *CHORIZO*

This classic, dry-cured *chorizo* hangs close to the heart of every Spaniard. It's the *pimentón*-and-garlic-spiked ambassador most associated with the *embutidos* of Spain.

Nearly every Spanish region has its own *chorizo*; this one is closely related to the very famous and IGP-protected *chorizo* of Cantimpalos, a small municipality located in Castilla y León. *Chorizo* there is found in three sizes: *sarta,* the smallest, cured around 20 days; *achorizado,* the classic U-shaped next size up, with a curing time of around 24 days; and *cular,* the largest, which is stuffed in a pork bung and needs to be cured around 40 days.

1   Place the meats, *tocino,* and grinder parts in the freezer for 30 minutes to par-freeze before attempting to grind. In a small mixing bowl, combine the T-SPX culture and the distilled water, making a slurry. Set aside for a minimum of 10 minutes at room temperature to bloom.

2   Using a mortar and pestle, crush together the garlic and salt to form an *ajosal.* If desired, you can finish the *ajosal* in a food processor fitted with the "S" blade.

3   In a mixing bowl, combine the *ajosal,* dextrose, sugar, and curing salt. Divide the mixture in half.

4   Add half of the *ajosal* mixture to the *cabacero* meat and *tocino.* Toss together. In a separate bowl, mix the *papada* with the other half of the *ajosal* mixture. Set the mixtures aside in the refrigerator to chill as you set up the grinder.

Since this style of *chorizo* can come in different casing sizes, note that the grind will vary depending on the size you are shooting for.

TO MAKE *CHORIZO SARTA* OR *ACHORIZADO*:

1   Fill a large bowl with ice. Place a smaller bowl inside the ice-filled bowl. Grind the *cabecero* meat and *tocino* mixture once through a medium-coarse (3⁄8 inch [9.5 mm]) die into the smaller bowl. Be careful: The meat mixture is wet, so it may squirt and pop out of the grinder.

2   Grind the *papada* through a medium (1⁄4 inch [6 mm]) die and combine with the *cabecero* meat and *tocino* mixture for mixing.

YIELD

Sarta: 8–10 loops of sausage per 2.2 pounds (1 kg);
Achorizado: 3–4 loops of sausage per 2.2 pounds (1 kg);
Cular: 1 *cular* per 2.2 pounds (1 kg)

INGREDIENTS	*CHARCUTIER'S* PERCENTAGE
per 2.2 pounds (1 kg) of the following blend of diced meats: 40% *cabecero* (*coppa*/head of the pork loin), 40% *papada* (pork jowl), and 20% *tocino* (pork fat)	**100%**
1⁄4 ounce (10 g) T-SPX culture (see p. 94)	1%
1⁄2 cup (100 mL) distilled water	
1⁄3 ounce (10 g) minced garlic	1%
3⁄4 ounce (24 g) kosher salt	2.4%
1⁄8 ounce (3 g) dextrose	3%
1⁄8 ounce (3 g) granulated sugar	3%
1⁄8 ounce (2.4 g) Instacure #2 or DQ #2 curing salt mix (see p. 83)	24%
1⁄4 cup (50 mL) dry white wine, such as Verdejo, chilled	
2⁄3 ounce (20 g) *pimentón dulce*	2%
1⁄8 ounce (2 g) dried oregano	2%

For *Sarta:* 2 feet (60 cm) 1–1 1⁄3-inch (25–35-mm) hog casings, soaked, or more as needed; for *Achorizado:* 2 feet (60 cm) 1 1⁄4–2-inch (30–50-mm) hog casings, soaked, or more as needed; for *Cular:* 1 pork bung, soaked

CONTINUED ON NEXT PAGE

CANTIMPALOS-STYLE *CHORIZO* CONTINUED FROM PREVIOUS PAGE

TO MAKE *CHORIZO CULAR:*

1   Fill a large bowl with ice. Place a smaller bowl inside the ice-filled bowl. Grind the *cabecero* meat and *tocino* mixture once through a coarse (¾ inch [19 mm]) die into the smaller bowl. Be careful: The meat mixture is wet, so it may squirt and pop out of the grinder.

2   Grind the *papada* through a medium-coarse (⅜ inch [9.5 mm]) die and combine with the *cabecero* meat and *tocino* mixture for mixing.

TO FINISH THE *CHORIZO:*

1   In a small mixing bowl, combine the wine, *pimentón,* and oregano, making a slurry. Keep the bowl containing the slurry chilled until ready to use.

2   Place the ground meats in the bowl of a stand mixer fitted with the paddle attachment (or you can just mix in a mixing bowl with a wooden spoon). Begin mixing on low speed. As the mixer runs, pour the wine slurry into the bowl in a steady stream.

3   Continue mixing on medium speed for 1 to 2 minutes, until the wine slurry has been fully incorporated into the mixture, a white residue forms on the sides of the bowl, and the mixture firms up.

4   Reduce the mixer speed to low and add the T-SPX slurry. Continue mixing for 1 minute, until the T-SPX slurry is fully incorporated into the mixture. Place the bowl containing the ground meat mixture in the refrigerator to keep it cold until you are ready to stuff the sausage into casings.

5   To make *chorizo sarta* or *achorizada:* Stuff the *sarta* or *achorizado* into the casings and tie into 12-inch (30-cm) loops, ending both with a butterfly knot series and a loop for hanging.

To make *chorizo cular:* Stuff and tie the *cular* as you would a *morcón* or *sobrasada,* using string to tie it off with a butterfly knot and bubble knot series and ending with a loop at one end for hanging (this type of sausage is not prepared as links; see p. 90 in Chapter 3 for illustrations and directions).

Using a sterile pin or sausage pricker, prick each sausage several times. Weigh each sausage to obtain a green weight; record the weights and tag each sausage with its green weight.

6   Ferment the sausages in a drying chamber set at 65°F to 80°F (18.3°C to 26.6°C) and 85% to 90% relative humidity for 2 to 3 days. Check the pH of the meat (see p. 95) to ensure that the level has dropped below 5.3 before the third day of drying.

7   Hang the sausages in a drying chamber set at 54°F to 60°F (12°C to 16°C) and 80% to 85% relative humidity for 1 to 2 months, until the sausages have lost about 35% of their green weight. (For the *sarta* and *achorizado,* this will take about 1 month; for the *cular,* it will take 1½ to 2 months.) At that point, the sausages will be ready to consume.

NOTES: You can ferment the sausages either before or after stuffing. It's really a matter of preference, since the meat firms up during the fermentation process and makes stuffing a little easier. On the other hand, it might be more efficient for you to make and stuff the sausages all in 1 day.

Bactoferm T-SPX is a slow-acting fermentation agent, so don't be surprised if you don't hit a pH of 5.3 after 2 days of drying time. The package directions provide more information; be sure to read them carefully.

RIOJANO-STYLE *CHORIZO*

Most Spaniards generally agree that other than their own regional *chorizo,* of course, some of the best *chorizos* in Spain are found in La Rioja. The Riojano *chorizo* is IGP-protected, even though it really doesn't deviate too much from the Cantimpalos-Style *Chorizo* (see recipe on p. 299). The major difference is that in La Rioja, the lean portion of the *masa* is mostly from the flavorful and expensive *lomo* (loin) section of the pig. That, combined with the unique microclimate in the region, makes for a very unique and delicious *embutido.*

1   Place the meats, *tocino,* and grinder parts in the freezer for 30 minutes to par-freeze before attempting to grind. In a small mixing bowl, combine the T-SPX culture and the distilled water, making a slurry. Set aside for a minimum of 10 minutes at room temperature to bloom.

2   Using a mortar and pestle, crush together the garlic and salt to form an *ajosal.* If desired, you can finish the *ajosal* in a food processor fitted with the "S" blade.

3   In a mixing bowl, combine the *ajosal,* dextrose, sugar, and curing salt. Divide the mixture in half.

4   Add half of the *ajosal* mixture to the *lomo* meat and *tocino.* Toss together. In a separate bowl, mix the *panceta* with the other half of the *ajosal* mixture. Set the mixtures aside in the refrigerator to chill as you set up the grinder.

5   Fill a large bowl with ice. Place a smaller bowl inside the ice-filled bowl. Grind the *lomo* meat and *tocino* mixture once through a medium-coarse (3⁄8 inch [9.5 mm]) die into the smaller bowl. Be careful: The meat mixture is wet, so it may squirt and pop out of the grinder.

6   Grind the *panceta* through a medium (1⁄4 inch [6 mm]) die and combine with the *lomo* meat and *tocino* mixture for mixing.

7   In a small mixing bowl, combine the wine and *pimentón,* making a slurry. Keep the bowl containing the slurry chilled until ready to use.

YIELD

3–4 loops of sausage per 2.2 pounds (1 kg)

INGREDIENTS	*CHARCUTIER'S* PERCENTAGE
per 2.2 pounds (1 kg) of the following blend of diced meats: 60% *lomo* (pork loin), 20% *panceta* (pork belly), and 20% *tocino* (pork fat)	**100%**
1⁄4 ounce (10 g) T-SPX culture (see p. 94)	**1%**
1⁄2 cup (100 mL) distilled water	
1⁄3 ounce (10 g) minced garlic	**1%**
3⁄4 ounce (24 g) kosher salt	**2.4%**
1⁄8 ounce (3 g) dextrose	**.3%**
1⁄8 ounce (3 g) granulated sugar	**.3%**
1⁄8 ounce (2.4 g) Instacure #2 or DQ #2 curing salt mix (see p. 83)	**.24%**
1⁄4 cup (50 mL) dry white wine, such as Verdejo, chilled	
2⁄3 ounce (20 g) *pimentón dulce*	**2%**
2 feet (60 cm) 11⁄4–11⁄2-inch (32–36-mm) hog casings, soaked, or more as needed	

8   Place the ground meats in the bowl of a stand mixer fitted with the paddle attachment (or you can just mix in a mixing bowl with a wooden spoon). Begin mixing on low speed. As the mixer runs, pour the wine slurry into the bowl in a steady stream.

9   Continue mixing on medium speed for 1 to 2 minutes, until the wine slurry has been fully incorporated into the mixture, a white residue forms on the sides of the bowl, and the mixture firms up.

10 Reduce the mixer speed to low and add the T-SPX slurry. Continue mixing for 1 minute, until the T-SPX slurry is fully incorporated into the mixture. Place the bowl containing the ground meat mixture in the refrigerator to keep it cold until you are ready to stuff the sausage into casings.

11 Stuff the mixture into the casings and tie into 12-inch (30-cm) loops, ending with a butterfly knot series and a loop for hanging (see Chapter 3 for illustrations and directions). Using a sterile pin or sausage pricker, prick each sausage several times. Weigh each sausage loop to obtain a green weight; record the weights and tag each sausage with its green weight.

12 Ferment the sausages in a drying chamber set at 65°F to 80°F (18.3°C to 26.6°C) and 85% to 90% relative humidity for 2 to 3 days. Check the pH of the meat (see p. 95) to ensure that the level has dropped below 5.3 before the third day of drying.

13 Hang the sausages in a drying chamber set at 54°F to 60°F (12°C to 16°C) and 80% to 85% relative humidity for 1 to 2 months (depending on their size), until the sausages have lost about 35% of their green weight. At that point, the sausages will be ready to consume.

NOTES: You can ferment the sausages either before or after stuffing. It's really a matter of preference, since the meat firms up during the fermentation process and makes stuffing a little easier. On the other hand, it might be more efficient for you to make and stuff the sausages all in 1 day.

Bactoferm T-SPX is a slow-acting fermentation agent, so don't be surprised if you don't hit a pH of 5.3 after 2 days of drying time. The package directions provide more information; be sure to read them carefully.

For spicier *chorizo,* instead of ⅔ ounce (20 g) *pimentón dulce,* use ⅓ ounce (10 g) of *pimentón dulce* and ⅓ ounce (10 g) of *pimentón picante.*

SOBRASADA

The tropical humidity in the Balearic Islands has long made it impossible to produce a dry-cured *chorizo* similar to ones found on the mainland. So when Sicilian maritime traders arrived in the sixteenth century (the word *"sobrasada"* stems from the old Italian word *"soppressa,"* or "compressed"), they shared their meat-curing techniques with the local populace. Just like that, a spreadable, *pâté*-like Majorcan star sausage was born.

If you've ever tried the Calabrian sausage *n'duja* (Chris Cosentino's *salumi* mecca Boccalone in San Francisco makes a killer version), the similarities are obvious—these sausages are closely linked by the proximity of their home countries and the travels of early merchant traders.

So here's the bad news: *Sobrasada de Mallorca*, or any *charcuterie* from Mallorca for that matter (there's actually a whole family of Mallorcan *embutidos* that are made from the meat of the Mallorcan black pigs), is simply not available in the United States. The sausage's manufacture is geographically protected by the European Union, and US customs can't handle the idea of a safe-for-consumption sausage made on a tropical island. But this recipe will provide your fix without costing you thousands for a plane ticket. It also happens to be one of the best *embutidos* you'll ever eat, so make a lot and spread the love—preferably onto some grilled bread.

For an even more authentically Majorcan dish, drizzle your little *sobrasada montadito* with some honey.

YIELD

2 *Sobrasadas* per 2.2 pounds (1 kg)

INGREDIENTS	CHARCUTIER'S PERCENTAGE
per 2.2 pounds (1 kg) of the following blend of diced meats: 30% *cabecero* (*coppa*/head of the pork loin), 30% *papada* (pork jowl), and 40% *tocino* (pork fat)	**100%**
¼ ounce (10 g) T-SPX culture (see p. 94)	**1%**
½ cup (100 mL) distilled water	
1 ounce (28 g) kosher salt	**2.8%**
⅛ ounce (3 g) dextrose	**.3%**
⅛ ounce (3 g) granulated sugar	**.3%**
⅛ ounce (2.4 g) Instacure #2 or DQ #2 curing salt mix (see p. 83)	**.24%**
¼ cup (50 mL) dry white wine, such as Verdejo, chilled	
2 ounces (60 g) *pimentón dulce*	**6%**
½ tablespoon (3 g) freshly ground black pepper	**.3%**
2 feet (60 cm) 2½-inch (65-mm) beef middles, rinsed and soaked, or more as needed or 1 *ciego de cerdo* (pork bung)	

1   Place the meats and *tocino* (in separate bowls) and grinder parts in the freezer for 30 minutes to par-freeze before attempting to grind. In a small mixing bowl, combine the T-SPX culture and the distilled water, making a slurry. Set aside for a minimum of 10 minutes at room temperature to bloom.

2   Toss the *cabecero* and *papada* meats with the salt, dextrose, sugar, and curing salt and set aside in the refrigerator to chill before grinding.

3   Fill a large bowl with ice. Place a smaller bowl inside the ice-filled bowl. Grind the *cabecero* and *papada* meats once through a medium-coarse (⅜ inch [9.5 mm]) die into the smaller bowl. Be careful: The meat mixture is wet, so it may squirt and pop out of the grinder. Return the ground meat to the freezer while grinding the rest of the ingredients.

4   Grind the *tocino* through the same die into a separate bowl set inside the ice-filled bowl. Return the ground *tocino* to the freezer while grinding the rest of the ingredients.

5   Grind the *cabecero* and *papada* meats again through a medium (¼ inch [6 mm]) die.

6   Grind the *tocino* again through an extra-small (⅛ inch [3 mm]) die. Combine the ground meats and *tocino* in the same bowl for mixing.

7   In a small mixing bowl, combine the wine, *pimentón,* and black pepper, making a slurry. Keep the bowl containing the slurry chilled until ready to use.

8   Place the ground meats in the bowl of a stand mixer fitted with the paddle attachment (or you can just mix in a mixing bowl with a wooden spoon). Begin mixing on low speed. As the mixer runs, pour the *pimentón* slurry into the bowl in a steady stream.

9   Continue mixing on medium speed for 1 to 2 minutes, until the *pimentón* slurry has been fully incorporated into the mixture, a white residue forms on the sides of the bowl, and the mixture firms up.

10 Reduce the mixer speed to low and add the T-SPX slurry. Continue mixing for 1 minute, until the T-SPX slurry is fully incorporated into the mixture. Place the bowl containing the ground meat mixture in the refrigerator to keep it cold until you are ready to stuff the sausage into casings.

11 Tightly stuff the mixture into the middles or *ciego* and use string to tie them off with a butterfly knot and bubble knot series for the middles or tie them like a *morcón* for the *ciego,* ending with a loop at one end for hanging (this type of sausage is not prepared as links; see Chapter 3 for illustrations and directions). Using a sterile pin or sausage pricker, prick each *Sobrasada* several times. Weigh each *Sobrasada* to obtain a green weight; record the weights and tag each *Sobrasada* with its green weight.

12 Ferment the sausages in a drying chamber set at 68°F (20°C) and 85% to 90% relative humidity for 2 to 3 days. Check the pH of the meat (see p. 95) to ensure that the level has dropped below 5.3 before the third day of drying.

13 Hang the sausages in a drying chamber set at 54°F to 60°F (12°C to 16°C) and 80% to 85% relative humidity for 2 to 6 months (depending on their size), until the sausages have lost about 35% of their green weight. At that point, the sausages will be ready to consume.

NOTES: You can ferment the sausages either before or after stuffing. It's really a matter of preference, since the meat firms up during the fermentation process and makes stuffing a little easier. On the other hand, it might be more efficient for you to make and stuff the sausages all in 1 day.

Bactoferm T-SPX is a slow-acting fermentation agent, so don't be surprised if you don't hit a pH of 5.3 after 2 days of drying time. The package directions provide more information; be sure to read them carefully.

For spicier *Sobrasada,* instead of 2 ounces (60 g) *pimentón dulce,* use 1⅓ ounces (40 g) of *pimentón dulce* and ⅔ ounce (20 g) of *pimentón picante.*

PAMPLONA-STYLE *CHORIZO*

You may not know it, but you've already heard of *El Encierro.*

Here in America, we know it as the running of the bulls, a Spanish pastime with origins in Spain's northeastern lands. The most famous run occurs in Pamplona during the Festival of San Fermín (thanks to Ernest Hemingway's writing and Billy Crystal getting his ass gored in the movie *City Slickers).*

But riddle me this: What do you think a modest Spanish town like Pamplona does with so much leftover beef? The whole point of running the bulls into a bull ring is for them to eventually meet the business end of a rifle or sword—which means that after the fun is over, there will be quite a few 800-pound bull carcasses to deal with. Over time, this *chorizo*—one with a bit of a personality disorder, since it's made with beef and pork—became one of the answers for keeping all of that meat from spoiling.

1   Place the meats, *tocino,* and grinder parts in the freezer for 30 minutes to par-freeze before attempting to grind. In a small mixing bowl, combine the T-SPX culture and the distilled water, making a slurry. Set aside for a minimum of 10 minutes at room temperature to bloom.

2   Using a mortar and pestle, crush together the garlic and salt to form an *ajosal.* If desired, you can finish the *ajosal* in a food processor fitted with the "S" blade.

3   Toss together the meats and *tocino* with the *ajosal,* dextrose, sugar, and curing salt. Set aside in the refrigerator to chill before grinding.

4   Fill a large bowl with ice. Place a smaller bowl inside the ice-filled bowl. Grind the *cabecero* and brisket meats and *tocino* once through a medium-coarse (⅜ inch [9.5 mm]) die into the smaller bowl. Be careful: The meat mixture is wet, so it may squirt and pop out of the grinder. Return the ground meat to the freezer for 15 to 20 minutes, until it is par-frozen.

5   Grind the *cabecero*-brisket meats and *tocino* again through a medium-fine (³⁄₁₆ inch [5 mm]) die into the same small bowl inside the large ice-filled bowl.

YIELD

2–4 links of sausage per
2.2 pounds (1 kg)

INGREDIENTS	*CHARCUTIER'S* PERCENTAGE
per 2.2 pounds (1 kg) of the following blend of diced meats: 45% *cabecero* (*coppa*/head of the pork loin), 30% beef brisket or chuck, and 25% *tocino* (pork fat)	**100%**
¼ ounce (10 g) T-SPX culture (see p. 94)	**1%**
½ cup (100 mL) distilled water	
⅓ ounce (10 g) minced garlic	**1%**
¾ ounce (24 g) kosher salt	**2.4%**
⅛ ounce (3 g) dextrose	**.3%**
⅛ ounce (3 g) granulated sugar	**.3%**
⅛ ounce (2.4 g) Instacure #2 or DQ #2 curing salt mix (see p. 83)	**.24%**
¼ cup (50 mL) dry white wine, such as Verdejo, chilled	
⅔ ounce (20 g) *pimentón dulce*	**2%**
2 feet (60 cm) 2½-inch (65-mm) beef middles, rinsed and soaked, or more as needed	

6   In a small mixing bowl, combine the wine and *pimentón*, making a slurry. Keep the bowl containing the slurry chilled until ready to use.

7   Place the ground meats in the bowl of a stand mixer fitted with the paddle attachment (or you can just mix in a mixing bowl with a wooden spoon). Begin mixing on low speed. As the mixer runs, pour the wine slurry into the bowl in a steady stream.

8   Continue mixing on medium speed for 1 to 2 minutes, until the wine slurry has been fully incorporated into the mixture, a white residue forms on the sides of the bowl, the meat appears sticky, and the mixture firms up.

9   Reduce the mixer speed to low and add the T-SPX slurry. Continue mixing for 1 minute, until the T-SPX slurry is fully incorporated into the mixture. Place the bowl containing the ground meat mixture in the refrigerator to keep it cold until you are ready to stuff the sausage into casings.

10   Stuff the mixture into the middles and use string to tie them off with a butterfly knot and bubble knot series, ending with a loop at one end for hanging (this type of sausage is not prepared as links; see Chapter 3 for illustrations and directions). Using a sterile pin or sausage pricker, prick each sausage several times. Weigh each sausage to obtain a green weight; record the weights and tag each sausage with its green weight.

11   Ferment the sausages in a drying chamber set at 65°F to 80°F (18.3°C to 26.6°C) and 85% to 90% relative humidity for 2 to 3 days. Check the pH of the meat (see p. 95) to ensure that the level has dropped below 5.3 before the third day of drying.

12   Hang the sausages in a drying chamber set at 54°F to 60°F (12°C to 16°C) and 80% to 85% relative humidity for 2 to 4 months (depending on their size), until the sausages have lost about 35% of their green weight. At that point, the sausages will be ready to consume.

NOTES: Pamplona-Style *Chorizo* is either made as a finely ground *embutido* or sometimes with chunks of meat folded in. To make it chunky, you can separate ⅓ of the meat mixture before grinding and cut it into rough chunks by hand. Fold these chunks into the rest of the ground meat mixture and proceed as directed.

You can ferment the sausages either before or after stuffing. It's really a matter of preference, since the meat firms up during the fermentation process and makes stuffing a little easier. On the other hand, it might be more efficient for you to make and stuff the sausages all in 1 day.

Bactoferm T-SPX is a slow-acting fermentation agent, so don't be surprised if you don't hit a pH of 5.3 after 2 days of drying time. The package directions provide more information; be sure to read them carefully.

If you don't have access to beef middles, you can stuff this chorizo into hog casings and link them however you want. Bear in mind, however, that the drying time will be much shorter, since the diameter of the sausage will be smaller.

CHORIZO ASTURIANO

Asturian *chorizo* doesn't stray far from the basic *chorizo* recipe from Cantimpalos. The major difference lies in the fact that almost all *chorizos* found in the Asturian region get smoked to some extent. A lot of Asturians own sheds for smoking meats, and each will give you a different answer when asked how long the *chorizo* needs to hang in the smoking shed. It's really a function of temperature and the smokiness of the room—in those Asturian sheds, a small oak fire is left smoldering under the *chorizos,* so the smoking process is long and gentle. In fact, some of these "sheds" are in fact gigantic rooms with *chorizos* hanging 6 feet over the fire. If your smoking rig is smaller and allows for more rapid exposure, be aware that your sausages will need less time in there (see Chapter 3 for more information on smoking meat).

1   Place the meats, *tocino,* and grinder parts in the freezer for 30 minutes to par-freeze before attempting to grind. In a small mixing bowl, combine the F-RM-52 culture and the distilled water, making a slurry. Set aside for a minimum of 10 minutes at room temperature to bloom.

2   Using a mortar and pestle, crush together the garlic and salt to form an *ajosal.* If desired, you can finish the *ajosal* in a food processor fitted with the "S" blade.

3   In a mixing bowl, combine the *ajosal,* dextrose, sugar, and curing salt. Divide the mixture in half.

4   Add half of the *ajosal* mixture to the *cabecero* meat and *tocino.* Toss together. In a separate bowl, mix the *papada* with the other half of the *ajosal* mixture. Set aside both mixtures in the refrigerator to chill before grinding.

5   Fill a large bowl with ice. Place a smaller bowl inside the ice-filled bowl. Grind the *cabecero* meat and *tocino* once through a medium-coarse (⅜ inch [9.5 mm]) die into the smaller bowl. Be careful: The meat mixture is wet, so it may squirt and pop out of the grinder.

6   Grind the *papada* through a medium (¼ inch [6 mm]) die. Combine the ground *cabecero* and *papada* meats and *tocino* in the same bowl for mixing.

7   In a small mixing bowl, combine the wine and *pimentón,* making a slurry. Keep the bowl containing the slurry chilled until ready to use.

YIELD

3–4 loops or 6–8 links of sausage per 2.2 pounds (1 kg)

INGREDIENTS	CHARCUTIER'S PERCENTAGE
per 2.2 pounds (1 kg) of the following blend of diced meats: 40% *cabecero* (*coppa*/head of the pork loin), 40% *papada* (pork jowl), and 20% *tocino* (pork fat)	100%
¼ ounce (10 g) F-RM-52 culture (see p. 94)	1%
½ cup (100 mL) distilled water	
⅓ ounce (10 g) minced garlic	1%
¾ ounce (24 g) kosher salt	2.4%
⅛ ounce (3 g) dextrose	.3%
⅛ ounce (3 g) granulated sugar	.3%
⅛ ounce (2.4 g) Instacure #2 or DQ #2 curing salt mix (see p. 83)	.24%
¼ cup (50 mL) dry white wine, such as Verdejo, chilled	
⅔ ounce (20 g) *pimentón dulce*	2%
2 feet (60 cm) 1¼–1½-inch (32–36-mm) hog casings, soaked, or more as needed	

8   Place the ground meats in the bowl of a stand mixer fitted with the paddle attachment (or you can just mix in a mixing bowl with a wooden spoon). Begin mixing on low speed. As the mixer runs, pour the wine slurry into the bowl in a steady stream.

9   Continue mixing on medium speed for 1 to 2 minutes, until the wine slurry has been fully incorporated into the mixture, a white residue forms on the sides of the bowl, the meat appears sticky, and the mixture firms up.

10 Reduce the mixer speed to low and add the F-RM-52 slurry. Continue mixing for 1 minute, until the F-RM-52 slurry is fully incorporated into the mixture. Place the bowl containing the ground meat mixture in the refrigerator to keep it cold until you are ready to stuff the sausage into casings.

11 Stuff the mixture into the casings and tie into 12-inch (30-cm) loops or 6-inch (15-cm) links, ending with a butterfly knot series and a loop for hanging (see Chapter 3 for illustrations and directions). Using a sterile pin or sausage pricker, prick each sausage several times. Weigh each sausage to obtain a green weight; record the weights and tag each sausage with its green weight.

12 Ferment the sausages in a drying chamber set at 86°F (30°C) and 80% to 85% relative humidity for at least 2 days. During this time, check the pH of the meat (see p. 95) to ensure that the level has dropped below 5.3 by the third day of drying.

13 Stock a cold smoker with oak wood and bring the temperature to 90°F (30°C) and the relative humidity to 80% to 85%. Hang the *chorizos* in the smoking chamber and smoke for 7 to 10 days, or break it up into 6-hour-long sessions for 10 to 14 days, returning the *chorizos* to the drying chamber set at 54°F to 60°F (12°C to 16°C) and 80% to 85% relative humidity in between sessions.

14 Hang the sausages in a drying chamber set at 54°F to 60°F (12°C to 16°C) and 80% to 85% relative humidity for 1 to 2 months (depending on their size), until the sausages have lost about 35% of their green weight. At that point, the sausages will be ready to consume.

NOTES: You can ferment the sausages either before or after stuffing. It's really a matter of preference, since the meat firms up during the fermentation process and makes stuffing a little easier. On the other hand, it might be more efficient for you to make and stuff the sausages all in 1 day.

For spicier *chorizo,* instead of ⅔ ounce (20 g) *pimentón dulce,* use ⅓ ounce (10 g) of *pimentón dulce* and ⅓ ounce (10 g) of *pimentón picante.*

FUET

Given Spain's famed horse riding and training culture, it's not a stretch to see where this very simply seasoned sausage got its name: *Fuet* means "whip." The obvious implication is that this long, thin sausage is either representative of the horsemen's training tool or the Spaniards' penchant for leather-clad *dominas*.

I'm not judging you, Spain. Just pointing out the obvious...

Fuet is one of the cornerstone sausages in Catalonia and, like most Catalan *charcutería,* it resists the red-colored inclusion of *pimentón*. That's because this recipe predates Columbus's return to the Old World with a bounty of peppers. At the same time, *Fuet* showcases the Medici spices that link Catalan *charcuterie* to the *salumis* of Italy and *saucissons* of France.

1   Place the meats, *tocino,* and grinder parts in the freezer for 30 minutes to par-freeze before attempting to grind. In a small mixing bowl, combine the T-SPX culture and the distilled water, making a slurry. Set aside for a minimum of 10 minutes at room temperature to bloom.

2   Using a mortar and pestle, crush together the garlic and salt to form an *ajosal*. If desired, you can finish the *ajosal* in a food processor fitted with the "S" blade.

3   In a mixing bowl, combine the *ajosal,* dextrose, sugar, and curing salt. Divide the mixture in half.

4   Add half of the *ajosal* mixture to the *cabecero* meat and *tocino*. Toss together. In a separate bowl, mix the *magro* with the other half of the *ajosal* mixture. Set aside both mixtures in the refrigerator to chill before grinding.

5   Fill a large bowl with ice. Place a smaller bowl inside the ice-filled bowl. Grind the *cabecero* meat and *tocino* once through a medium-coarse (⅜ inch [9.5 mm]) die into the smaller bowl. Be careful: The meat mixture is wet, so it may squirt and pop out of the grinder. Return the ground meat to the refrigerator for now.

6   Grind the *magro* meat through a medium-coarse (⅜ inch [9.5 mm]) die into a separate small bowl inside the large ice-filled bowl. Return the ground meat to the freezer for 15 to 20 minutes, until it is par-frozen.

YIELD

6–10 links of sausage per 2.2 pounds (1 kg)

INGREDIENTS	CHARCUTIER'S PERCENTAGE
per 2.2 pounds (1 kg) of the following blend of diced meats: 40% *magro* (lean pork meat), 40% *cabecero* (*coppa*/head of the pork loin), and 20% *tocino* (pork fat)	100%
¼ ounce (10 g) T-SPX culture (see p. 94)	1%
½ cup (100 mL) distilled water	
⅓ ounce (10 g) minced garlic	1%
1 ounce (28 g) kosher salt	2.8%
⅛ ounce (3 g) dextrose	.3%
⅛ ounce (3 g) granulated sugar	.3%
⅛ ounce (2.4 g) Instacure #2 or DQ #2 curing salt mix (see p. 83)	.24%
¼ cup (50 mL) dry white wine, such as Verdejo, chilled	
⅛ ounce (5 g) ground white pepper	.5%
2 feet (60 cm) 1¼–1½-inch (32–36-mm) hog casings, soaked	

FUET CONTINUED FROM PREVIOUS PAGE

7   Grind the *magro* meat again through a medium (¼ inch [6 mm]) die into the same small bowl inside the large ice-filled bowl. Combine the *cabecero* and *magro* meats and the *tocino* in the same bowl for mixing.

8   In a small mixing bowl, combine the wine and white pepper, making a slurry. Keep the bowl containing the slurry chilled until ready to use.

9   Place the ground meats in the bowl of a stand mixer fitted with the paddle attachment (or you can just mix in a mixing bowl with a wooden spoon). Begin mixing on low speed. As the mixer runs, pour the wine slurry into the bowl in a steady stream.

10   Continue mixing on medium speed for 1 to 2 minutes, until the wine slurry has been fully incorporated into the mixture, a white residue forms on the sides of the bowl, the meat appears sticky, and the mixture firms up.

11   Reduce the mixer speed to low and add the T-SPX slurry. Continue mixing for 1 minute, until the T-SPX slurry is fully incorporated into the mixture. Place the bowl containing the ground meat mixture in the refrigerator to keep it cold until you are ready to stuff the sausage into casings.

12   Stuff the mixture into the casings and tie into very long 18-inch (46-cm) links, ending with a butterfly knot series and a loop for hanging (see Chapter 3 for illustrations and directions). Using a sterile pin or sausage pricker, prick each sausage several times. Weigh each sausage to obtain a green weight; record the weights and tag each sausage with its green weight.

13   Ferment the sausages in a drying chamber set at 65°F to 80°F (18.3°C to 26.6°C) and 85% to 90% relative humidity for 2 to 3 days. Check the pH of the meat (see p. 95) to ensure that the level has dropped below 5.3 before the third day of drying.

14   Hang the sausages in a drying chamber set at 54°F to 60°F (12°C to 16°C) and 80% to 85% relative humidity for 1 to 2 months (depending on their size), until the sausages have lost about 35% of their green weight. At that point, the sausages will be ready to consume.

NOTES: You can ferment the sausages either before or after stuffing. It's really a matter of preference, since the meat firms up during the fermentation process and makes stuffing a little easier. On the other hand, it might be more efficient for you to make and stuff the sausages all in 1 day.

Bactoferm T-SPX is a slow-acting fermentation agent, so don't be surprised if you don't hit a pH of 5.3 after 2 days of drying time. The package directions provide more information; be sure to read them carefully.

LOMO EMBUCHADO

Fernanda, one of the head *sabias* at our *matanzas* in Extremadura, has her own way of making *Lomo Embuchado*. Her eyes light up and her fingers dance as she slathers each loin in garlic-infused fat and then ties off the casings...it's like she knows exactly how good each loin will be in 3 months, the time needed to transform that hunk of raw meat into one of my favorite *embutidos*.

So this is your fair warning: This is not a typical recipe for the *Lomo Embuchado* you find at tourist-infested *tapas* joints near the Plaza Major in Madrid. Fernanda's Extremeñan recipe is special because of its garlic- and paprika-infused pork fat bath. The resulting *lomo* is worth the extra steps, or so says Fernanda. And, at our *matanzas*, you do what the head *sabía* says or she holds the liquor hostage.

1   Place the Basic *Salazón* in a measuring cup. If using, whisk in the curing salt. Set aside.

2   Place the *lomo* on a work surface, such as a baking sheet. Rub the curing mixture into the meat, pressing hard and distributing it evenly over its entire surface.

3   Cover the *lomo* in plastic wrap. Place the *lomo* in a large baking dish and weigh it down with a few canned goods. Refrigerate the dish for 1 day per 2.2 pounds (1 kg) of meat, flipping it over daily to redistribute the cure. The *lomo* is ready when you notice that it has firmed up a little to the touch.

4   Remove the *lomo* from the cure and rinse it thoroughly under cold water. Wash the baking dish, dry it, and return the *lomo* to the dish. Refrigerate for 1 hour per day it spent in the curing mixture.

5   In a small skillet over low heat, warm the *manteca*. Add the garlic and cook for 15 minutes, until the garlic is softened and very fragrant but not browned. Remove from the heat and add the *pimentón*, creating an *adobo*.

6   Smear the *adobo* all over the *lomo* and set it aside for stuffing.

INGREDIENTS	CHARCUTIER'S PERCENTAGE
per 2.2 pounds (1 kg) of *lomo* (pork loin), trimmed of any visible silverskin	**100%**
1 ounce (25 g) Basic *Salazón* (see recipe on p. 111)	**2.5%**
4 ounces (50 g) *manteca* (pork fat)	**5%**
1¾ ounces (50 g) minced garlic	**5%**
2 tablespoons (10 g) *pimentón dulce*	**1%**
4 feet (120 cm) 2½-inch (65-mm) beef middles, rinsed and soaked	

OPTIONAL

⅛ ounce (3 g) Instacure #2 or DQ #2 curing salt mix	**.3%**

CONTINUED ON NEXT PAGE

LOMO EMBUCHADO CONTINUED FROM PREVIOUS PAGE

7 Stuff the *lomo* into the beef middle and use string to tie it off with a butterfly knot and bubble knot series as you would a *morcón* or *sobrasada*, ending with a loop at one end for hanging (this type of sausage is not prepared as links; see Chapter 3 for illustrations and directions). Using a sterile pin or sausage pricker, prick it several times. Weigh the *lomo* to obtain the green weight; record and tag it with its green weight.

8 Hang in a drying chamber set at 68°F (20°C) and 95% relative humidity for 2 days. Reset the drying chamber to 54°F to 60°F (12°C to 16°C) and 80% to 85% relative humidity and rehang the *lomo* in the chamber for 2 to 3 months (depending on size), until the sausage has lost about 32% to 35% of its green weight. At that point, it will be ready to consume.

NOTE: For spicier *lomo,* instead of 2 tablespoons (10 g) *pimentón dulce,* use 1 tablespoon (5 g) of *pimentón dulce* and 1 tablespoon (5 g) of *pimentón picante.*

TRUCOS DE LA COCINA: STUFFING WHOLE-MUSCLE *EMBUTIDOS*

FERNANDA HAS A cool little trick for getting the greased-up *lomo* into its casing: She ties a string around one end of the loin, and ties a spoon to the other end of the string. The spoon then goes into the casing, and deftly helps pull the loin into place. As Fernanda says with typical *sabía* wisdom: "It's even easier than putting on a condom."

SALCHICHÓN

Salchichón, saucisson, salami—embutidos of a feather from different parts of the world all hailing from the same Latin root word *salsus,* for "salted."

The Spanish representative, *Salchichón,* is most famously made in the Catalan town of Vic where IGP protection covers the local products. It's a pretty basic sausage made in a similar style to its French and Italian brothers—all include a common Medici-spice profile (think: pepper, nutmeg, white wine) and rely on good pork and the local microclimate to make for a subtly nuanced *embutido.*

1 Place the meats, *tocino,* and grinder parts in the freezer for 30 minutes to par-freeze before attempting to grind. In a small mixing bowl, combine the T-SPX culture and the distilled water, making a slurry. Set aside for a minimum of 10 minutes at room temperature to bloom.

2 Using a mortar and pestle, crush together the garlic and salt to form an *ajosal.* If desired, you can finish the *ajosal* in a food processor fitted with the "S" blade.

3 In a mixing bowl, combine the meats and *tocino* with the *ajosal,* dextrose, sugar, and curing salt. Set aside in the refrigerator to chill before you begin grinding.

4 Fill a large bowl with ice. Place a smaller bowl inside the ice-filled bowl. Grind the *cabecero* and *panceta* meats and *tocino* once through a medium-coarse (3/8 inch [9.5 mm]) die into the smaller bowl. Be careful: The meat mixture is wet, so it may squirt and pop out of the grinder. Return the ground meat to the freezer for 15 to 20 minutes, until it is par-frozen.

5 Grind the *cabecero* and *panceta* meats and *tocino* again through a medium-fine (3/16 inch [5 mm]) die into the same bowl over ice.

6 In a small mixing bowl, combine the wine, black pepper, white pepper, and nutmeg, making a slurry. Keep the bowl containing the slurry chilled until ready to use.

7 Place the ground meats in the bowl of a stand mixer fitted with the paddle attachment (or you can just mix in a mixing bowl with a wooden spoon). Begin mixing on low speed. As the mixer runs, pour the wine slurry into the bowl in a steady stream.

YIELD

2–4 links of sausage per 2.2 pounds (1 kg)

INGREDIENTS	*CHARCUTIER'S* PERCENTAGE
per 2.2 pounds (1 kg) of the following blend of diced meats: 40% *cabecero* (*coppa*/head of the pork loin), 40% *panceta* (pork belly), and 20% *tocino* (pork fat)	100%
¼ ounce (10 g) T-SPX culture (see p. 94)	1%
½ cup (100 mL) distilled water	
⅛ ounce (5 g) minced garlic	.5%
1 ounce (28 g) kosher salt	2.8%
⅛ ounce (3 g) dextrose	.3%
⅛ ounce (3 g) granulated sugar	.3%
⅛ ounce (2.4 g) Instacure #2 or DQ #2 curing salt mix (see p. 83)	.24%
¼ cup (50 mL) dry white wine, such as Verdejo, chilled	
⅛ ounce (3 g) freshly and coarsely ground black pepper	.3%
⅛ ounce (2 g) ground white pepper	.2%
⅛ ounce (1 g) freshly grated nutmeg	.1%
2 feet (60 cm) 2½-inch (65-mm) beef middles, rinsed and soaked, or more as needed	

CONTINUED ON NEXT PAGE

SALCHICHÓN CONTINUED FROM PREVIOUS PAGE

8   Continue mixing on medium speed for 1 to 2 minutes, until the wine slurry has been fully incorporated into the mixture, a white residue forms on the sides of the bowl, the meat appears sticky, and the mixture firms up.

9   Reduce the mixer speed to low and add the T-SPX slurry. Continue mixing for 1 minute, until the T-SPX slurry is fully incorporated into the mixture. Place the bowl containing the ground meat mixture in the refrigerator to keep it cold until you are ready to stuff the sausage into casings.

10   Stuff the mixture into the middles and use string to tie them off with a butterfly knot and bubble knot series, ending with a loop at one end for hanging. (This type of sausage is not prepared as links; see p. 90 in Chapter 3 for illustrations and directions.) Using a sterile pin or sausage pricker, prick each sausage several times. Weigh each sausage to obtain a green weight; record the weights and tag each sausage with its green weight.

11   Ferment the sausages in a drying chamber set at 65°F to 80°F (18.3°C to 26.6°C) and 85% to 90% relative humidity for 2 to 3 days. Check the pH of the meat (see p. 95) to ensure that the level has dropped below 5.3 before the third day of drying.

12   Hang the sausages in a drying chamber set at 54°F to 60°F (12°C to 16°C) and 80% to 85% relative humidity for 2 to 4 months (depending on their size), until the sausages have lost about 35% of their green weight. At that point, the sausages will be ready to consume.

NOTES: You can ferment the sausages either before or after stuffing. It's really a matter of preference, since the meat firms up during the fermentation process and makes stuffing a little easier. On the other hand, it might be more efficient for you to make and stuff the sausages all in 1 day.

Bactoferm T-SPX is a slow-acting fermentation agent, so don't be surprised if you don't hit a pH of 5.3 after 2 days of drying time. The package directions provide more information; be sure to read them carefully.

If you don't have access to beef middles, you can stuff this *chorizo* into hog casings and link them however you want. Bear in mind, however, that the drying time will be much shorter, since the diameter of the sausage will be smaller.

Salchichón (L) and Pamplona-style *chorizo* (R)

PATATERA

Spain reveres both the pig and the potato, so the evolution of the *patatera*—a potato-based *chorizo*—was inevitable. This *embutido* is a favorite in the northern reaches of Extremadura, where it is sliced and smashed onto a piece of bread for snacking and *tapas*.

In a *patatera,* some of the lean *magro* is replaced with cooked potatoes, which is a really good way to stretch recipes in tough times and also use perishable veggies before they spoil. The trick is to dry out the cooked potato really well before making the *masa,* as a wetter *masa* means a longer drying time.

At *matanzas,* the *sabias* dry out the potatoes by cooking them, placing them in a colander, and weighting them down to help press out the moisture.

1 In a large stockpot over medium–high heat, cover the potatoes with water and bring to a boil. Reduce the heat to medium and simmer for 30 to 40 minutes, until cooked through. Drain the potatoes.

2 Using a ricer or potato masher, mash the potatoes and set aside to cool to room temperature (the more steam that is released, the less moisture will remain).

3 Place the potatoes in a colander, cover with plastic wrap, and weigh down with a few canned goods. Place a plate under the colander to collect any moisture. Refrigerate overnight.

4 Place the meat, *tocino,* and grinder parts in the freezer for 30 minutes to par-freeze before attempting to grind. In a small mixing bowl, combine the T-SPX culture and the distilled water, making a slurry. Set aside for a minimum of 10 minutes at room temperature to bloom.

5 Using a mortar and pestle, crush together the garlic and salt to form an *ajosal.* If desired, you can finish the *ajosal* in a food processor fitted with the "S" blade.

6 In a mixing bowl, combine the meat and *tocino* with the *ajosal,* dextrose, sugar, and curing salt. Toss together and set aside in the refrigerator to chill before you begin grinding.

YIELD

3–4 loops of sausage per 2.2 pounds (1 kg)

INGREDIENTS	*CHARCUTIER'S* PERCENTAGE
per 2.2 pounds (1 kg) of the following blend of diced meats: 20% *cabecero* (*coppa*/head of the pork loin) and 80% *tocino* (pork fat)	**100%**
4.4 pounds (2 kg) russet potatoes, washed, peeled	**200%**
Water, to cover	
¼ ounce (10 g) T-SPX culture (see p. 94)	**1%**
½ cup (100 mL) distilled water	
⅓ ounce (10 g) minced garlic	**1%**
1 ounce (25 g) kosher salt	**2.5%**
⅛ ounce (3 g) dextrose	**.3%**
⅛ ounce (3 g) granulated sugar	**.3%**
⅛ ounce (2.4 g) Instacure #2 or DQ #2 curing salt mix (see p. 83)	**.24%**
¼ cup (50 mL) chilled water	
1 ounce (25 g) *pimentón dulce*	**2.5%**
2 feet (60 cm) 1¼–1½-inch (32–36-mm) hog casings, soaked	

CONTINUED ON NEXT PAGE

PATATERA CONTINUED FROM PREVIOUS PAGE

7 Fill a large bowl with ice. Place a smaller bowl inside the ice-filled bowl. Grind the *cabecero* meat and *tocino* once through a medium-coarse (⅜ inch [9.5 mm]) die into the smaller bowl. Be careful: The meat mixture is wet, so it may squirt and pop out of the grinder. Return the ground meat to the freezer for 15 to 20 minutes, until it is par-frozen.

8 Grind the *cabecero* meat and *tocino* again through a medium-fine (³⁄₁₆ inch [5 mm]) die into the same bowl over ice.

9 In a small mixing bowl, combine the chilled water and *pimentón*, making a slurry. Keep the bowl containing the slurry chilled until ready to use.

10 Place the ground meats in the bowl of a stand mixer fitted with the paddle attachment (or you can just mix in a mixing bowl with a wooden spoon). Begin mixing on low speed. As the mixer runs, pour the water slurry into the bowl in a steady stream.

11 Continue mixing on medium speed for 1 to 2 minutes, until the water slurry has been fully incorporated into the mixture, a white residue forms on the sides of the bowl, the meat appears sticky, and the mixture firms up.

12 Reduce the mixer speed to low and add the T-SPX slurry. Continue mixing for 1 minute, until the T-SPX slurry is fully incorporated into the mixture. Place the bowl containing the ground meat mixture in the refrigerator to keep it cold until you are ready to stuff the sausage into casings.

13 Stuff the mixture into the casings and tie into 12-inch (30-cm) loops, ending with a butterfly knot series and a loop for hanging (see p. 90 in Chapter 3 for illustrations and directions). Using a sterile pin or sausage pricker, prick each sausage several times. Weigh each sausage loop to obtain a green weight; record the weights and tag each sausage with its green weight.

14 Ferment the sausages in a drying chamber set at 65°F to 80°F (18.3°C to 26.6°C) and 85% to 90% relative humidity for 2 to 3 days. Check the pH of the meat (see p. 95) to ensure that the level has dropped below 5.3 before the third day of drying.

15 Hang the sausages in a drying chamber set at 54°F to 60°F (12°C to 16°C) and 80% to 85% relative humidity for 1 to 2 months (depending on their size), until the sausages have lost about 35% of their green weight. At that point, the sausages will be ready to consume.

NOTES: You can ferment the sausages either before or after stuffing. It's really a matter of preference, since the meat firms up during the fermentation process and makes stuffing a little easier. On the other hand, it might be more efficient for you to make and stuff the sausages all in 1 day.

Bactoferm T-SPX is a slow-acting fermentation agent, so don't be surprised if you don't hit a pH of 5.3 after 2 days of drying time. The package directions provide more information; be sure to read them carefully.

For spicier *Patatera,* instead of 1 ounce (25 g) of *pimentón dulce,* use ½ ounce (15 g) of *pimentón dulce* and ⅓ ounce (10 g) of *pimentón picante.*

Just as you would when making gnocchi, rice the potatoes while they are hot. Doing so yields the best texture for the sausage.

TRUCOS DE LA COCINA: MORCILLA DE PATATERA

PATATERA'S OTHER NAME—*morcilla de patatera*—is a point of confusion for many living outside of the northern *pueblos* of Extremadura. That's because the word *morcilla* in the rest of Spain usually indicates that blood is included in the *embutido,* but that's not the case here.

Here's the deal: Extremadura is an area of Spain very much entrenched in the old ways. Folks there are only vaguely aware of email, recipes never change, and *morcilla* has always referred to any kind of sausage, blood or no blood. So here—and only here—*morcilla* can mean any type of sausage.

CHORIZO DE CALABAZA

While cooking at Adolfo's in Toledo, I was fortunate to walk to work every day through the city's *Casco Viejo*. I made a point of stopping at a local butcher shop almost every day to see what meats and *embutidos* the old guys in back were playing with. One day during the colder months of autumn, I saw a pumpkin sausage hanging in the display window and talked to the guys about it. Turns out, it's a classic combo they kick around when the season is right. Much like the *Patatera,* this sausage uses a starchy, plant-based friend—in this case pumpkin—as the binding agent for the sausage. And since *pimentón* and pumpkin make for a great flavor combo, this sausage is a delicious take on the classic *chorizo*.

The moral of this story: Become friends with your local butchers, because cool old guys make good stuff.

1   In a large stockpot over medium heat, cover the pumpkin with water and bring to a boil. Reduce the heat to medium and simmer for 30 to 40 minutes, until cooked through. Drain the pumpkin.

2   Using a ricer or potato masher, mash the pumpkin and set aside to cool to room temperature (the more steam that is released, the less moisture will remain).

3   Place the pumpkin in a colander, cover with plastic wrap, and weigh down with a few canned goods. Place a plate under the colander to collect any moisture. Refrigerate overnight.

4   Place the meat, *tocino,* and grinder parts in the freezer for 30 minutes to par-freeze before attempting to grind. In a small mixing bowl, combine the T-SPX culture and the distilled water, making a slurry. Set aside for a minimum of 10 minutes at room temperature to bloom.

5   Using a mortar and pestle, crush together the garlic and salt to form an *ajosal.* If desired, you can finish the *ajosal* in a food processor fitted with the "S" blade.

6   In a mixing bowl, combine the meat and *tocino* with the *ajosal,* dextrose, sugar, and curing salt. Toss together and refrigerate momentarily as you quickly set up the grinder.

YIELD

3–4 loops of sausage per 2.2 pounds (1 kg)

INGREDIENTS	CHARCUTIER'S PERCENTAGE
per 2.2 pounds (1 kg) of the following blend of diced meats: 20% *cabecero* (*coppa*/head of the pork loin) and 80% *tocino* (pork fat)	100%
4.4 pounds (2 kg) diced fresh pumpkin	200%
Water, to cover	
¼ ounce (10 g) T-SPX culture (see p. 94)	1%
½ cup (100 mL) distilled water	
⅓ ounce (10 g) minced garlic	1%
1 ounce (25 g) kosher salt	2.5%
⅛ ounce (3 g) dextrose	.3%
⅛ ounce (3 g) granulated sugar	.3%
⅛ ounce (2.4 g) Instacure #2 or DQ #2 curing salt mix (see p. 83)	.24%
¼ cup (50 mL) dry white wine, such as Verdejo, chilled	
1 ounce (25 g) *pimentón dulce*	2.5%
⅛ ounce (5 g) freshly ground black pepper	.5%
½ teaspoon (1 g) whole allspice, toasted and ground	.1%
½ teaspoon (1 g) freshly grated nutmeg	.1%
2 feet (60 cm) 1¼–1½-inch (32–36-mm) hog casings, soaked, or more as needed	

7   Fill a large bowl with ice. Place a smaller bowl inside the ice-filled bowl. Grind the *cabecero* meat and *tocino* once through a medium-coarse (³⁄₈ inch [9.5 mm]) die into the smaller bowl. Be careful: The meat mixture is wet, so it may squirt and pop out of the grinder. Return the ground meat to the freezer for 15 to 20 minutes, until it is par-frozen.

8   Grind the *cabecero* meat and *tocino* again through a medium-fine (³⁄₁₆ inch [5 mm]) die into the same bowl over ice.

9   In a small mixing bowl, combine the wine, *pimentón,* black pepper, allspice, and nutmeg, making a slurry. Keep the bowl containing the slurry chilled until ready to use.

10 Place the ground meats in the bowl of a stand mixer fitted with the paddle attachment (or you can just mix in a mixing bowl with a wooden spoon). Begin mixing on low speed. As the mixer runs, pour the wine slurry into the bowl in a steady stream.

11 Continue mixing on medium speed for 1 to 2 minutes, until the wine slurry has been fully incorporated into the mixture, a white residue forms on the sides of the bowl, the meat appears sticky, and the mixture firms up.

12 Reduce the mixer speed to low and add the T-SPX slurry. Continue mixing for 1 minute, until the T-SPX slurry is fully incorporated into the mixture. Place the bowl containing the ground meat mixture in the refrigerator to keep it cold until you are ready to stuff the sausage into casings.

13 Stuff the mixture into the casings and tie into 12-inch (30-cm) loops, ending with a butterfly knot series and a loop for hanging (see Chapter 3 for illustrations and directions). Using a sterile pin or sausage pricker, prick each sausage several times. Weigh each sausage loop to obtain a green weight; record the weights and tag each sausage with its green weight.

14 Ferment the sausages in a drying chamber set at 65°F to 80°F (18.3°C to 26.6°C) and 85% to 90% relative humidity for 2 to 3 days. Check the pH of the meat (see p. 95) to ensure that the level has dropped below 5.3 before the third day of drying.

15 Hang the sausages in a drying chamber set at 54°F to 60°F (12°C to 16°C) and 80% to 85% relative humidity for 1 to 2 months (depending on their size), until the sausages have lost about 35% of their green weight. At that point, the sausages will be ready to consume.

NOTES: You can ferment the sausages either before or after stuffing. It's really a matter of preference, since the meat firms up during the fermentation process and makes stuffing a little easier. On the other hand, it might be more efficient for you to make and stuff the sausages all in 1 day.

Bactoferm T-SPX is a slow-acting fermentation agent, so don't be surprised if you don't hit a pH of 5.3 after 2 days of drying time. The package directions provide more information; be sure to read them carefully.

For spicier *Chorizo de Calabaza,* instead of 1 ounce (25 g) of *pimentón dulce,* use ½ ounce (15 g) of *pimentón dulce* and ⅓ ounce (10 g) of *pimentón picante.*

Just as you would when making gnocchi, rice the pumpkin while it is hot. Doing so yields the best texture for the sausage.

MORCILLA ACHORIZADA

Morcilla Achorizada, also known as *Morcilla de Jaén,* is very near and dear to my heart. In fact, it's one of the reasons that I started my journey down the path of *charcutería.* And it's all thanks to a fortuitous meeting with the coolest bartender in Madrid: *jamón* slicer, ex-boxer, and tattooed *jefe* Manolo Delgado.

If you ask him, Manolo isn't shy about saying that he was "born in Madrid, but created in the bars of Sol," where he has toiled for over 16 years. And it was there, in his domain behind the counter at La Soberbia, that our group of cooks on a scholarship tour with the Spanish Institute for Foreign Trade (*Instituto Español de Comercio Exterior;* ICEX) first stumbled into his bar and consecrated our friendship over glasses of Patxaran and a plate of this amazing *embutido.* Manolo served us the *morcilla* as a sort of pressed Panini, thinly sliced and sandwiched between warm, charred bread slices that he drizzled liberally with a garlicky olive oil. It was the perfect foil for our drinking, and one of the reasons I make a pilgrimage back to La Soberbia to see Manolo every time I return to Spain.

The trick with this *morcilla* is cooking the potatoes and onions into a flavorful *sofrito* and a formless mush. The *sofrito* gives the sausage its unique texture.

YIELD

3 to 4 loops of sausage per
2.2 pounds (1 kg)

INGREDIENTS	CHARCUTIER'S PERCENTAGE
per 2.2 pounds (1 kg) of the following blend of diced meats: 40% *cabecero* (*coppa*/head of the pork loin), 40% *papada* (pork jowl), and 20% *tocino* (pork fat)	**100%**

FOR THE POTATOES:

1.1	pounds (500 g) russet potatoes, peeled......................	**50%**
	Water, to cover	

FOR THE *SOFRITO:*

10½	ounces (300 g) *manteca* (pork lard)................................	**30%**
18	ounces (500 g) minced white onions	**50%**
1¾	ounces (50 g) kosher salt..............	**5%**
1	fresh bay leaf	

FOR THE RICE:

18	ounces (500 g) uncooked bomba* rice............................	**50%**
2	quarts (2 L) water	
⅓	ounce (10 g) kosher salt	**1%**
2	cups (500 mL) *sangre* (pork blood), whisked to prevent coagulation.................	**50%**

continued

* Bomba rice is a type of rice from Spain typically used for making paella. It is unique because it absorbs 3 times its volume in liquid.

TO MAKE THE POTATOES:

1   In a large stockpot over medium–high heat, cover the potatoes with water and bring to a boil. Reduce the heat to medium and simmer for 30 to 40 minutes, until cooked through. Drain the potatoes.

2   Using a ricer or potato masher, mash the potatoes and set aside to cool to room temperature (the more steam that is released, the less moisture will remain).

3   Place the potatoes in a colander, cover with plastic wrap, and weigh down with a few canned goods. Place a plate under the colander to collect any moisture. Refrigerate overnight.

TO MAKE THE *SOFRITO*:

1   In a large saucepan, warm the *manteca* over medium heat until rippling but not smoking. Add the minced onions, salt, and bay leaf. Reduce the heat to low and allow the onions to sweat for 30 to 40 minutes, until they are cooked through but not browned.

TO MAKE THE RICE:

1   To the saucepan containing the *sofrito,* add the rice. Sauté for 3 to 5 minutes, until the rice is warmed and coated with the *sofrito.*

2   Raise the heat to medium–high and add the water and salt. Cover the pan and bring it to a boil. Reduce the heat to low and simmer, covered, for 30 minutes, stirring frequently.

3   Remove the lid and cook for another 20 to 30 minutes, until the rice is a mushy porridge and individual grains cannot be distinguished.

INGREDIENTS	CHARCUTIER'S PERCENTAGE
FOR THE *MASA*:	
¼ ounce (10 g) T-SPX culture (see p. 94)	**1%**
½ cup (100 mL) distilled water	
1 ounce (28 g) kosher salt	**2.8%**
⅓ ounce (10 g) minced garlic	**1%**
⅛ ounce (3 g) dextrose	**3%**
⅛ ounce (3 g) granulated sugar	**3%**
⅛ ounce (2.4 g) Instacure #2 or DQ #2 curing salt mix (see p. 83)	**24%**
1 ounce (25 g) *pimentón dulce*	**2.5%**
¼ cup (50 mL) dry white wine, such as Verdejo, chilled	
⅛ ounce (2 g) dried oregano	**2%**
2 feet (60 cm) 1¼–1½-inch (32–36-mm) hog casings, soaked, or more as needed	

TO FINISH THE DISH:

1   Add the potatoes to the *sofrito* and rice mixture and cook for 10 to 20 minutes on low heat, until the mixture is dry and combined into a homogenous mass. Discard the bay leaf and cool the mixture for 10 minutes.

2   Add the *sangre* to the warm potato–rice mixture. The mixture should change color from red to reddish-black as it warms through. Set aside.

3   Place the meats, *tocino,* and grinder parts in the freezer for 30 minutes to par-freeze before attempting to grind. In a small mixing bowl, combine the T-SPX culture and the distilled water, making a slurry. Set aside for a minimum of 10 minutes at room temperature to bloom.

4   Using a mortar and pestle, crush together the salt and garlic to form an *ajosal.* If desired, you can finish the *ajosal* in a food processor fitted with the "S" blade.

5   In a mixing bowl, combine the *ajosal,* dextrose, sugar, and curing salt. Divide the mixture in half.

6   Add half of the *ajosal* mixture to the *cabecero* meat and *tocino.* Toss together. In a separate bowl, mix the *papada* with the other half of the *ajosal* mixture. Set aside both mixtures in the refrigerator to chill before you begin grinding.

CONTINUED ON NEXT PAGE

MORCILLA ACHORIZADA CONTINUED FROM PREVIOUS PAGE

7 Fill a large bowl with ice. Place a smaller bowl inside the ice-filled bowl. Grind the *cabecero* meat and *tocino* once through a medium-coarse (⅜ inch [9.5 mm]) die into the smaller bowl. Be careful: The meat mixture is wet, so it may squirt and pop out of the grinder.

8 Grind the *papada* through a medium (¼ inch [6 mm]) die and combine with the *cabecero* meat for mixing. Return the ground meats to the freezer for 15 to 20 minutes, until they are par-frozen.

9 In a small mixing bowl, combine the wine, *pimentón,* and oregano, making a slurry. Keep the bowl containing the slurry chilled until ready to use.

10 Place the ground meats in the bowl of a stand mixer fitted with the paddle attachment (or you can just mix in a mixing bowl with a wooden spoon). Begin mixing on low speed. As the mixer runs, pour the wine slurry into the bowl in a steady stream.

11 Continue mixing on medium speed for 1 to 2 minutes, until the wine slurry has been fully incorporated into the mixture, a white residue forms on the sides of the bowl, the meat appears sticky, and the mixture firms up.

12 Reduce the mixer speed to low and add the T-SPX slurry. Continue mixing for 1 minute, until the T-SPX slurry is fully incorporated into the mixture.

13 Increase the speed to medium and add the rice–potato mixture. Continue mixing for 1 minute, until the mixtures are fully blended. Place the bowl containing the ground meat mixture in the refrigerator to keep it cold until you are ready to stuff the sausage into casings.

14 Stuff the mixture into the casings and tie into 12-inch (30-cm) loops, ending with a butterfly knot series and a loop for hanging (see Chapter 3 for illustrations and directions). Using a sterile pin or sausage pricker, prick each sausage several times. Weigh each sausage loop to obtain a green weight; record the weights and tag each sausage with its green weight.

15 Ferment the sausages in a drying chamber set at 65°F to 80°F (18.3°C to 26.6°C) and 85% to 90% relative humidity for 2 to 3 days. Check the pH of the meat (see p. 95) to ensure that the level has dropped below 5.3 before the third day of drying.

16 Stock a cold smoker with oak wood and bring the temperature to 90°F (30°C) and the relative humidity to 80% to 85%. Hang the *morcillas* in the smoking chamber and smoke for 2 to 4 days, or break it up into 6-hour-long sessions for 4 to 8 days, returning the *morcillas* to the drying chamber set at 54°F to 60°F (12°C to 16°C) and 80% to 85% relative humidity in between sessions.

17 Hang the sausages in a drying chamber set at 54°F to 60°F (12°C to 16°C) and 80% to 85% relative humidity for 1 to 2 months (depending on their size), until the sausages have lost about 35% of their green weight. At that point, the sausages will be ready to consume.

NOTES: You can ferment the sausages either before or after stuffing. It's really a matter of preference, since the meat firms up during the fermentation process and makes stuffing a little easier. On the other hand, it might be more efficient for you to make and stuff the sausages all in 1 day.

Bactoferm T-SPX is a slow-acting fermentation agent, so don't be surprised if you don't hit a pH of 5.3 after 2 days of drying time. The package directions provide more information; be sure to read them carefully.

Just as you would when making gnocchi, rice the potatoes while they are hot. Doing so yields the best texture for the sausage.

SABADIEGO

Whenever you break down a pig, there's going to be lots of leftover meat, trim, and gnarly bits of offal. That's where the story of the *Sabadiego* begins.

With a lineage that dates back to almost the eighth century, the *Sabadiego* has a fascinating history in Spanish culinary lore. It's long been presumed to be the poorest of the poor *embutidos* a person could eat—a sign of personal penitence and humility via eating habits.

Specifically, the name *"sabadiego"* means "sausage of Saturday," since that was the day that it was traditionally consumed. In the Catholic tradition of old Spain, Fridays were meatless, so most folks were craving meat by Saturday. But on a Saturday, the only meat you could consume would be a type not considered "luxurious"—for example, *Jamón,* or in the case of sausage, *Sabadiego.*

As with so many old-fashioned culinary customs, the *Sabadiego* fell out of favor over the years when meat got cheaper and more plentiful. Before losing their cultural heritage, however, a group of men from the town of Noreña in Asturias founded the *Orden del Sabadiego* in 1988 with a mission to reclaim the tradition.

The guys do a bunch of charity work in and around the area and get to dress up in some pretty fly Harry Potter-esque capes, but a regular day means getting together in the Order's chapter house, cooking *Sabadiego*-based dinners (*fabada* and other stews are a common usage for this *embutido),* and doing what guys do in private clubs: play cards, smoke, and let the liquor flow. The only taboo subjects in the clubhouse are politics or *fútbol,* which is probably a good idea considering sharp cooking implements are almost always within arm's reach.

YIELD

3 to 4 loops of sausage per 2.2 pounds (1 kg)

INGREDIENTS	CHARCUTIER'S PERCENTAGE
per 2.2 pounds (1 kg) of the following blend of diced meats: 40% *cabecero* (*coppa*/head of the pork loin), 40% *papada* (pork jowl), and 20% *tocino* (pork fat)	**100%**
18 ounces (500 g) *manteca* (pork lard)	**50%**
4.4 pounds (2 kg) minced white onions	**200%**
2¾ ounces (78 g) kosher salt, divided	**7.8%**
2 fresh bay leaves	
2 cups (500 mL) *sangre* (pork blood), whisked to prevent coagulation	**50%**
¼ ounce (10 g) T-SPX culture (see p. 94)	**1%**
½ cup (100 mL) distilled water	
⅔ ounce (20 g) minced garlic, divided	**2%**
⅛ ounce (3 g) dextrose	**.3%**
⅛ ounce (3 g) granulated sugar	**.3%**
⅛ ounce (2.4 g) Instacure #2 or DQ #2 curing salt mix (see p. 83)	**.24%**
¼ cup (50 mL) chilled water	
1 ounce (25 g) *pimentón dulce*	**2.5%**
⅛ ounce (2 g) dried oregano	**.2%**
2 feet (60 cm) 1¼–1½-inch (32–36-mm) hog casings, soaked, or more as needed	

CONTINUED ON NEXT PAGE

SABADIEGO CONTINUED FROM PREVIOUS PAGE

1   In a large saucepan, warm the *manteca* over medium heat until rippling but not smoking. Add the minced onions, 1¾ ounces (50 g) of the salt, and the bay leaves. Reduce the heat to low and allow the onions to sweat for 30 to 40 minutes, until they are cooked through but not browned. Discard the bay leaves and cool the mixture for 10 minutes.

2   Add the *sangre* to the saucepan. The mixture should change color from red to reddish-black as it warms through. Set aside.

3   Place the meats, *tocino,* and grinder parts in the freezer for 30 minutes to par-freeze before attempting to grind. In a small mixing bowl, combine the T-SPX culture and the distilled water, making a slurry. Set aside for a minimum of 10 minutes at room temperature to bloom.

4   Using a mortar and pestle, crush together the garlic and the remaining 1 ounce (28 g) of the salt to form an *ajosal.* If desired, you can finish the *ajosal* in a food processor fitted with the "S" blade.

5   In a mixing bowl, combine the *ajosal,* dextrose, sugar, and curing salt. Divide the mixture in half.

6   Add half of the *ajosal* mixture to the *cabecero* meat and *tocino.* Toss together. In a separate bowl, mix the *papada* with the other half of the *ajosal* mixture. Set aside both mixtures in the refrigerator to chill before you begin grinding.

7   Fill a large bowl with ice. Place a smaller bowl inside the ice-filled bowl. Grind the *cabecero* meat and *tocino* once through a medium-coarse (⅜ inch [9.5 mm]) die into the smaller bowl. Be careful: The meat mixture is wet, so it may squirt and pop out of the grinder.

8   Grind the *papada* through a medium (¼ inch [6 mm]) die and combine with the *cabecero* meat for mixing.

9   In a small mixing bowl, combine the chilled water, *pimentón,* and oregano, making a slurry. Keep the bowl containing the slurry chilled until ready to use.

10   Place the ground meats in the bowl of a stand mixer fitted with the paddle attachment (or you can just mix in a mixing bowl with a wooden spoon). Begin mixing on low speed. As the mixer runs, pour the water slurry into the bowl in a steady stream.

11   Continue mixing on medium speed for 1 to 2 minutes, until the water slurry has been fully incorporated into the mixture, a white residue forms on the sides of the bowl, the meat appears sticky, and the mixture firms up.

12   Reduce the mixer speed to low and add the T-SPX slurry. Continue mixing for 1 minute, until the T-SPX slurry is fully incorporated into the mixture. Place the bowl containing the ground meat mixture in the refrigerator to keep it cold until you are ready to stuff the sausage into casings.

13   Stuff the mixture into the casings and tie into 12-inch (30-cm) loops, ending with a butterfly knot series and a loop for hanging (see Chapter 3 for illustrations and directions). Using a sterile pin or sausage pricker, prick each sausage several times. Weigh each sausage to obtain a green weight; record the weights and tag each sausage with its green weight.

14   Ferment the sausages in a drying chamber set at 65°F to 80°F (18.3°C to 26.6°C) and 85% to 90% relative humidity for 2 to 3 days. Check the pH of the meat (see p. 95) to ensure that the level has dropped below 5.3 before the third day of drying.

15   Stock a cold smoker with oak wood and bring the temperature to 90°F (30°C) and the relative humidity to 80% to 85%. Hang the *Sabadiegos* in the smoking chamber and smoke for 7 to 10 days, or break it up into 6-hour-long sessions for 10 to 14 days, returning the *Sabadiegos* to the drying chamber set at 54°F to 60°F (12°C to 16°C) and 80% to 85% relative humidity in between sessions.

16   Hang the sausages in a drying chamber set at 54°F to 60°F (12°C to 16°C) and 80% to 85% relative humidity for 1 to 2 months (depending on their size), until the sausages have lost about 35% of their green weight. At that point, the sausages will be ready to consume.

NOTES: You can ferment the sausages either before or after stuffing. It's really a matter of preference, since the meat firms up during the fermentation process and makes stuffing a little easier. On the other hand, it might be more efficient for you to make and stuff the sausages all in 1 day.

Bactoferm T-SPX is a slow-acting fermentation agent, so don't be surprised if you don't hit a pH of 5.3 after 2 days of drying time. The package directions provide more information; be sure to read them carefully.

To make a more traditional *Sabadiego,* replace 5% of the meat with soaked tripe pieces (aka chitterlings). Cook as indicated for *Bull Catalan* (see recipe on p. 280), until the meat is soft enough to grind with the *papada*.

MORCÓN

What happens when a savvy *sabía* sees *chorizo masa* and some lean trim from butchering pigs kicking around? *Morcón,* that's what. Both get stuffed into a *ciego,* tied up, and hung up, allowing nature to do her magical curing act.

This *embutido* is a big-time favorite throughout Spain—especially in Extremadura and northern Andalucía. It's a bit leaner than *chorizo,* as there's no back fat added into the mix of already lean meats. While the flavors and curing methodology are very similar to the basic *chorizo masa,* the textures involved are quite different and unique to the *Morcón.*

1   Place the meats and grinder parts in the freezer for 30 minutes to par-freeze before attempting to grind. In a small mixing bowl, combine the T-SPX culture and the distilled water, making a slurry. Set aside for a minimum of 10 minutes at room temperature to bloom.

2   Using a mortar and pestle, crush together the salt and garlic to form an *ajosal.* If desired, you can finish the *ajosal* in a food processor fitted with the "S" blade.

3   In a mixing bowl, combine the *ajosal,* dextrose, sugar, and curing salt. Divide the mixture in half.

4   Add half of the *ajosal* mixture to the *cabecero* meat. Toss together. In a separate bowl, mix the *papada* and *magro* with the other half of the *ajosal* mixture. Set aside both mixtures in the refrigerator to chill before you begin grinding.

5   Fill a large bowl with ice. Place a smaller bowl inside the ice-filled bowl. Grind the *papada* and *magro* meats once through a medium-coarse (3⁄8 inch [9.5 mm]) die into the smaller bowl. Be careful: The meat mixture is wet, so it may squirt and pop out of the grinder.

6   Combine the ground meats with the diced *cabecero* meat for mixing.

7   In a small mixing bowl, combine the wine, *pimentón,* and black pepper, making a slurry. Keep the bowl containing the slurry chilled until ready to use.

8   Place the ground meat mixture in the bowl of a stand mixer fitted with the paddle attachment (or you can just mix in a mixing bowl with a wooden spoon). Begin mixing on low speed. As the mixer runs, pour the wine slurry into the bowl in a steady stream.

YIELD

2 *morcón* per 2.2 pounds (1 kg)

INGREDIENTS	CHARCUTIER'S PERCENTAGE
per 2.2 pounds (1 kg) of the following blend of diced meats: 40% *papada* (pork jowl), 30% *cabecero* (*coppa*/head of the pork loin), and 30% *magro* (lean pork meat)	**100%**
1⁄4 ounce (10 g) T-SPX culture (see p. 94)	**1%**
1⁄2 cup (100 mL) distilled water	
1 ounce (28 g) kosher salt	**2.8%**
1⁄3 ounce (10 g) minced garlic	**1%**
1⁄8 ounce (3 g) dextrose	**.3%**
1⁄8 ounce (3 g) granulated sugar	**.3%**
1⁄8 ounce (2.4 g) Instacure #2 or DQ #2 curing salt mix (see p. 83)	**.24%**
1⁄4 cup (50 mL) dry white wine, such as Verdejo, chilled	
1 ounce (25 g) *pimentón dulce*	**2.5%**
1⁄8 ounce (5 g) freshly ground black pepper	**.5%**
1 *ciego de cerdo* (pork bung), soaked	

CONTINUED ON NEXT PAGE

MORCÓN CONTINUED FROM PREVIOUS PAGE

9   Continue mixing on medium speed for 1 to 2 minutes, until the wine slurry has been fully incorporated into the mixture, a white residue forms on the sides of the bowl, the meat appears sticky, and the mixture firms up.

10   Reduce the mixer speed to low and add the T-SPX slurry. Continue mixing for 1 minute, until the T-SPX slurry is fully incorporated into the mixture. Place the bowl containing the ground meat mixture in the refrigerator to keep it cold until you are ready to stuff the sausage into casings.

11   Tightly stuff the mixture into the *ciego* and use string to tie it off with a butterfly knot and bubble knot series, ending with a loop at one end for hanging (this type of sausage is not prepared as links; see Chapter 3 for illustrations and directions). Using a sterile pin or sausage pricker, prick the *Morcón* several times. Weigh the *Morcón* to obtain a green weight; record and tag each *Morcón* with the weight.

12   Ferment the *Morcón* in a drying chamber set at 65°F to 80°F (18.3°C to 26.6°C) and 85% to 90% relative humidity for 2 to 3 days. Check the pH of the meat (see p. 95) to ensure that the level has dropped below 5.3 before the third day of drying.

13   Hang the *Morcón* in a drying chamber set at 54°F to 60°F (12°C to 16°C) and 80% to 85% relative humidity for 3 to 4 months (depending on their size), until the *Morcón* has lost about 35% of its green weight. At that point, the *Morcón* will be ready to consume.

NOTES: You can ferment the sausages either before or after stuffing. It's really a matter of preference, since the meat firms up during the fermentation process and makes stuffing a little easier. On the other hand, it might be more efficient for you to make and stuff the sausages all in 1 day.

Bactoferm T-SPX is a slow-acting fermentation agent, so don't be surprised if you don't hit a pH of 5.3 after 2 days of drying time. The package directions provide more information; be sure to read them carefully.

For a spicier *Morcón,* instead of 1 ounce (25 g) of *pimentón dulce,* use ½ ounce (15 g) of *pimentón dulce* and ⅓ ounce (10 g) of *pimentón picante.*

CURED
EMBUTIDOS
TRADITIONAL AND MODERN RECIPES

COCIDO MADRILEÑO

If you are a Spaniard and nursing a severe Sunday-morning hangover, there's only one cure to bring you back from the dead: *Cocido Madrileño,* the ultimate chicken soup-based restorative.

But, as most Spaniards will tell you, the best *cocido* in Madrid isn't made in a restaurant. Oh, sure, you can get a great *cocido* at La Bola, Malacatín, or some of the other joints in town that are famous for serving *cocido* on an everyday basis. But the average Spaniard will tell you that his or her mom makes the best in all of Spain.

For my money, the best is made by Elaine Hill, the mom of Saul Aparicio-Hill, our Spanish Institute for Foreign Trade (*Instituto Español de Comercio Exterior;* ICEX) tour guide and the coolest guy in Spain. While she may not be a native Spaniard, Elaine's *cocido* puts any chef-inspired version I've ever had to shame.

YIELD

4–6 entrée servings (and lots of leftovers)

INGREDIENTS

9	ounces (250 g) dried chickpeas
1	tablespoon (20 g) baking soda
	Cold water, to cover
2.2	pounds (1 kg) chicken bones, feet, or wings
2.2	pounds (1 kg) beef or pork bones
1	*jamón* bone
1	stewing chicken
4	Cantimpalos-Style *Chorizo* sausages (see recipe on p. 299)
4	*Morcilla de Arroz* or *Morcilla de Cebolla* sausages (see recipes on pp. 258 and 260)
12	ounces (300 g) *Panceta Curada* or *Jamón* (see recipes on pp. 122 and 143)
12	ounces (300 g) *tocino fresco* (salted pork back fat)
12	ounces (300 g) beef *morcillo* (shank) or *pecho* (brisket), thoroughly rinsed
6	medium russet potatoes, peeled and pierced with a fork
4	medium carrots, peeled and trimmed
4	small turnips, peeled and trimmed
2	medium zucchini or other squash, cleaned and trimmed
2	ribs celery, cleaned and trimmed
2	leeks, cleaned and trimmed
	Kosher salt, as needed
	Extra virgin olive oil, as needed
4	cloves garlic, peeled and sliced *Godfather* thin
2	heads Savoy cabbage, trimmed and chiffonaded

continued

Cocido is an ancient dish typically served in three stages (called *vuelcos*, meaning to "turn out"). The first *vuelco* is the rich broth and some thin noodles; followed by the second *vuelco,* the veggies and chickpeas; and last, the third *vuelco,* the meats, including *chorizo, morcilla, jamón, tocino,* and other meaty friends.

The *cocido* meal is typically served as a Sunday family lunch or dinner with lots of intentional leftovers. The leftovers then get consumed over the next few days, generally in a *ropa vieja* on Day 2 and *croquettas* made from the remaining meats on Day 3.

One important note: When serving Elaine's recipe, including her *Tomate Frito* (see recipe on p. 407) is an absolute must. I'm not kidding when I say it is one of the best condiments you will ever try, especially for serving with an intense, meaty dish like *cocido.*

INGREDIENTS *(continued)*

1	tablespoon (7 g) *pimentón dulce*
1	cup (235 mL) *Tomate Frito* (see recipe on p. 407)
1	whole clove garlic, peeled and destemmed
3	tablespoons (10 g) dried oregano
2	tablespoons (12 g) cumin seed, toasted and ground
¼	cup (50 mL) red wine vinegar
8	ounces (250 g) *fideos* or vermicelli pasta

1   In a large mixing bowl, combine the chickpeas and baking soda and cover with the water. Stir to combine and soak overnight at room temperature.

2   In a large stockpot, barely cover the chicken bones, feet, or wings; beef or pork bones; and *jamón* bones with water. Place over medium–high heat and bring to a boil to purge the bones of any blood or impurities.

3   Drain the water from the stockpot and cover again with cold water. Return to medium–high heat and bring to a boil. Reduce the heat to low and simmer for 3 hours, skimming off any foam or impurities.

4   Drain the chickpeas, place them in cheesecloth, and tie the cheesecloth closed, creating a bag. Place the chickpeas in the cheesecloth in the stockpot and continue to simmer for 1 hour. Remove from the heat.

5   Strain the broth into another stockpot. Remove and discard the bones. Transfer the chickpeas from the cheesecloth to a small mixing bowl and set aside.

6   Place the chicken, sausages, *panceta, tocino,* and *morcillo* in the stockpot. Cover with the broth and place over medium–high heat. Bring to a vigorous boil. Reduce the heat to a high simmer and skim off any foam or impurities but leaving the fat in the broth since it will emulsify into the broth (see p. 335 for more information about this technique).

7   Simmer on low heat until the meats are cooked through: Remove the chicken after 45 to 60 minutes, the sausages after 1 hour, the *tocino* and the *panceta* after 1½ hours, and the *morcillo* after 2 hours, or when it shreds easily. As you remove each meat from the broth, set it aside on a serving platter.

8   Preheat the oven to 250°F (120°C). Once the meats are cool enough to handle, cut or shred them into bite-sized pieces. Return the pieces to the serving platter, splash them with a little broth, cover with foil, and place in the oven to keep warm.

9   Place the potatoes, carrots, turnips, squash, celery, and leeks in the broth and cook until softened. Remove the squash after 10 to 15 minutes, the leeks after 15 to 20 minutes, and the turnips, carrots, and potatoes after 20 to 25 minutes. As you remove each vegetable from the broth, set it aside on a separate serving platter. Cut the veggies into large chunks, return them to the serving platter, splash them with a little broth, cover with foil, and place in the oven to keep warm.

CONTINUED ON NEXT PAGE

COCIDO MADRILEÑO

CONTINUED FROM PREVIOUS PAGE

10   Raise the heat to medium–high and bring the broth to a boil. Continue boiling the broth for 10 to 15 minutes, until the broth's volume has been reduced by ¼ and the fat has emulsified. Taste and season with the salt. Transfer the broth to a large bowl and set aside to hold warm.

11   In a large sauté pan, warm a thin layer of the oil over medium–high heat until rippling but not smoking. Add the sliced garlic and sauté for 2 minutes, until the garlic is just starting to brown. Add the cabbage, season again with the salt, and sauté for 5 minutes, until the cabbage starts to wilt. Stir to keep the cabbage and garlic moving in the pan (and to prevent the garlic from burning). Add the *pimentón* and sauté for 5 minutes, periodically stirring, until the cabbage's volume has been reduced by ½ but it still retains a little crunch. Remove from the heat. Taste and reseason with the salt. Transfer to a large platter, cover with foil, and place in the oven to keep warm.

12   Place the *Tomate Frito* in the sauté pan and return the pan to medium–high heat. Bring to a boil and reduce the heat to medium. Simmer for 7 to 10 minutes, until its volume is reduced by ½.

13   Finely grate the whole garlic clove directly over the sauté pan and add the oregano and cumin. Stir and cook for 1 minute. Remove from the heat and stir in the red wine vinegar. Transfer to a serving bowl and place the bowl in the oven to keep warm.

14   In a large saucepan over medium heat, warm 2 cups (500 mL) of the broth per person (trust me, they'll want seconds—especially if they are hung over!). When the broth reaches a simmer, add the pasta and cook for 4 to 6 minutes, until just *al dente*. Remove from the heat. Taste and reseason with the salt as necessary.

15   Transfer the broth and pasta into bowls and serve as the first course with the *Tomate Frito*.

16   Remove the meat and vegetable platters from the oven. Serve with the reserved chickpeas. Serve the remaining courses when ready, vegetables first and then the meats. Continue serving until those in attendance beg for mercy.

NOTES: *Cocido* recipes are intended to provide leftovers—stock you can freeze and meats to use in other dishes. So unless you are feeding a big group, this recipe will provide for many meals to come.

If you are a confident cook, you can speed up the cooking process by separating the broth into multiple pots and cooking everything at the same time. At the end, simply combine all the various broths together to meld their flavors. That's restaurant-level efficiency, Chef.

TRUCOS DE LA COCINA: SPANISH BOILED BROTHS

THE SPANIARDS HAVE a peculiar way of simmering their broth for stews and *potajes* that runs counter to pretty much every way you've ever heard of making a stock or broth. The first time I saw the technique at work was while making a family meal of *Cocido Madrileño* at Adolfo Muñoz's restaurant in Toledo. We started by simmering a whole beef shank with a chicken, some *tocino,* and other goodies and—like a good and diligent cook—I kept the heat low and skimmed the fat off the top of the liquid. A cloudy broth is a bad broth, right? Wrong! I learned that lesson quickly when my *jefe de la cocina,* Sergio, bellowed at me from across the kitchen in his Manchego-inflected accent:

"¡Oye! ¿Qué haces, Americano?"

His knowing smile said, "You've got a lot to learn, *amigo.*" Sergio then walked over, ceremoniously dumped the skimmed fat back into the pot, and kicked the heat up to a rolling boil where it stayed for much of the cooking process.

Here's the deal: As Sergio graciously explained to me, fat is intentionally emulsified into the liquid for many of Spain's *potaje* preparations. It provides creaminess (historically, it also provided valuable calories) in the final broth, so the fat is almost never skimmed off and the broth is almost always kept at a boil to facilitate the emulsification process.

The Spaniards aren't alone in using this technique. Japanese *tonkotsu*-style ramen uses a similar broth, where stock is infused with pork fat and marrow. The trick to getting a perfect milky-white broth is to ensure that all of the blood and particulates are purged from the bones and meats before cooking.

CHORIZO EN SIDRA

I get all excited whenever a *tapas* bar has *Chorizo en Sidra* on the menu. There's just something about this delicious Asturian dish of *chorizo* braised in apple cider that, for me, is the epitome of what pork and apples can be. It breaks my heart, however, when instead I get a hunk of sausage cooked until its dried and desiccated death, the sauce a pool of sludgy yuck reminiscent of a Jersey Shore hookup: greasy and unappealing when sober.

So consider this recipe a reclamation of sorts—an opportunity to rescue the classic *tapa* from a life of over-braised servitude. The trick is to gently cook the *chorizo* until it is just done, making sure that the sausage retains lots of juicy fat.

1   In a medium sauté pan over medium heat, warm the oil until rippling but not smoking. Add the apples and cook for 4 to 6 minutes, until lightly caramelized all over. Remove from the pan, transfer to a bowl, and set aside.

2   Add the sausages to the pan and sauté for 2 minutes, until browned.

3   Add the raisins and cider to the pan and simmer for 4 to 6 minutes, until the liquid has reduced by ½ and the raisins are plump. Return the browned apples to the pan, stir, and cook for 1 minute, until warmed through. Season to taste with the salt and cider vinegar.

4   Add the dark leafy greens (if using) and cook for just a minute in the warm sauce, until wilting.

5   Remove from the heat and transfer to a serving bowl.

NOTE: Kale is a very popular ingredient at the moment, and it goes really well in this dish. I like to throw a handful into the sauté right at the end of cooking. It wilts a little but still adds great crunch to the finished dish.

YIELD

4–6 *tapa* servings

INGREDIENTS

¼   cup (60 mL) extra virgin olive oil

1   medium Granny Smith or other tart apple

4   *Chorizo Asturiano* sausages (see recipe on p. 308), cut into ½-inch (13-mm) rounds

1½   ounces (40 g) golden raisins, soaked in cider vinegar for 10 minutes

1   cup (240 mL) sweet hard cider

    Kosher salt, to taste

    Cider vinegar, to taste

OPTIONAL

½   ounce (15 g) chiffonaded dark leafy greens, such as kale or spinach

CROQUETAS DE JAMÓN

This recipe comes thanks to my friendship with Paras Shah, a great chef from Queens, New York. Paras and I shared a few Spanish adventures after we both won the 2009 scholarship from the Spanish Institute for Foreign Trade (*Instituto Español de Comercio Exterior;* ICEX) and were the lone Americans on the tour. "P," as we came to know him, is hands down one of the most gifted cooks I know, and his pedigree working at some of the best restaurants in New York was only furthered by his time cooking at Restaurante Echaurren—home of these famous *croquetas*—as well as at the famed El Bulli prior to its closing.

Echaurren is a unique, Michelin-starred hotel and restaurant in the one-horse town of Ezcaray, La Rioja. Though they are pretty far from the rest of the world up there, Echaurren is famous throughout Spain for these classic *croquetas.* I'm not exaggerating when I say people drive for miles, especially around the holiday season, for these little balls of fried, *jamón*-laced béchamel.

It's a comfort-food thing, but don't take my word for it: Here, in his own words, is Paras on those famous *croquetas:*

When I first stepped foot into Ezcaray, La Rioja, I certainly realized that I wasn't in Kansas anymore—nor in Queens for that matter; I was miles away from everything that I had ever known, food-wise as well as in lifestyle.

Upon entering Echaurren, I felt the tradition that seemed to ooze out of the old wooden beams. All I had known about traditional Spanish food at that point was the fairly watered-down fare that appealed to the tourist crowd in Madrid. I was about to delve fist deep into tradition...

I walked into the kitchen and began learning what La Rioja was all about: lamb, *chorizo*, potatoes, *potajes,* and those brilliant *croquetas.*

The *croquetas* at Echaurren embody the tradition of the restaurant; they're built off of a béchamel-based *masa* that needed to be tended to like a newborn.

After the *masa* cooled, it was with a deft hand that you had to form the *croquetas*—gently but confidently *"con carino sin miedo"* breading them. Then, they would take that hot oil bath for a couple of minutes until golden brown, with a shell that magically holds together as if the individual grains of the bread crumbs are protecting the unctuous treasure inside.

Then you bite into one and in that one moment, all is revealed about making true, traditional Spanish food: simple technique, loving preparation, and absolute joy to eat.

1   In a large saucepan over medium–high heat, place the chicken breast. Cover with the stock and bring to a boil. Reduce the heat to medium and simmer for 20 minutes, until the chicken breast is fully cooked. Remove from the heat and allow the chicken to come to room temperature in the liquid.

2   Remove the chicken from the stock, reserving the stock in a measuring cup, and very finely mince the chicken (or pulverize it in the bowl of a food processor fitted with the "S" blade). Place the minced chicken in a bowl and refrigerate until ready to use.

3   Return the saucepan to medium heat and add the butter. Once the butter has melted, add the onion, season with the salt, and cook for 20 to 25 minutes, until the onion has softened. Add the *jamón* and continue cooking for 10 minutes, until it has softened.

4   Add the reserved chicken to the saucepan and stir to incorporate. Whisk in the flour and stir until a thick paste that looks like a roux forms. Continue cooking for 5 to 7 minutes, until the flour takes on a nutty smell and light golden color.

5   To the saucepan, slowly add the milk, 1 cup (500 mL) at a time, while whisking constantly. The end result will be a fluid, but thick, béchamel sauce. Raise the heat to medium–high and bring to a boil, whisking constantly. Reduce the heat to medium and simmer for 10 more minutes or until the starchiness is cooked out of the sauce.

6   To the saucepan, add the hardboiled eggs and reserved stock. Whisk to combine and continue simmering for 5 minutes. Remove from the heat.

7   Line a baking sheet with parchment paper. Pour the contents of the saucepan—the *masa*—onto the baking sheet and refrigerate for 4 hours to overnight, until it has fully cooled and set up.

8   In a large sauté pan over medium–high heat, heat the oil to 375°F (190°C). Line a baking sheet with paper towels.

9   Place two large, shallow bowls on the counter. In one, place the bread crumbs, and in the other, place the beaten egg yolks. Set a platter next to the bowls.

YIELD

8–10 *tapa* servings

INGREDIENTS

1   (6–8-ounce [170–230-g]) chicken breast

3½ cups (100 g) gelatinous beef or chicken stock or 1 cup (250 mL) demi-glace

5⅔ ounces (160 g) unsalted butter

1   medium yellow onion, cut into small dice (preferably in a food processor)

    Kosher salt, as needed

1¾ ounces (50 g) *Jamón* (see recipe on p. 143), diced

7   ounces (200 g) unbleached all-purpose flour

2   quarts (2 L) whole milk

3   large hardboiled eggs, minced

2   cups (500 mL) virgin olive oil, plus more as needed, for frying

1   cup (108 g) finely ground bread crumbs

6   large egg yolks, beaten

CONTINUED ON NEXT PAGE

CROQUETAS DE JAMÓN CONTINUED FROM PREVIOUS PAGE

10 Using an ice-cream scoop or similar deep spoon, scoop up some of the *masa*. Roll it into a cylinder in your hands, forming a *croqueta*. Dip the *croqueta* into the bread crumbs and roll it until it is completely covered. Dip the *croqueta* into the beaten eggs and roll it until it is thoroughly coated. Return the *croqueta* to the bread crumbs for a second coating. Set aside the *croqueta* on the platter. Repeat the process until all the *masa* has been used.

11 Taking care not to crowd the pan, fry each of the *croquetas* for 6 to 8 minutes, until they are golden brown and cooked through. As each is cooked, remove it to the prepared baking sheet to drain. Remove from the heat.

12 Serve the *croquetas* quickly, after they've drained for just a few minutes. The hotter you serve them, the more gooey and delicious the béchamel will be.

VARIATIONS

To make *Chorizo Croquetas:* Substitute *chorizo* for the *jamón*. You can also omit the hardboiled eggs if they're not your thing, but otherwise continue as above.

To make Lobster *Croquetas:* Substitute deshelled, cooked lobster meat for the *jamón*. If you want, you can also substitute a reduced lobster or shellfish broth for the beef or chicken broth.

To make *Pringue Croquetas:* Substitute the cooked meat from a stew or *cocido* for the *jamón*. Meats like *morcilla* and *panceta* work well.

EMPANADA GALLEGA

Empanadas—along with *calzones* and other stuffed breads—owe their heritage to the conquests of the Moors and the influence of Indian culinary traditions (in this case, *samosas)* that have traveled westward over the centuries. The *empanada* is thoroughly entrenched in the soul of the Galician people; it's one of their most famous dishes. You'll find it everywhere there, usually filled with a tomato-heavy *sofrito* and other ingredients, ranging from meats to seafood to veggies.

My friend Margarita Torres-Vidal, a local chef who specializes in seafood and all things Gallego (including raising supermodel-grade daughters), made this for lunch one day on a whim. Margarita happened to have some *mejillones* on hand (this isn't uncommon, since Galicia has great seafood) and wanted to teach the visiting *Americanos* how to make a proper *empanada*.

Not surprisingly, it was delicious in many ways, but especially because Margarita insisted on letting the *empanada* cool before eating. As it cools, the pie sinks a little, and the crust absorbs all the flavors of the *sofrito* and mussels. This dish begs for a light, acidic Galician white wine, like an Albariño or Verdejo.

1   In the bowl of a stand mixer fitted with the whisk attachment, mix together the milk, warm water, yeast, honey, sugar, butter, and *manteca* on low speed for 1 minute. Set aside at room temperature for 10 minutes, until the yeast blooms and gets foamy.

2   Replace the whisk attachment with the dough-hook attachment. Add the flour to the bowl and knead on low speed for 7 to 10 minutes, until the dough forms a not-too-sticky ball. (Note: If you prefer doing things by hand, you can instead use the "volcano" technique. On a clean work surface, pile the flour into a mound. Make a large depression in the center. Pour the milk mixture into the depression and, using your hand or a fork, stir very slowly, gradually incorporating more and more flour into the dough.) If the dough is too sticky, add a little flour; if it is not sticky enough, add a little more water.

3   Add the salt and continue kneading on low speed for 5 to 10 minutes, until the *masa* is smooth, elastic, and bounces back immediately when pressed with a finger.

YIELD

6–8 entrée or 8–12 *tapa* servings

INGREDIENTS	BAKER'S PERCENTAGE
per 21 ounces (600 g) of bread flour...	**100%**
⅔ cup (150 g) whole milk, warmed	**25%**
⅔ cup (150 g) warm water, plus more as needed	**25%**
1 (⅓ ounce [10 g]) packet instant or active dry yeast	**1.6%**
1 tablespoon (20 g) honey..............	**3.3%**
1 tablespoon (15 g) granulated sugar	**2.5%**
3½ ounces (100 g) unsalted butter......	**17%**
3½ ounces (100 g) *manteca* (pork lard), plus more for greasing.........	**17%**
1 ounce (25 g) kosher salt, plus more to taste...........................	**4%**
1½ quarts (1.5 L) water	
2.2 pounds (1 kg) Prince Edward Island mussels, cleaned and debearded..................................	**170%**
2 cups (500 mL) Basic *Sofrito* (see recipe on p. 418)	
4 *Chorizo Asturiano* sausages (see recipe on p. 308), cut into small dice	
2 large whole eggs, beaten	

CONTINUED ON NEXT PAGE

EMPANADA GALLEGA CONTINUED FROM PREVIOUS PAGE

4   Using some *manteca,* lightly grease a large mixing bowl. Place the dough ball in the bowl and cover with a damp cloth. Let it rise in a warm place for 30 minutes, until it roughly doubles in size.

5   Place a colander inside a large bowl. In a medium stockpot over medium–high heat, bring the water to a boil.

6   Using a sieve, immerse ⅓ of the mussels in the boiling water for 1 to 3 minutes, until they open. As soon as each opens, remove it from the boiling water and place in the colander. Repeat until all of the mussels have been cooked. Remove from the heat.

7   Remove each mussel from its shell. Discard the shells and return the mussels to the colander. Place the colander and bowl in the refrigerator to cool for 7 to 10 minutes.

8   In a large saucepan, prepare the Basic *Sofrito* with the *Chorizo Asturiano,* also adding in the reserved mussel liquor from the bowl. Remove from the heat.

9   Add the cooled mussels to the *sofrito*. Taste and season as necessary with the salt.

10   Preheat the oven to 400°F (200°C). Grease a 9 × 12-inch (22.5 × 30-cm) baking dish with some *manteca*.

11   Remove the dough from the bowl. Punch it down and divide it in half. Form each half into a ball. Return one of the balls to the bowl and re-cover it with the damp cloth.

12   On a clean work surface, roll the dough into a ¼-inch-thick (6-mm-thick) 10 × 13-inch (25 × 33-cm) rectangle. Lay the dough in the prepared baking dish, allowing any extra dough to drape over the sides.

13   Pour the contents of the saucepan into the baking dish. Using a spatula, smooth and spread it evenly in the dish.

14   Remove the remaining dough from the bowl and roll it into a rectangle of the same size. Place it on top of the *empanada* filling and crimp together the seams of the two sheets of dough, completely sealing it.

15   Brush the top of the *empanada* with the eggs. Place the dish in the oven and bake for 30 to 45 minutes, until the crust is golden brown. Remove from the oven and let cool for 15 to 20 minutes before serving.

NOTE: Traditionally, *empanadas* are left to cool slightly. Once the filling cools, the dough settles and the *empanada* depresses a little in the middle.

MIGAS MANCHEGAS

While cooking for Adolfo in Toledo, I learned a lot about this dish. It's one of Adolfo's specialties, and he makes it better than anyone else I have ever met. *Migas* is a peasant dish from La Mancha that blends seasoned bread, cured meats, and whatever veggies are on hand. It was just the thing for poor shepherds who needed a quick, high-calorie meal while working in the pastures.

Since Adolfo owns a vineyard, he always throws a handful of grapes into his *migas*. You should, too…it's a brilliant idea, one of the many reasons why he's The Man of La Mancha.

1   Place the grated bread crumbs and water in a small mixing bowl and soak for 2 hours to overnight.

2   In a medium saucepan over medium–high heat, warm the oil until it is rippling but not smoking. Add the garlic. Stir well and cook for 4 to 6 minutes, until the garlic starts to turn golden brown and smells toasty.

3   Add the sausages and *panceta* to the saucepan. Season with the salt and black pepper, stir, and cook for 5 minutes, until the sausages and *panceta* release their fat, turning the mixture golden red.

4   Add 4 chopped Piquillo *Confit* peppers and the liquid to the saucepan. Cook for 5 minutes, until everything is warmed through.

5   Add the soaked bread crumbs to the saucepan and toss well. Continue tossing and cooking for 15 minutes, until the bread crumbs start to toast and turn golden brown.

6   Remove from the heat. Add the grapes and toss well. Taste and reseason with the salt, black pepper, and finishing oil.

7   Transfer to a serving bowl. Top with the remaining strips of Piquillo *Confit* peppers and serve warm.

NOTE: This recipe is made to be served alongside a red wine from Castilla-La Mancha. Of course, I highly recommend the Pago de Alma wines Adolfo makes on his estate—but you're going to have to visit him to get some.

YIELD

4–6 entrée servings

INGREDIENTS

1   (23-ounce [650 g]) day-old baguette, finely grated (see *Trucos de la Cocina* on the next page)

1½ cups (355 mL) water

½   cup (120 mL) extra virgin olive oil

12   cloves garlic, lightly crushed but skins intact

4   Cantimpalos-Style or Riojano-Style *Chorizo* sausages (see recipes on pp. 299 and 302), cut into small dice

½   pound (225 g) *panceta* (cured pork belly), cut into small dice

Kosher salt, to taste

Freshly ground black pepper, to taste

8   Piquillo *Confit* peppers (see recipe on p. 408), divided

1   (12-ounce [300 g]) bunch champagne or other small grapes

Extra virgin olive oil, as needed, for finishing

TRUCOS DE LA COCINA: PAN DE CRUZ

PAN DE CRUZ is an ancient bread meant for shepherds. Its dry, dense outer crust keeps the bread's interior fresh for almost a week, making it the perfect bread for *migas*. This IGP-protected food is more important for what it isn't than for what it is. It isn't salted, it isn't terribly flavorful (despite what my friends in La Mancha say), and it isn't easy to find outside of La Mancha.

Instead of *Pan de Cruz*, use a crappy supermarket baguette. Leave the bread out for a day to get dry. Grate it using the big holes of a box grater, and you will have a very close approximation of the crispiness of *Pan de Cruz*.

Make sure you stay away from any bread that's dense, wheaty, moist, or chewy. This recipe demands a dry, airy white bread that will soak up the delicious rendered pork fat.

CALLOS DE CARMEN

If you don't like stinky foods, now's the time to turn the page.

We are going to be talking about delicious tripe, which the Spaniards lovingly call *callos*. But I'm not talking about just any old *callos* recipe: This is the best tripe dish I have ever come across, anywhere, and it's a family recipe belonging to one of the *sabias* from Extremadura, Carmen. Carmen learned the recipe from her mother, who likely learned it from her mother. It's sticky, spicy, and one of the best surprise dishes that will be put in front of you in the midst of the *matanzas* that Carmen presides over.

Callos, however, take a little lovin' to get to a point where you would want to actually put them in your mouth and chew. So here are some things to consider for handling and cooking your *callos:* You can substitute beef *callos* for the pork *callos,* but pork tripe, which needless to say is what we used at the *matanzas,* is typically fattier and thus better suited. Also: *Callos* smell bad. It's a fact of life. The technique in the recipe is the best way to get rid of the sweet stank of *callos,* though nothing is a substitute for soaking them with citrus peels for a day.

Tripe in the United States typically comes pretty clean already, but make sure to rinse it well anyway—just to make triple sure that you get rid of any nasty flotsam or jetsam. Trust me: You can *never* get your *callos* clean enough.

TO MAKE THE *CALDO:*

1   In a large stockpot, place all of the *caldo* ingredients except the salt. Cover with cold water.

2   Place the stockpot over medium–high heat. Bring to a boil and then reduce the heat to medium. Simmer the *caldo* for 2 hours, skimming off and discarding any impurities or scum that rise to the surface. Remove from the heat. Strain into a large mixing bowl and set aside. Reserve the ham bone but discard all of the other solids. Season to taste with the salt.

YIELD

6–8 entrée servings

INGREDIENTS

FOR THE *CALDO:*

1	ham bone
6	medium carrots, peeled and cut into large dice
3	medium leeks, cleaned and cut into large dice
3	medium Roma tomatoes, cut in half
1	head garlic, split in half
3	fresh bay leaves
3	whole cloves
15	black peppercorns
	Kosher salt, to taste

FOR THE *CALLOS:*

2.2	pounds (1 kg) pork or beef tripe, "dressed," rinsed, and cut into 3-inch (7.5-cm) squares
3	pork trotters, rinsed and split
4	Cantimpalos-Style or Riojano-Style *Chorizo* sausages (see recipes on pp. 299 and 302)
4	*Morcilla de Arroz* or *Morcilla de Cebolla* sausages (see recipes on pp. 258 and 260)
3½	ounces (100 g) *Jamón* (see recipe on p. 143), diced

continued

CONTINUED ON NEXT PAGE

CALLOS DE CARMEN

CONTINUED FROM PREVIOUS PAGE

TO MAKE THE *CALLOS*:

1   In a large stockpot, place the tripe and trotters. Cover with cold water. Place the stockpot over medium–high heat. Bring to a boil. Strain the water through a colander and reserve the tripe and trotters. Repeat the process a second time. Remove from the heat.

2   Return the tripe and trotters to the stockpot. Cover with the reserved *caldo* and add the ham bone. Place the stockpot over medium–high heat. Bring to a boil and then reduce the heat to medium. Simmer for 2 hours, until the trotters are starting to break apart and the tripe has softened.

3   Add the sausages and *jamón* to the stockpot. Cook for 1 hour, until the trotter is falling apart and you are able to easily poke a hole in the tripe. Remove from the heat and remove the meats and the *callos* from the *caldo,* keeping the trotters separate from the rest of the meats. Set the meats aside to cool. Place the *callos* in a bowl. Reserve at least 2 cups (500 mL) of the *caldo* to use as gelatinous stock for *Croquettas* (see recipe on p. 338), *Cocido Madrileño* (see recipe on p. 332), or other recipes that need flavorful stock. Remove and discard the ham bone.

4   Once they are cool enough to handle, dice the sausages and *jamón*, pull the meat from the trotters, and add to the bowl with the *callos*.

TO FINISH THE DISH:

1   In a large saucepan, prepare the Basic *Sofrito,* using olive oil as the fat and including the garlic, bay leaf, tomato paste, and chile pepper options from the recipe.

2   Add the flour and *pimentones* to the saucepan. Allow the *sofrito* to simmer for 1 minute, taking care not to burn the *pimentones*. Deglaze the pan with the sherry and cook for 1 minute, until bubbling. Add the *Tomate Frito* and 1 cup (250 mL) of the reserved *caldo* and cook for 1 minute.

3   Add the meats and *callos* to the saucepan and cook for 15 minutes, until the dish is heated through. Remove from the heat and discard the bay leaf. Add more stock as needed. Taste and season with the salt and black pepper.

INGREDIENTS *continued*

FOR THE FINISHED DISH:

1   cup (240 mL) Basic *Sofrito* (see recipe on p. 418)

2   tablespoons (15 g) unbleached all-purpose flour

2   tablespoons (14 g) *pimentón dulce*

1   tablespoon (7 g) *pimentón picante*

¼   cup (50 mL) amontillado sherry, chilled

½   cup (100 mL) *Tomate Frito* (see recipe on p. 407)

Kosher salt, to taste

Freshly ground black pepper, to taste

NOTE: This dish is usually eaten communally, so your best bet is to serve the *callos* in a big earthenware bowl and with some rustic bread for dunking.

LACÓN CON GRELOS

Every region in Spain has a favorite vegetable, fruit, or legume that the locals look forward to as a herald of seasonal changes. In La Rioja, the chill of winter brings about a hankering for a classic *potaje* made from the *caparrone* bean. In Navarra, tiny, very expensive peas called *lagrimas* make their debut in early spring. Summer means a wealth of figs in the Moorish-inspired kitchens of the south. And in fall, in Galicia, the local turnip greens, *grelos,* are harvested from October through January.

Grelos are featured prominently in the hearty, stew-heavy regions of the north. Paired with *Lacón* in this recipe, they'll call to mind Southern soul-food collard greens.

1   Soak the *Lacón* in water overnight to remove some of its salt.

2   Drain the *Lacón.* Place it in a large stockpot. Cover the *Lacón* by 1 inch (2.5 cm) with the water. Place the stockpot over medium–high heat. Bring to a boil and reduce the heat to medium. Simmer for 1 hour, until the *Lacón* shreds easily.

3   Add the sausages and potatoes to the stockpot. Continue cooking for 20 minutes, until the potatoes are almost cooked.

4   Add the *grelos* to the stockpot. Cook for 10 minutes, until the stems have softened.

5   Remove the meats from the pot and place them in a medium mixing bowl to cool. Meanwhile, taste the *caldo* and season with the oil, salt, black pepper, and *pimentón.* Remove from the heat.

6   After the meats have cooled, cut them into bite-sized pieces and place them on a platter.

7   Ladle the *caldo* into bowls. Serve warm with the meats.

NOTE: Try *Lacón con Grelos* served with a rustic bread and an acidic Gallego wine, like Albariño.

YIELD

6–8 entrée servings

INGREDIENTS

1.65   pounds (750 g) *Lacón Cocido* (see recipe on p. 151)

Cold water, to cover

3   Cantimpalos-Style or Riojano-Style *Chorizo* sausages (see recipes on pp. 299 and 302)

3   medium Yukon gold potatoes

1   bunch *grelos* (turnip greens)

Extra virgin olive oil, as needed

Kosher salt, to taste

Freshly ground black pepper, to taste

Pimentón dulce, to taste

LENTEJAS CASTELLANAS

In the past, lentils and other legumes have gotten many a poor Spanish family through hard times—in fact, they are often sarcastically referred to as "poor man's meat." Lentils in particular provide a nutritious base for many of the famous *potajes* that make up the *comida casera* from the Manchego countryside, and this lentil recipe is especially well known in La Mancha, where it's a popular lunchtime meal.

Since lentils don't require a lot of time for preparation and everyone loves them, they are the perfect foil for restaurant family meals like ours at Adolfo.

One note for this dish: Once lentils are cooked and ready to serve, acid is your friend. That might mean splashing a good dose of sherry vinegar into the pot, but *mi amiga* Janet Mendel, a Spanish cookbook author, recommends serving some pickled onions as a garnish. Whatever you do, make sure something acidic gets in there to break up the richness of the dish.

1   In a large saucepan, prepare the Basic *Sofrito,* using olive oil as the fat and doubling the amount of garlic called for in the recipe.

2   Add the *pimentones* to the saucepan. Allow the *sofrito* to simmer for 1 minute, taking care not to burn the *pimentones.* Remove from the heat and set aside.

3   In a large stockpot, combine the lentils, *panceta,* sausages, oil, bay leaves, carrots, and garlic. Barely cover with cold water. Place the stockpot over medium–high heat. Bring to a boil and then reduce the heat to medium–low. Cook for 20 to 25 minutes, until the lentils are just starting to soften. (From this point forward, do not stir the stockpot. Doing so will break up the lentils. Instead, just give the stockpot a shake from time to time.)

4   Return the saucepan containing the *sofrito* to medium heat. Cook for 3 to 5 minutes, until warmed through.

5   Add the *sofrito* and the potatoes to the stockpot, shaking the pot to incorporate the *sofrito* and settle the potatoes. Cook for 30 minutes, until the potatoes and lentils are cooked through. Remove and discard the bay leaves, carrots, and garlic. Taste and season with the salt, black pepper, and sherry vinegar. Remove from the heat. Transfer to a large serving bowl and serve warm.

NOTE: Serve *Lentejas Castellanas* with some rustic bread for dunking and some *encurtidos,* like pickled onions or peppers.

YIELD

6–8 entrée servings

INGREDIENTS

1   cup (240 mL) Basic *Sofrito* (see recipe on p. 418)

⅔   ounce (21 g) *pimentón dulce*

⅓   ounce (10 g) *pimentón picante*

18   ounces (500 g) lentils, such as Green Castellanas or Puy

9   ounces (250 g) *panceta* (cured pork belly)

3   Cantimpalos-Style or Riojano-Style *Chorizo* sausages (see recipes on pp. 299 and 302)

3   *Morcilla de Cebolla* sausages (see recipe on p. 260)

½   cup (100 mL) extra virgin olive oil

2   fresh bay leaves

2   medium carrots, peeled and trimmed

1   head garlic, split in half

4   medium russet potatoes, peeled and "cracked" into medium dice (see p. 161 for more on this technique)

½   cup (100 mL) dry white wine, such as Verdejo

Kosher salt, to taste

Freshly ground black pepper, to taste

Sherry vinegar, as needed

BOCADILLO DE CHORIZO PAMPLONA

Don't get me wrong: Toledo is a great little town. But let's just say that if you're an *Americano* line cook who's just gotten off work and is hungry for a snack...well, this ain't New York, kid.

But if you know where to look, you can find gold—like I did many nights at my favorite little Toledano *bocadillo* stand. A *bocadillo* is just about anything that can be stuffed into a baguette-like roll, and this little Toledo shop was the kind of late-night place that caters to the inebriated. Thus, the portions are big, greasy, and satisfying.

Bocadillo de Chorizo Pamplona was my sanctuary after a long day of feeding the masses: a bunch of deliciousness including onions caramelized to the point of becoming a warm marmalade and a soft goat cheese. For extra points, bake your own bread, but there are plenty of bakeries around that make decent baguettes.

1 On the bottom half of 1 of the *barras,* spread ¼ of the cheese. Top with ¼ of the *chorizo* slices. Spoon ¼ of the *mermalada* on top of the *chorizo*. Place the other half of the *barra* on top.

2 Repeat with the rest of the *barras* until all 4 *bocadillos* are prepared. Serve.

YIELD

4 entrée servings

INGREDIENTS

4   medium *barras* (small baguettes of bread), sliced lengthwise

9   ounces (250 g) soft Capricho de Cabra, Miticrema, or other Spanish goat or sheep cheese

1   large Pamplona-Style *Chorizo* sausage (see recipe on p. 306), thinly sliced

½   cup (100 g) *Mermelada de Cebolla* (see recipe on p. 401)

CHAPTER

9

PÂTÉS Y TERRINAS

**C**OLD FOODS PRESENT
a challenge since, generally speaking, it's a lot easier to make
hot foods delicious. You see, hot foods have the advantages of
aroma, volatile seasonings, and complex textures to help things
along, whereas cold foods typically need to be seasoned aggres-
sively to compensate for their lack of aromatic appeal.

And herein lies the trials and the tribulations of great *pâté*,
the Rodney Dangerfield of the *charcuterie* world, which has
only recently begun to get the mainstream respect it so
rightfully deserves.

Pâtés and *terrinas* (terrines) really aren't that far removed from some types of sausage: there's ground meat and maybe some liver or other offal goodies mixed with fat and spices. All of it is either coarsely or finely ground or emulsified into a homogenous mass. It's only when the mixture gets stuffed into a mold, baked, chilled, and sliced that this preparation becomes substantially different from an *embutido*.

These preparations used to be all the rage: Dating back to ancient Greece and continuing on through to medieval times, the French Revolution, and most of the *haute cuisine* movement, *pâtés* and terrines were celebrated not only for their refinement but also for their convenience of preparation and service (you could bake several at once, chill, and slice at your pleasure). Also, since they were highly caloric and easily transportable, *pâtés* were a perfect meal for the working class or for roving armies.

But for this very same reason—their high calorie and fat content—*pâtés* and terrines got a bad rap and fell out of favor during the end of the twentieth century. In fact, pretty much the only survivor of the technique during that time (at least on American tables) was terrine's distant cousin, the meatloaf.

In recent years, however, our *charcuterie* renaissance has made *pâté* popular again. Culinary globalization, a renewed focus on classical culinary techniques, craft butchery, and different methods of fermentation and preservation have led us to a point where *pâtés* and terrines can now be found at the humblest of restaurants—or just about every other hipster joint in Williamsburg.

In this chapter, you will find some of the *pâtés* and terrines that made their way across the French border and influenced the gastronomy of Spain. Typically, you will find three different types—(1) coarse *pâtés;* (2) smoother, puréed types, akin to *mousselines;* and (3) liver terrines.

Much like their French counterparts, the coarse terrines of Spain have varying degrees of liver and a somewhat rustic appearance. Their smoother cousins, however, are typically a suspension of fat and meat that allows them to be spread on toasted bread. This is also true of liver terrines, like the ever-popular *foie gras* terrine and chicken liver *pâtés*.

TRUCOS DE LA COCINA: PÂTÉ TECHNIQUE

MAKING A *PÂTÉ* is not that different from making an *embutido,* in that you want to keep everything cold throughout the grinding process. So keeping bowls and blades chilled, par-freezing the meat, and progressive grinding using different dies will all help ensure that your *pâté* has the best chance for a successful texture and the least chance of splitting.

Also, as with some *embutidos,* you should definitely take the opportunity to taste the *masa* before finishing the terrine. Unlike the process for *embutidos,* however, you'll need to poach a bit of it in simmering water instead of sautéing it. Sautéing a *pâté prueba* will break the mix and won't taste like the finished *pâté*. To poach the *prueba,* just put a little bit in some plastic wrap and simmer it in warm water for 5 to 10 minutes, until cooked.

PÂTÉ DE JABALÍ

Since the *Ibérico* pig is descended from wild boars, it should come as no surprise that wild boars are a common sight in Spain's countryside. So, when it comes time to hunt down a few of those troublemakers, the people of Cuenca, a small town in La Mancha, make this *pâté* with some of the meat.

This *pâté* is what French *charcutiers* would call *"grand-mère"* style, which just means that there's a lot of liver in the mix and that the binding agent (in *charcuterie*-speak, the *panade)* is bread-based. The Spaniards don't really codify things in this fashion, however, so I only bring it up to give my *charcuterie* nerds out there a frame of reference.

The main thing you should know about this *pâté*—or for any *pâté* in which you caramelize the ingredients—is that you'll need to achieve deep browning to get as much flavor as possible into the *pâté*. This means you want an *über*-hot pan and meat that's been dried thoroughly (by patting down with some paper towels). Otherwise, you'll steam the liver, ruin this killer *pâté,* and somewhere, a fairy will die.

Once you've successfully made this *pâté,* you should serve it with lots of pungent, acidic goodies. Pickles, mustards, and pieces of good, crusty bread are all great ideas, as are glasses of amontillado sherry, some beers, and a group of friends to share your efforts and do the dishes.

1   In a small mixing bowl, combine the brandy, 1 ounce (25 g) of the salt, the brown sugar, the black pepper, the thyme, and the bay leaves. Stir and divide into two medium mixing bowls.

2   Place the *hígado* pieces in the first bowl and toss well. Place the *aguja, papada,* and *panceta* meats in the second bowl and toss well. Cover both bowls with plastic wrap and refrigerate overnight. In a small mixing bowl, combine the raisins and sherry. Set aside and allow the fruits to macerate and plump at room temperature overnight.

3   Line a bowl with paper towels. Remove the *hígado* from the marinade. Strain the marinade into a measuring cup and reserve; discard the solids. Pat the *hígado* dry thoroughly.

4   In a large saucepan over medium–high heat, melt the butter until foaming but not brown. Place the *hígado* pieces in the saucepan, without crowding them, and sear for 4 to 6 minutes, until heavily browned on one side. Flip the pieces

YIELD

Approximately 1 terrine per 2.2 pounds (1 kg) of meat

INGREDIENTS	*CHARCUTIER'S* PERCENTAGE
per 2.2 pounds (1 kg) of the following blend of large-diced meats: 40% *hígado* (pork liver); 30% *aguja* (wild boar or pork collar), *panceta* (wild boar or pork belly), or *paleta* (wild boar or pork shoulder); 15% *papada* (wild boar or pork jowl); and 15% *panceta* (wild boar or pork belly)	**100%**
½ cup (100 mL) brandy	
2 ounces (55 g) kosher salt, divided	**5.5%**
½ ounce (15 g) brown sugar	**1.5%**
⅛ ounce (5 g) freshly ground black pepper	**.5%**
5 sprigs fresh thyme	
2 fresh bay leaves	
2¾ ounces (80 g) golden raisins	
½ cup (120 mL) Pedro Ximénez (PX) sherry	
1¾ ounces (50 g) unsalted butter	
5 ounces (150 g) minced yellow onions	
2 slices bread, crusts removed, torn into pieces	
¾ cup (150 mL) heavy cream	
2 whole large eggs, beaten	
⅛ ounce (2 g) TCM #1 or DQ #1 curing salt mix (see p. 83)	**.2%**
2 teaspoons (4 g) Four-Spice Blend (see recipe on p. 409)	**.4%**
1 teaspoon (2 g) freshly grated nutmeg	**.2%**
1 teaspoon (2 g) whole cloves, toasted and ground	**.2%**

and sear for 2 to 4 minutes on the opposite side, taking care not to cook the pieces past a medium-rare point. Remove from the heat.

5   Place the *hígado* pieces in the prepared bowl to drain. Blot them dry thoroughly. Reserve the melted butter and fond in the saucepan.

6   Return the saucepan to medium heat. Add the onions to the pan and season with the remaining 1 ounce (25 g) of the salt. Sweat the onions for 10 minutes, until they are soft but have not taken on color. (You can add a splash of water to the pan as needed to keep the onions or the fond from burning.)

7   Continue cooking the onions for 10 to 20 minutes, until they are deeply caramelized, making sure to not burn the fond along the way.

8   Carefully add the reserved brandy marinade from the *hígado* to the pan and flambé. Once the flames die down, stir up any browned bits and simmer for 8 to 10 minutes, until the liquid has nearly evaporated. Remove from the heat and transfer the onions to the bowl containing the *hígado*. Chill the mixture in the refrigerator for 1 to 2 hours, until it is cold.

9   In a large mixing bowl, combine the bread, cream, eggs, curing salt, Four-Spice Blend, nutmeg, and cloves. Toss to combine, creating a *panade*.

10   Fill a large bowl with ice, and place a smaller bowl inside the ice-filled bowl. Transfer the meats, *hígado*, and onions to the bowl containing the *panade*. (Reserve the brandy marinade for the *aguja*, *papada*, and *panceta* meats.) Toss to combine.

11   Grind the meat mixture once through a medium-coarse (⅜ inch [9.5 mm]) die into the smaller bowl. Be careful: The meat mixture is wet, so it may squirt and pop out of the grinder. Add the macerated raisins and sherry to the ground meat mixture.

12   Place the ground meat mixture in the bowl of a stand mixer fitted with the paddle attachment (or you can just mix in a mixing bowl with a sturdy spoon). Begin mixing on low speed. As the mixer runs for 1 minute, pour the reserved brandy marinade for the *aguja*, *papada*, and *panceta* meats into the bowl in a steady stream.

13   Continue mixing on medium speed for 1 to 2 minutes, until the marinade has been fully incorporated into the mixture, a white residue forms on the sides of the bowl, and the *masa* begins to come together.

14   Moisten the inside of a 2-quart (2-L) terrine mold with some water. Line the mold with plastic wrap, allowing the wrap to hang over the sides of the mold.

15   Preheat the oven to 300°F (150°C).

16   Using an overhand motion, fling the *masa* in the bowl into the mold. Pat it down with a rubber spatula, as this will help eliminate any air that may be trapped in the mold. Once the mold is filled, fold the excess plastic wrap over the top of the *masa*. Wrap the mold with foil and cover with its lid.

17   Place the mold into a deep roasting pan and place the pan in the preheated oven. Pour hot tap water into the pan until it reaches halfway up the side of the wrapped mold.

18   Bake for 60 to 90 minutes, until the center of the *pâté* reaches 155°F (68°C). Remove from the oven. Remove the mold from the roasting pan and immediately place it in a second roasting pan filled with ice water. Place a few cans on the mold to weigh it down and let it rest in the ice bath for 60 minutes, or until the outside is cold to the touch. Transfer the weighted *pâté* to the refrigerator for at least 8 hours, until thoroughly chilled.

19   Gently unmold the *pâté* and slice into servings. Serve cold.

NOTE: This *pâté* is particularly delicious served with grilled bread and some *encurtidos*.

JUAN MARI ARZAK'S
*PASTEL DE CABRACHO**

Anyone who knows about the gastronomic history of Spain knows the story of Chef Juan Mari Arzak. Chef Arzak is largely responsible for laying the foundation of the New Basque cuisine movement and for much of modern Spanish gastronomy at the Michelin level.

During the 1970s and 1980s, as France basked in the glow of the *nouvelle cuisine* movement, the *nouveau*-Frenchie ideals of lighter, refined sauces migrated into Basque country. Chef Arzak and a group of his contemporaries known collectively as "The Gang of Twelve" decided to incorporate these ideals into traditional Basque cooking. Several three-Michelin-starred restaurants later, here we are.

Back in the mid-'70s, this recipe from Chef Arzak for a *mousseline*-based fish terrine became a signature dish of the *nouvelle cuisine Basque.* Today, you'll find it widely available throughout northern Spain as one of the most oft-imitated forms of seafood *charcuterie.*

This dish is a cross between a terrine and what the Spanish call a *pastel* (cake). Its main ingredient is scorpion fish, a quite affordable type of fish that's always been popular in northern Spain but is just now becoming widely available in North America because of overfishing of more popular varieties. The Spanish enjoy this fish for its white, meaty flesh, but if you can't find it, then halibut or a similar fish makes a great substitute.

This terrine is typically served with lots of *crostini* and pairs great with a hard *sidra* or a *txakoli.*

YIELD

Approximately 1 terrine per 1.1 pounds (500 g) of fish

INGREDIENTS	*CHARCUTIER'S* PERCENTAGE
per 1.1 pounds (500 g) of scorpion fish, skinned and deboned	**100%**
¼ cup (50 mL) extra virgin olive oil	
¼ cup (50 mL) white wine	
2 medium white onions, peeled, destemmed, and cut into large dice	
2 medium carrots, peeled and cut into small dice	
2 leeks, cleaned and cut into large dice	
2 fresh bay leaves	
1¼ ounces (35 g) kosher salt, plus more to taste	**7%**
10 black peppercorns	
1 cup (250 mL) heavy cream	
8 whole large eggs, beaten	
1 cup (235 mL) *Tomate Frito* (see recipe on p. 407)	
Freshly ground black pepper, to taste	
Unsalted butter, as needed	
Fresh bread crumbs, as needed	

1   In a large stockpot over medium heat, combine the oil, wine, onions, carrots, leeks, bay leaves, salt, and peppercorns. Stir to combine, making a *court bouillon.* Cook for 15 to 20 minutes, until the *court bouillon* reaches 170°F (77°C).

2   Add the scorpion fish to the stockpot and poach for 5 to 7 minutes, until it is just cooked and warmed throughout. Remove from the heat.

3   Remove the fish from the stockpot and reserve ½ cup (100 mL) of the *court bouillon;* discard the rest. Set the fish aside in a strainer to cool to room temperature.

4   Preheat the oven to 430°F (225°C). Shred the fish into small pieces, ensuring that no bones are left. Place in a small bowl and set aside.

5   In a medium saucepan over low heat, warm the cream for 3 to 4 minutes, until it just comes to a simmer. Remove from the heat and pour the cream into a medium mixing bowl.

6   Whisk the reserved *court bouillon,* eggs, and *Tomate Frito* into the bowl until thoroughly incorporated. Gently fold the reserved fish into the mixture, creating a *pastel,* and season with the salt and black pepper.

7   Grease a large baking dish or 2-quart (2-L) terrine mold with the butter. Sprinkle with the bread crumbs until the entire pan is coated. Pour the *pastel* into the baking dish and cover with foil.

8   Place the baking dish into a deep roasting pan and place the pan in the preheated oven. Pour hot tap water into the pan until it reaches halfway up the side of the baking dish.

9   Bake for 60 to 70 minutes, until a skewer inserted into the center of the *pastel* comes out clean. Remove from the oven and allow to cool to room temperature. Remove the mold from the roasting pan and immediately place it in a second roasting pan filled with ice water. Let the *pastel* rest in the ice bath for 60 minutes, or until the outside is cold to the touch.

10 Transfer the *pastel* to the refrigerator for at least 8 hours, until it is thoroughly chilled. Slice into servings. Serve cold.

*A special thank you to Chef Juan Mari for both lighting the path of our modern cuisine movement and for providing this recipe.

NOTE: Juan Mari serves this *pastel* in slices alongside a salad and some mayonnaise or *alioli.*

MANTECA COLORÁ

Manteca Colorá is basically an Andalucían take on *rilletes,* those fancy, Frenchie, fat-covered cooked meats meant to be stored and aged under a layer of fat for enjoyment all winter long.

This version is all about pork and heady Moorish spices, and it's a great way to use up whatever lean meat trim that you have on hand from various *charcuterie* projects.

Make sure that you watch the cooking time, since you don't want to poach the meat so long that it gets dried out. You want creamy, dreamy spreadable goodness, not something that winds up dry and mealy like cat food.

1   Preheat the oven to 250°F (121°C).

2   In a large mixing bowl, combine the meat and *tocino.* Add the *Moruno* Spice and season with the salt. Set aside.

3   In a large ovenproof saucepan, warm the *manteca* over medium heat until rippling but not smoking. Add the garlic. Toast the garlic in the fat for 3 to 5 minutes, until golden. Remove and discard the garlic.

4   Add the pork to the saucepan in a single layer. Sear for 4 to 6 minutes on one side, until it is golden brown. Turn the meat over and repeat, searing it until golden. Remove the pieces of pork to a plate as they are done and repeat in batches until all the pork has been seared.

5   Return the pieces of pork to the saucepan and add the sherry. Remove from the heat.

6   Cover the saucepan tightly with the lid or foil and place the pan in the oven. Bake for 3 to 4 hours, until the pork shreds easily. Remove from the oven.

7   Remove the pork from the saucepan, reserving the braising liquid. Using two forks, finely shred the pork and place in a mixing bowl. Season the pork to taste with the *Moruno* Spice and salt.

8   Transfer the seasoned pork to individual ramekins. Top each serving with the reserved braising liquid until each serving is covered entirely.

9   Transfer the ramekins to the refrigerator for at least 8 hours, until it is thoroughly chilled. Serve cold.

YIELD

4 to 6 (4-fluid-ounce [120-mL]) ramekins per 2.2 pounds (1 kg) of meat

INGREDIENTS	CHARCUTIER'S PERCENTAGE
per 2.2 pounds (1 kg) of *magro* (lean pork meat), cut into large chunks	**100%**
7 ounces (200 g) *tocino* (pork fat), chilled and cut into large dice	**20%**
2 tablespoons (15 g) *Moruno* Spice (see recipe on p. 188), plus more to taste	**1.5%**
2 ounces (55 g) kosher salt, plus more to taste	**5.5%**
3½ ounces (100 g) *manteca* (pork lard), plus more for greasing	**10%**
5 cloves garlic, unpeeled and lightly smashed	
½ cup (100 mL) amontillado sherry	

NOTES: The *Manteca Colorá* will keep a few weeks in the refrigerator, because the fat acts as a barrier against oxygen and bacteria. In fact, it gets better with a little age and ripening.

Like any fatty spread, *Manteca Colorá* is best served with some grilled bread and pickles or alongside a nice, acidic salad.

PÂTÉ DE CERDO

Whereas *Pâté de Jabalí* (see recipe on p. 358) would be considered in French *charcuterie* circles a *grand-mère* style of *pâté* for including lots of liver, this pork *pâté* would be considered more *campagne* style, as it's a little more refined.

Jean-Luc Figueras, a Michelin-starred chef from Barcelona, made this recipe at one of our *matanzas* in Extremadura. To be honest, he came up with it spontaneously after some serious cajoling on the part of the *sabias.* They gave him a hard time about earning his keep, for a couple of good reasons: (1) He's a very well-known chef in Spain and (2) he was partaking of much of the wine that day. So, with a little smile, he gave them a look and said, "Watch this."

That was when the fun started. Jean-Luc grabbed some of the pork livers, some jowl meat, and some of the copious amounts of *Ibérico* fat that we'd accumulated. He combined that with a few shallots and a bunch of caul fat he referred to as his *condom prehistorico* before grinding and packing the mixture into a terrine mold.

A short time later, over a lunch of this classic *pâté,* Jean-Luc had definitely "earned his keep" and then some—he was welcome to all the wine he could consume for the rest of the *matanza.*

To the *charcutier* go the spoils.

1   In a small mixing bowl, combine the brandy, the wine, 1 ounce (25 g) of the salt, the brown sugar, the black pepper, the thyme, and the bay leaf. Stir and divide into two medium mixing bowls. Add the milk to one of the bowls and stir to combine.

2   Place the *hígado* pieces in the bowl containing the milk and toss well. Place the *aguja, papada,* and *panceta* pieces in the second bowl and toss well. Cover both bowls with plastic wrap and refrigerate overnight.

3   In a large mixing bowl, combine the remaining 1 ounce (25 g) of the salt, the cream, the eggs, the parsley, the shallots, the garlic, the flour, the Four-Spice Blend, the nutmeg, and the cloves. Stir well, creating a *panade,* and set aside.

4   Remove and discard the thyme and bay leaf from the marinade.

YIELD

Approximately 1 terrine per 2.2 pounds (1 kg) of meat

INGREDIENTS	CHARCUTIER'S PERCENTAGE
per 2.2 pounds (1 kg) of the following blend of large-diced meats: 20% *hígado* (pork liver), 35% *aguja* (pork collar), 35% *papada* (pork jowl), and 10% *panceta* (pork belly)	**100%**
¼ cup (50 mL) brandy	
¼ cup (50 mL) white wine	
2 ounces (55 g) kosher salt, divided	**5.5%**
⅓ ounce (10 g) brown sugar	**1%**
⅛ ounce (5 g) freshly ground black pepper	**.5%**
5 sprigs fresh thyme	
1 fresh bay leaf	
½ cup (100 mL) whole milk	
¾ cup (150 mL) heavy cream	
2 whole large eggs, beaten	
2 ounces (60 g) minced fresh flat-leaf parsley	**6%**
1¾ ounces (50 g) minced shallots	**5%**
1 ounce (25 g) minced garlic	**2.5%**
2 tablespoons (20 g) unbleached all-purpose flour	**2%**
2 teaspoons (4 g) Four-Spice Blend (see recipe on p. 409)	**.4%**
1 teaspoon (2 g) freshly grated nutmeg	**.2%**
2 (.5 g) whole cloves, toasted and ground	**.05%**
Caul fat, soaked, as needed (see Notes)	

CONTINUED ON NEXT PAGE

PÂTÉ DE CERDO CONTINUED FROM PREVIOUS PAGE

5   Fill a large bowl with ice, and place a smaller bowl inside the ice-filled bowl. Grind the meat mixture once through a medium-coarse (⅜ inch [9.5 mm]) die into the smaller bowl. Be careful: The meat mixture is wet, so it may squirt and pop out of the grinder.

6   Grind the *hígado* mixture once through a medium-fine (3/16 inch [5 mm]) die into the same bowl containing the ground meats.

7   Transfer the ground meats and *hígado* to the bowl containing the *panade*. Stir to combine.

8   Place the ground meat and *hígado* mixture in the bowl of a stand mixer fitted with the paddle attachment (or you can just mix in a mixing bowl with a sturdy spoon). Begin mixing on medium speed for 1 to 2 minutes, until a white residue forms on the sides of the bowl and the *masa* begins to come together.

9   Moisten the inside of a terrine mold with some water. Line the mold with plastic wrap, allowing the wrap to hang over the sides of the mold.

10   Preheat the oven to 300°F (150°C).

11   Using an overhand motion, fling the *masa* in the bowl into the mold. Pat it down with a rubber spatula, as this will help eliminate any air that may be trapped in the mold. Once the mold is filled, fold the excess plastic wrap over the top of the *masa*. Wrap the mold with foil and cover with its lid.

12   Place the mold into a deep roasting pan and place the pan in the preheated oven. Pour hot tap water into the pan until it reaches halfway up the side of the wrapped mold.

13   Bake for 60 to 90 minutes, until the center of the *pâté* reaches 155°F (68°C). Remove the mold from the roasting pan and immediately place it in a second roasting pan filled with ice water. Place a few cans on the mold to weigh it down and let it rest in the ice bath for 60 minutes, or until the outside is cold to the touch. Transfer the weighted *pâté* to the refrigerator for at least 8 hours, until it is thoroughly chilled.

14   Gently unmold the *pâté* and slice into servings. Serve cold.

NOTES: This *pâté* is particularly delicious served with grilled bread and some *encurtidos*.

If you can't track down any caul fat, some thick-sliced bacon will also do the trick to grease the terrine.

Opposite page: *Pâté de Jabalí* (L) and *Pâté de Cerdo* (R)

CACHUELA

Since most *matanzas* occur in the wee hours of the morning, the first things that you eat will always be caloric, fatty, caffeinated, and/or alcoholic.

At the Rocamador *matanzas*, this means a breakfast of very strong coffee fortified with Chinchón served alongside a plate full of fried eggs, *migas*, and *Cachuela*, a chunky *pâté* made from lots of offal and sometimes blood. Loaded with lots of fat and organ meat, this is exactly the sort of meal you need to steel yourself for the work to come in breaking down hundreds of pounds of piggie.

The trick in making *Cachuela* is in cooking the livers and kidneys until they are just barely done, and then blitzing them in the food processor while they're still hot.

1 In a large sauté pan over medium–high heat, warm the *manteca* until rippling but not smoking. Add the garlic cloves and sauté for 5 to 8 minutes, until they start to turn deep golden brown. Remove and discard the garlic.

2 Season the *asadura* with the salt. Add it to the sauté pan with the bay leaf and sear the *asadura* on 1 side for 1 minute, until a crust forms. Flip the pieces and sear for another 2 to 3 minutes on the opposite side, until cooked through. Remove the *asadura* from the sauté pan and transfer to a platter. Cover with foil.

3 In the same sauté pan containing the *manteca*, sear the *hígado* pieces, taking care not to crowd them, for 4 to 6 minutes, until heavily browned on one side. Flip the pieces and sear for 2 to 4 minutes on the opposite side, taking care not to cook the pieces past a medium-rare point. Remove the *hígado* pieces from the sauté pan and transfer to the same platter containing the *asadura*. Re-cover with the foil.

4 Pour the contents of the sauté pan into the bowl of a food processor fitted with the "S" blade or a blender. Add the reserved *asadura* and *hígado* pieces, *pimentón*, cumin seed, Four-Spice Blend, and orange zest. Pulse until the purée is as smooth as you prefer (see Note).

5 Taste the purée and reseason as necessary with the salt, black pepper, cumin, and Four-Spice Blend. Transfer the *Cachuela* to individual ramekins or a terrine mold.

6 Transfer the ramekins or terrine mold to the refrigerator for at least 8 hours, until thoroughly chilled. Serve cold.

YIELD

1 terrine or 4 to 6 (4-fluid-ounce [120-mL]) ramekins per 1.1 pounds (500 g) of meat

INGREDIENTS	CHARCUTIER'S PERCENTAGE
per 1.1 pounds (500 g) of trimmed, large-diced, and very dry *asadura* (pork offal), such as kidneys or hearts, or *magro* (lean pork meat)	**100%**
1.1 pounds (500 g) *manteca* (pork lard)	100%
5 cloves garlic, unpeeled and lightly smashed	
1¼ ounces (35 g) kosher salt, plus more to taste	7%
1 fresh bay leaf	
1.1 pounds (500 g) *hígado* (pork liver), trimmed, cubed, and very dry	100%
1 tablespoon (7 g) *pimentón dulce*	
1 teaspoon (2 g) freshly ground black pepper, plus more as needed	
1 teaspoon (2 g) cumin seed, toasted and ground, plus more as needed	
2 teaspoons (4 g) Four-Spice Blend (see recipe on p. 409), plus more as needed	
Zest of ½ orange	

NOTE: *Cachuela* is normally left chunky, so you get different textures. If you feel like blending the *cachuela* all the way to a smooth purée and then passing it through a sieve, making a *mousseline,* you have what the *sabias* call *zurrapa*. It is essentially the *cachuela* in a puréed form, and it's generally eaten in the same manner as a breakfast/snack during *matanzas*.

PÂTÉ DE HÍGADO DE POLLO

I saw chicken liver *pâtés* almost everywhere I worked and lived in Spain. The stuff is a universally loved and delicious artery clogger.

In Madrid, little shops sell it laced with everything from port wine to chunks of black truffles. In the *pinotxo* bars of the north, you'll often find it paired with grilled local mushrooms and other vegetables. And at Daní Garcia's Michelin-starred restaurant Calima, in the south, we used a similar *pâté* as garnish for a rabbit dish.

This recipe, a sort of amalgam of different techniques that I've picked up, is a super-smooth *pâté* with lots of flavor, especially from its secret ingredient: vanilla bean, which adds a really nice aromatic quality to this otherwise heavy dish.

The gelée on top is a classic finishing component for this *pâté* that has an important purpose other than flavor: Without something on top of the *pâté,* it will oxidize and turn a little gray over time. If you can't find some PX sherry to make it with, consider using a cheaper amontillado or oloroso.

1   In a large mixing bowl, combine the *hígados,* cream, 5 of the thyme sprigs, bay leaves, vanilla bean, and Four-Spice Blend. Cover with plastic wrap and refrigerate overnight.

2   Remove and discard the bay leaves, thyme, and vanilla bean. Drain the *hígados,* reserving the marinade in the refrigerator. Dry them very well and season with the salt.

3   In a large sauté pan over medium–high heat, warm 1½ ounces (40 g) of the duck fat until rippling but not smoking. Add the *hígados* to the pan without crowding them and sear on 1 side for 2 minutes, until a crust forms.

4   Flip the *hígados* and sear for 30 seconds. Remove the *hígados* from the pan and place them in a blender (they should be slightly undercooked).

5   Return the sauté pan to medium–high heat. Add the remaining 1½ ounces (40 g) of the duck fat and the remaining 5 thyme sprigs, which should crackle a little. Add the onions to the pan, season with the salt, and toss well. Cook for 15 to 20 minutes, until the onions start to turn light brown and caramelize.

YIELD

1 terrine or 4 to 6 (4-fluid-ounce [120-mL]) ramekins per 1.1 pounds (500 g) of *hígados de pollo*

INGREDIENTS	*CHARCUTIER'S* PERCENTAGE
per 2.2 pounds (1 kg) of *hígados de pollo* (chicken livers), deveined and cleaned of any green bile spots............................ **100%**	

1	cup (250 mL) heavy cream	
10	fresh sprigs thyme, divided	
2	fresh bay leaves	
½	vanilla bean, split	
2	teaspoons (4 g) Four-Spice Blend (see recipe on p. 409)	
2	ounces (55 g) kosher salt, plus more to taste........................ **5.5%**	
3	ounces (85 g) duck fat or unsalted butter, divided	
2	medium yellow onions, minced	
5	cloves garlic, peeled, destemmed, and crushed	
½	cup (100 mL) amontillado sherry, chilled	
14	ounces (400 g) unsalted butter, softened	
	Freshly ground black pepper, to taste	
¼	cup (50 mL) water	
2	teaspoons (4 g) unflavored powdered gelatin	
½	cup (100 mL) Pedro Ximénez (PX) sherry	

CONTINUED ON NEXT PAGE

PÂTÉ DE HÍGADO DE POLLO CONTINUED FROM PREVIOUS PAGE

6   Add the garlic to the sauté pan and continue cooking for 20 minutes, until the onions are deeply caramelized. If the fond or onions starts to darken too quickly, add a little water to the pan and keep cooking.

7   Carefully add the amontillado sherry to the sauté pan. You should get some flambé action; shake the pan until the flame dies. Cook for 1 minute, cooking off some of the alcohol. Remove and discard the thyme. Transfer the contents of the sauté pan to a blender.

8   Return the sauté pan to medium–high heat. Add the reserved liver marinade and simmer for 6 to 8 minutes, until the cream is reduced by ½. Remove from the heat. Transfer the hot cream to the blender.

9   Process the mixture in the blender on high speed for 1 minute, until it is very smooth (it's good to process the liver mixture while it's hot, as that helps you achieve a smoother purée). Add the butter bit by bit while processing, until all the butter is incorporated. Taste the *pâté* and reseason with the salt and black pepper as necessary.

10 Using a *chinois* or other fine sieve, strain the *pâté* into a serving bowl or individual ramekins. Place the *pâté(s)* onto a baking sheet and set them aside for the moment.

11 Place the water in a small saucepan over medium heat and then sprinkle the gelatin on the water's surface. Warm until the gelatin just melts. Add the PX sherry and whisk to combine. Remove from the heat.

12 Spoon the warm gelée over the *pâté* until it is completely covered.

13 Refrigerate at least 8 hours or overnight to set. Serve cold.

NOTE: Pretty much everyone serves this alongside some grilled bread—it's a very rich first course akin to *foie gras,* which is why some pickled goodies help a lot to break up the richness.

CHAPTER FEATURE

EDUARDO SOUSA'S FOIE GRAS

NO BOOK ON *charcuterie* is complete without discussing the sworn enemy of PETA and your cardiologist: decadent, deliciously fatty duck liver known in Spanish and all other languages as *foie gras.*

As a culinary delicacy, *foie gras* is alive and well in Spain—much as it is in the United States, despite some farcical attempts to regulate its production and consumption in my home state (I'm looking squarely at you, California).

More importantly, however, the world has taken notice of Spanish production methods for *foie*—especially those of farmer Eduardo Sousa, which are sometimes characterized as more humane because they do not involve the controversial tube-feeding method known as *gavage*.

Specifically, the work being done by Eduardo and his family at La Patería de Sousa, a small farm and *dehesa* in Extremadura, has been hailed by many experts—even (gasp!) some French ones—as groundbreaking and has earned the small company numerous culinary and innovation awards.

Essentially, Eduardo's operation provides a large meadow for the geese to play in. He lets them eat, hang out, and do what gaggles of geese do: naturally gorge themselves in preparation for the winter. In fact, they are free to come and go as they please, but since the area is safe and the geese are happy, they choose to stay. Come the winter harvest, the geese meet their maker, and that's the interesting part—the livers from

Eduardo's geese are surprisingly similar to birds raised elsewhere with the *gavage* method of tube-feeding.

For the record, I'm a very partisan individual on the subject of *foie gras*. I've visited a number of our farms in the United States, and I have observed the tube-assisted *gavage* feeding method with my own eyes (watching ducks and geese literally line up to be fed by their handlers). While this was enough to convince me that the *gavage* process is not a harmful one to the animal's well-being (if it were painful, the animals would flee from their feedings), I still hope that there's a compromise to be had between those opposed to *foie gras* and those concerned with the personal freedoms that these laws obviously encroach upon. Here's hoping, at the very least, that products like Eduardo Sousa's *foie gras* are one step down that road to compromise…

FOIE SALADA EN TORCHON

Maybe you don't hear about it every day, but the *foie torchon* is damn near as integral a part of fine dining as immersion circulators and Adderall-enhanced line cooks with tattoo sleeves.

That's because—aside from the fact that *foie* is a luxury item that most mortals are afraid of screwing around with—making a *torchon* is representative of a popular culinary truism. This mantra, which was first intoned by the great chef Fernand Point, now serves as a battle cry for perfectionist cooks in blue aprons all over the world, and it goes a little something like this: "Perfection is a lot of little things done well."

A perfectly executed *torchon* lays bare the attention and care the cook put into each step, and a poorly executed *torchon* shows the sins that plagued its creator throughout (often haste is high on the list). There's nowhere to hide with a *torchon,* for technique and patience are everything when creating the right texture and flavor with so few ingredients.

But here's the thing: No cook I know who's worth his or her Nenox knives backs down from that kind of challenge. We thrive on walking a daily tightrope of nerves, adrenaline, and reality-altering substances, and the stakes are that much higher when you're talking about an ingredient that can't be recycled into a family meal if it gets fucked up.

So your challenge, dear reader, should you choose to accept it, is this recipe: a basic *torchon* of *foie gras* that's both a large step in the direction of *charcuterie* mastery and a killer snack.

Good luck, and may the *foie* be with you.

1   Place the *foie* in a large mixing bowl and cover with the sparkling water. (I prefer Vichy Catalan in this recipe, since it's a little bit saltier than other sparkling waters.) Chill in the refrigerator overnight.

2   Remove the *foie* from the water and dry it well. Place it on a plate and allow it to come to room temperature (it's easier to handle when it's warmer; from this point forward, don't worry too much about breaking up the *foie,* as it's basically like Play-Doh).

3   Gently pull the lobes of the *foie* apart, trying your best to avoid tearing the veins. Starting at the bottom of the first lobe, locate the major vein. Using a butter knife, a small palette knife, or your finger, lift and pull the vein until you reach the groups of smaller veins that spiral out from the top of it. As you are deveining the *foie,* make sure to also pinch out any green bile spots or discoloration.

YIELD

2 *torchons* per 2.2 pounds (1 kg) of *foie gras*

INGREDIENTS	CHARCUTIER'S PERCENTAGE
per 2.2 pounds (1 kg) of *foie gras* (goose livers), grade B, room temperature	**100%**
Sparkling water, such as Vichy Catalan, to cover	
½ ounce (15 g) kosher salt	**1.5%**
⅛ ounce (5 g) granulated sugar	**.5%**
⅛ ounce (2 g) TCM #1 or DQ #1 curing salt mix (see p. 83)	**.2%**
1½ teaspoons (3 g) Four-Spice Blend (see recipe on p. 409)	**.3%**
¼ cup (50 mL) high-quality rum or brandy	
Coarse sea salt or rock salt, to cover	

CONTINUED ON NEXT PAGE

FOIE SALADA EN TORCHON CONTINUED FROM PREVIOUS PAGE

4   Flatten out the lobe of the *foie* along its natural breaking point (in the center). At this point, you should be able to clear any smaller veins. Set the lobe aside on a plate and repeat the process with the other lobe. Continue until both lobes have been deveined and flattened.

5   In a large, shallow bowl, combine the salt, sugar, curing salt, and Four-Spice Blend, creating a cure. Place the *foie* pieces into the bowl with the cure and toss to combine. Sprinkle with the rum.

6   Transfer the *foie* to a large baking dish. Pack down the *foie* by covering it with plastic wrap and pressing down on the wrap. The *foie* should be flattened and compacted slightly. Place in the refrigerator to cure for 1 to 3 days.

7   Remove the *foie* from the cure and allow it to rise to room temperature.

8   Place a large piece of plastic wrap on the counter (if you don't have any plastic wrap, stack a few layers of parchment paper). Lay ½ of the *foie* on the plastic wrap and reserve the other ½ in the refrigerator.

9   Form the *foie* into a rough rectangle. Create a tight log by rolling the *foie* up like a jelly roll and twisting the ends of the plastic wrap. Set the *foie* log aside in the refrigerator for 10 minutes to chill.

10   While the *foie* is chilling, cut and stack 2 identical strips of cheesecloth that are each a few inches wider than the *foie* log and a few feet long. Place the *foie* log on the bottom of the cheesecloth's long side (in other words, the cheesecloth should be oriented with the top/bottom as the long sides of the rectangle, and right/left as the short sides). Wrap the cheesecloth tightly around the log. Then, twist the ends to shape the log into a compact cylinder.

11   Tighten and tie off both ends of the *foie* with a tight knot and wrap a long piece of twine around one end. Cinch and tie this string while twisting the other end, creating a sort of corkscrew that will tighten the *foie* with each pass.

12   Continue this process of cinching and tying until the *foie* is a tightly compact log about 2 inches (5 cm) in diameter. You will know the log is tight enough when the pressure from cinching the twine causes some of the *foie* to seep through the cheesecloth. Place the log—now a *torchon*—in the baking dish. Repeat the process with the other *foie* ½, and then refrigerate both *torchons* for at least 4 hours, until cold.

13   Pour a thick layer of the coarse salt into a foodsafe container large enough to bury the *torchons*. Nestle the *torchons*—still in their cheesecloth wrappers—into the salt. Cover with enough of the coarse salt to bury them completely, then cover the dish with plastic wrap and allow the *foie* to sit in the salt, undisturbed, for 3 days in the refrigerator.

14   Remove the *foie torchon* from the salt and wipe off any excess salt still clinging to it. At this point, the *foie* is ready to be served, or you can hang it in a drying chamber for 1 to 3 more days to dry it a little further.

NOTE: If you are super anal retentive about not having any speckling on your *foie* due to the dark spices in the cure, double the amount of Four-Spice Blend specified in the recipe but don't add it to the cure. Instead, mix it with the salt in which the *torchon* is buried. It will permeate the *foie* with flavor without marking it up.

TRUCOS DE LA COCINA: THE FOIE LASSO

MAKING A *TORCHON* of *foie* is a rite of passage for any young cook earning his or her stripes in the culinary arts. It's one of those Mr. Miyagi-esque lessons we go through to learn patience, attention to detail, and appreciation for craftsmanship.

Like a lot of cooks before me, after managing to roll up a decent *torchon* or two, I learned some tricks for getting quicker and tighter torques to my *foie*. The best of those tricks was the lasso technique, in which you wrap and cinch twine around the end of the *torchon,* methodically tightening the *foie* into a state of perfect compression.

To give credit where it's due, I first came across this method while talking to Paras Shah, a great chef and old friend from Queens, New York, who also participated in the 2009 ICEX culinary scholarship tour of Spain. He learned it during his time working at Momofuku Noodle Bar under Chef David Chang, so lots of respect to the team there for their technically sound methodology.

I outline the basics of the technique in the *Foie Salada en Torchon* recipe, but here's the gist of it: Once your *foie* is rolled up, loop a piece of twine around one end (the lassoing part) and start winding the string around the end while tightening with each wrap. You will be left with a *torchon* that is super tight, compact, and devoid of any air holes or other imperfections.

Practice, practice chef...

CHAPTER

10

GUARNICIÓNES Y SALSAS

N THIS CHAPTER,

you will meet some of *charcuterie's* best friends: Pickles, jams, spices, and sauces that support rich and delicious cured foods by providing their perfect complements: They are sour, sweet, and acidic partners.

In the first part of this chapter, you'll see some of the unique pickles typically served in Spain. Not only are many of these common *tapas* on their own—things like pickled veggies, olives, fish, and other acidic treats—but these same snacks also typically complement most plates of *charcutería* as well.

Likewise, Spain has a bit of a sweet tooth when it comes to *charcuterie* garnishes, and no one does sweet jams and *mermeladas* like the nuns who make and sell these specialties from their *conventos* across the Spanish countryside; those recipes are found in the second part of this chapter.

Finally, I discuss some of the basic sauces and base recipes that come from the home kitchens of Spain. These are the basic recipes that most *sabias* and restaurant kitchens have on hand to add flourish to everything from meat-curing projects to more mundane cooking efforts, like soups, stews, and other dishes.

ACEITUNAS

Who doesn't love marinated olives? But when was the last time that a restaurant put a plate of them in front of you that made you stop and say: "Wow! These are different and delicious!"

I can tell you when that happened for me—it was this recipe. I was cooking at Calima and our chef de cuisine, the aptly named David Olivas (*olivas* is, ironically, another name for olives) was prepping a catered party for the weekend. He was making a bowl of these olives from his family's recipe and offered them around for the cooks to try.

At first, I thought it was a joke or that David had screwed up: All of the olives were smashed. But it turns out that this only made them better! Since the entire olive—inside and out—was exposed to the marinade, the olives absorbed the flavors more readily, and they were really delicious.

As you might expect, the olives were so good that we ate almost all of them...and let's just say that David was less than thrilled that he had to make another batch for the party. Such are the hazards of making great food, so I recommend you make a double batch.

The olives keep for a while and only get better with time.

1   Using a peeler or paring knife, remove the zest of the orange and lemon in strips. Set the zest aside in a small bowl and juice the remaining fruit, reserving the juice in a large bowl.

2   To the same large bowl, add the garlic, oil, sherry vinegar, honey, thyme, rosemary, bay leaves, pickled pepper, and salt. Stir to combine. Taste the marinade and reseason as necessary.

3   Place in the refrigerator at least overnight; the dish is best when the olives have marinated for a few days.

YIELD

Tapas for 8–10 people per 2.2 pounds (1 kg)

INGREDIENTS

per 2.2 pounds (1 kg) of lightly smashed Spanish olives, such as arbequina, gordal, or manzanilla

1	medium orange
1	medium lemon
10	whole cloves garlic, peeled
1	cup (250 mL) Spanish olive oil, such as piqual or arbequina
½	cup (100 mL) sherry vinegar
1	tablespoon (20 g) rosemary, thyme, or orange blossom honey
4	fresh sprigs thyme
4	fresh sprigs rosemary
2	fresh bay leaves
2	Cured Guindillas (see recipe on p. 383) or other spicy chile pepper, such as Fresno
1	ounce (20 g) kosher salt, plus more to taste

CHAPTER FEATURE

ENCURTIDOS

WAY BEFORE PICKLES became a culinary cliché on the American food scene, Spaniards had been preserving veggies and fruits for centuries in spicy, vinegary solutions called *encurtidos*.

Somewhat loosely tied to the history of *tapas*—which, depending on whom you ask, is all about the need to keep flies out of your sherry or are a means of complying with ancient laws about having to serve food with alcoholic drinks—some cities in Spain have whole economies based around IGP-protected pickles that are either a fundamental part of the area's cuisine or a major component in their cooking.

In the next few pages, you'll read about a series of *encurtido* recipes I've picked up over my travels. All of them keep for a good while in the fridge, or you can process them for canning and thus indefinite shelf storage (see p. 130 for details).

CURED GUINDILLAS

Guindilla is the name for both the fresh and pickled versions of a skinny, green, hot pepper from the north of Spain (in Basque they are called *piparras)*.

At mealtime, the jar of pickled peppers is traditionally plopped in the middle of a table, and the peppers are eaten between spoonfuls of hearty dishes like *alubias negras,* because the acid offsets the dishes' heaviness and the chilies add as much or as little heat as you want.

You are sometimes able to find these peppers growing here in the United States, but if not, go for a medium-hot cousin like banana, Hungarian, cubanelle, or wax peppers. Just remember that Spaniards are total wimps when it comes to heat, so keep the peppers on the mild side if you're feeding them to anyone from Spain—unless you want to see them run around the room holding their tongue screaming *"¡Me pica!"*

Yes, I've seen it happen...

1   Place the chilies in a jar or container that just holds them and cover with the water. Retain the chilies in the jar and pour the water into a measuring cup. Note the amount. Discard half the water, and add an equal quantity of the vinegar. Set aside.

2   In a medium saucepan over medium–high heat, place the sugar and 1 tablespoon of the water. Do not stir; instead, swirl the pan lightly and heat the mixture for 4 to 6 minutes, until a light amber caramel forms.

3   Add the reserved water and vinegar mixture and the salt to the saucepan. Reduce the heat to medium and cook for 6 to 10 minutes, until the mixture reaches a simmer and all of the sugar and salt have dissolved. Taste the pickling liquid and reseason as necessary.

4   In a large foodsafe plastic container, combine the peppers and the hot pickling liquid. Weigh down the peppers as needed to keep them submerged in the liquid, and allow the liquid to come to room temperature. Seal the container.

5   Store in the refrigerator for 2 to 4 days, or until pickled. Serve cold.

YIELD

3–5 pickled peppers

INGREDIENTS

per 1 cup (250 mL) of liquid

3–5   yellow chile peppers, such as guindillas/piparras, cubanelles, banana, or Hungarian wax, pricked a few times with a toothpick or fork

Water, as needed

White wine vinegar, as needed

2   tablespoons (25 g) granulated sugar, or more as needed

1   tablespoon (10 g) kosher salt, or more as needed

AJO DULCE

Whenever I do a big *charcuterie* board, this is my pickled weapon of choice. These little beauties always get people talking, as the garlic cloves wind up looking like deep brown jewels coated in a sweet–sour poaching liquid. (Whatever you do, don't throw out this liquid—it's great as a salad dressing or marinade!)

Apple cider vinegar works really well here, but in a pinch I've used everything from rice wine vinegar to red wine vinegar. Just make sure to really get a good caramel when you're cooking the sugar and water—it's the trick to get the color and deep flavor in the pickle.

1   Place the garlic in a jar or container that just holds them and cover with the water. Retain the garlic in the jar and pour the water into a measuring cup. Note the amount. Discard half the water, and add an equal quantity of the vinegar. Set aside.

2   In a medium saucepan over medium–high heat, place the sugar and 1 tablespoon of the water. Do not stir; instead, swirl the pan lightly and heat the mixture for 4 to 6 minutes, until a light amber caramel forms.

3   Add the garlic cloves and salt to the saucepan and stir to coat the garlic with the caramel. Cook for 2 minutes, until the caramel turns slightly darker.

4   Add the reserved water and vinegar mixture to the saucepan. Reduce the heat to medium and simmer for 15 to 20 minutes, until the liquid reduces to a syrupy consistency and the garlic is soft. Taste the pickling liquid and reseason as necessary. Remove from the heat and set aside to cool to room temperature.

5   Transfer the mixture to a large foodsafe plastic container. Weigh down the garlic cloves as needed to keep them submerged in the liquid. Seal the container.

6   Store in the refrigerator for 2 to 4 days, or until pickled. Serve cold.

YIELD

10–20 cloves of pickled garlic

INGREDIENTS

per 1 cup (250 mL) of pickling liquid

1   head garlic, cloves peeled and destemmed

Water, as needed

Apple cider vinegar, as needed

1¾   ounces (50 g) granulated sugar, or more as needed

2   tablespoons (20 g) kosher salt, or more as needed

AJOS ENCURTIDOS

Pickled cloves of garlic are probably the most famous and universal *encurtido* in Spain. You'll find them everywhere, from *tapas* bars to store shelves.

The previous pickled garlic recipe is on the sweeter side, but this one is much more savory and needs a long bath in the pickling liquid to soften the cloves and take the edge off the garlic flavor. When I serve these cloves, I typically mix them with other pickled veggies, since Americans don't have quite the tolerance that Spaniards have for popping whole cloves of garlic in their mouth—pickled or not.

1   Fill a bowl with ice and water and set aside.

2   To a medium saucepan over medium–high heat, add the garlic and cover with the cold water. Bring the water to a boil and then immediately remove from the heat. Let the garlic steep in the water for 5 minutes (this removes some of its harshness).

3   Drain the garlic into a colander. Plunge the colander into the ice bath, stopping the cooking process.

4   Place the garlic in a jar or container that just holds them and cover with the water. Retain the garlic in the jar and pour the water into a measuring cup. Note the amount. Discard half the water, and add an equal quantity of the vinegar. Set aside.

5   In a medium saucepan over medium–high heat, place the garlic, reserved water and vinegar mixture, the salt, and the sugar. Cook for 8 to 10 minutes, until the salt and sugar have dissolved. Remove from the heat, taste the mixture, and reseason as necessary.

6   Transfer the mixture to a large foodsafe plastic container. Weigh down the garlic cloves as needed to keep them submerged in the liquid. Seal the container.

7   Store in the refrigerator for 1 week, until pickled. Serve cold.

YIELD

10–20 cloves of pickled garlic

INGREDIENTS

per 1 cup (250 mL) of pickling liquid

1	head garlic, cloves peeled and destemmed
	Cold water, as needed
	Apple cider vinegar, as needed
⅔	ounce (20 g) kosher salt
⅓	ounce (10 g) granulated sugar

ALMAGRO-STYLE EGGPLANT

When I first arrived in Toledo to cook for Adolfo, I was shocked to find a welcome basket waiting for me in my apartment— that's just how cool Adolfo and his family are.

Nestled in the basket, alongside other local culinary delicacies like *turrón* and some bottles of wine, was a jar of freaky pickled baby eggplants bobbing around in a hazy red liquid. Later, I learned that these little eggplants were a famous snack from the nearby town of Almagro and, when I mustered the courage to crack the lid and try one, they didn't last long.

Eggplant came to Spain via Syria from Asia, where the eggplant is a native species. Given that eggplant is not native to Spain, and considering the cumin and fennel used in the pickling process, it's not hard to see a Moorish influence at work in this recipe. These very special and unique pickles—an IGP-protected food—are made from a particular type of eggplant that grows around Almagro. The eggplants, which are harvested when they are still very small, young, and tender, get brined for a bit, blanched, and then skewered with fennel and bell pepper before eventually finding their home in a spicy, *pimentón*-infused pickling liquid.

Obviously, you won't be finding miniature Almagro eggplant at your favorite hipster farmer's veggie stall at the Green Market anytime soon. But there's good news: Miniature Indian eggplant works here, as does an heirloom variety called Fairytale eggplant. Look for it both at your favorite hipster stand at the farmers' market or at ethnic produce stands.

YIELD

2.2 pounds (1 kg) pickled eggplant

INGREDIENTS

per 2.2 pounds (1 kg) of miniature eggplant, such as Fairytale heirloom, stems on

FOR THE EGGPLANT:

Cold water, as needed

Kosher salt, as needed

FOR THE PICKLING LIQUID:

1 medium yellow onion, peeled, destemmed, and halved

2 medium red bell peppers, deseeded and cut into large dice

1 head fennel, cored, cut into thirds (tops reserved), and sliced into small skewer-sized julienne

10 whole cloves garlic, peeled and destemmed

2 cups (500 mL) water

1 ounce (25 g) kosher salt

⅓ ounce (10 g) cumin seed, toasted

⅓ ounce (10 g) fennel seed, toasted

1 ounce (25 g) *pimentón picante*

1 ounce (25 g) *pimentón agridulce*

2 cups (500 mL) sherry vinegar

¼ cup (50 mL) *Tomate Frito* (see recipe on p. 407)

2 tablespoons (25 mL) extra virgin olive oil

1 fresh bay leaf

TO PREPARE THE EGGPLANTS:

1   Place the eggplants in a jar or container that just holds them and cover with the water. Retain the eggplants in the jar and pour the water into a measuring cup. Note the amount. Add 1¾ ounces (50 g) of the salt per cup of water and stir until dissolved. Pour the water into the jar containing the eggplants and soak at room temperature for at least 3 hours (this will remove some of the eggplants' natural bitterness).

2   For each eggplant, make a lengthwise cut from the bottom of the vegetable up to its middle, leaving the top half intact.

3   Fill a bowl with ice and water and set aside.

4   Place the scored eggplants into a large saucepan and cover with cold water. Place the saucepan over medium–high heat and bring to a boil. Immediately remove from the heat. Let the eggplants poach in the water for 2 minutes.

5   Drain the eggplants into a colander and plunge the colander into the ice bath, stopping the cooking process. Set the eggplants aside for the moment.

TO MAKE THE PICKLING LIQUID:

1   In a large saucepan, cover the onion, red bell peppers, and fennel bulb pieces with the water. Set aside.

2   Using a mortar and pestle, crush together the garlic, salt, cumin seed, and fennel seed to form an *ajosal*. If desired, you can finish the *ajosal* in a food processor fitted with the "S" blade.

3   Place the saucepan over medium–high heat and add the *ajosal,* both *pimentones,* vinegar, *Tomate Frito,* oil, and bay leaf. Bring to a boil and then immediately remove from the heat. Taste the pickling liquid, reseason as necessary, and set aside for the moment.

4   Pull out some of the red pepper pieces and stuff them into the scored areas of the eggplants. Then use the branches of the fennel tops as you would a skewer or toothpick to close up the eggplant.

5   In a large foodsafe plastic container, combine the stuffed eggplants, the warm pickling liquid, and the remaining pieces of onion and fennel. Weigh down the eggplants as needed to keep them submerged in the liquid. Let the liquid cool to room temperature and seal the container.

6   Store in the refrigerator for 2 to 4 days, or until pickled. Serve cold.

CURED *CEBOLLITAS*

These little pickled onions are great for *charcuterie* boards and salads alike. They are crunchy, vinegary, and a nice counterpoint to rich dressings and meats.

When you put the beets into the marinade you will get a bright scarlet pickled onion with a hint of the beet's sweetness—really cool stuff for adding color to dishes. If you aren't into the red color, just omit the beets.

1   Prepare the onions by slicing off their tips (the parts opposite the root ends).

2   Fill a bowl with ice and water and set aside.

3   Bring a small saucepan of water to a rolling boil. Add the onions to the boiling water and cook for 3 to 4 minutes, until they soften slightly. Drain the onions into a colander and plunge the colander into the ice bath, stopping the cooking process. Rinse the saucepan for reuse.

4   Slip the onions out of their skins via the slits you created at the onion tips.

5   Place the onions in a jar or container that just holds them and cover with the water. Retain the onions in the jar and pour the water into a measuring cup. Note the amount. Discard half the water, and add an equal quantity of the vinegar. Set aside.

6   Return the saucepan to medium–high heat. Add the sugar and 1 tablespoon of the water. Do not stir; instead, swirl the pan lightly and heat the mixture for 4 to 6 minutes, until a light amber caramel forms.

7   Add the onions, beet, and salt to the saucepan and stir to coat with the caramel. Cook for 2 to 3 minutes more, until the caramel turns slightly darker.

8   Add the reserved water and vinegar mixture, the bay leaf, the star anise, and the peppercorns to the saucepan. Reduce the heat to medium and simmer for 3 to 5 minutes more, until the sugar and salt have dissolved. Remove from the heat and set aside to cool to room temperature.

9   Taste the cooled mixture and reseason as necessary. Transfer the mixture to a large foodsafe plastic container. Weigh down the onions as needed to keep them submerged in the liquid. Seal the container.

10 Store in the refrigerator for 2 to 4 days, or until pickled. Serve cold.

YIELD

15–20 pickled pearl onions

INGREDIENTS

per 1 cup (250 mL) of pickling liquid

4½ ounces (120 g) pearl onions (should be around 20)

Water, as needed

Red wine vinegar, as needed

⅓   ounce (10 g) granulated sugar, or more as needed

1   medium red beet, peeled and quartered

1   ounce (25 g) kosher salt, or more as needed

1   fresh bay leaf

½   star anise

⅛   ounce (5 g) black peppercorns

CURED *PEPINILLOS*

I don't know exactly when Frenchie cornichon pickles crossed the border, but they have, for sure—you can find them pretty much anywhere in Spain. This is a basic recipe for cornichons, which are basically miniature sweetish cucumber pickles that are typically made using Fin de Meaux cucumbers.

Cornichons are ubiquitous on *charcuterie* boards, but substituting tiny Persian cukes is a fair trade (just bear in mind that it will take more time to pickle them, since they're bigger).

TO PREPARE THE CUCUMBERS:

1   Place the cucumbers in a bowl and toss them with the salt. Set aside at room temperature for at least 1 hour, allowing the salt to draw the water out of the cucumbers. Rinse and drain the cucumbers. Set aside for the moment.

TO MAKE THE PICKLING LIQUID:

1   In a medium saucepan over medium–high heat, place the vinegar, hot water, sugar, peppercorns, mustard seed, salt, cloves, onion, tarragon, and bay leaves. Cook for 4 to 6 minutes, until the salt and sugar have dissolved.

2   Remove from the heat, taste, and reseason as necessary.

3   In a large foodsafe plastic container, combine the cucumbers and the hot pickling liquid. Weigh down the cucumbers as needed to keep them submerged in the liquid. Seal the container.

4   Store in the refrigerator for 2 weeks, until pickled. Serve cold.

YIELD

2.2 pounds (1 kg) pickled cucumbers

INGREDIENTS

per 2.2 pounds (1 kg) of baby cucumbers, such as Fin de Meaux, rinsed

FOR THE CUCUMBERS:

1¼ ounces (35 g) kosher salt

FOR THE PICKLING LIQUID:

2   cups (500 mL) white wine vinegar

2   cups (500 mL) hot tap water

⅓   ounce (10 g) granulated sugar, or more as needed

1   ounce (25 g) black peppercorns, toasted

1   ounce (25 g) whole mustard seed, toasted

1   ounce (25 g) kosher salt, or more as needed

3   whole cloves, toasted

1   medium white onion, peeled and julienned

5   sprigs fresh tarragon

2   fresh bay leaves

HUEVOS DE CORDONIZ

You will most likely find pickled quail eggs perched on top of *pintxos* in San Sebastián or Navarra. Much more often than not, they're a garnish on a *tapa,* rather than a *tapa* on their own.

I fell in love with this recipe from Pamplona for pickled quail eggs in a *pimentón*-spiked pickling liquid. The red liquid tints the white of the egg and also imparts the *pimentón's* smoky goodness. Just make sure to use real-deal *pimentón de la Vera,* which is way smokier than standard paprika.

TO PREPARE THE QUAIL EGGS:

1   Place the quail eggs in a large saucepan and cover with the water. Add the vinegar and salt. Place the saucepan over medium–high heat and bring to a rolling boil.

2   Immediately remove from the heat. Allow the saucepan to sit, covered, for 3 minutes. Meanwhile, fill a bowl with ice and water and set aside.

3   Drain the quail eggs into a colander and plunge the colander into the ice bath, stopping the cooking process. Refrigerate the eggs for at least an hour, until thoroughly chilled.

4   Crack and peel the eggs. Discard the shells and return the eggs to the refrigerator.

TO MAKE THE PICKLING LIQUID:

1   In a large saucepan, combine all the pickling liquid ingredients. Bring to a boil. Remove from the heat, taste the pickling liquid and reseason as necessary, and set aside to cool to room temperature.

2   Transfer the eggs and pickling liquid to a large foodsafe plastic container. Weigh down the eggs as needed to keep them submerged in the liquid. Seal the container.

3   Store in the refrigerator for 2 to 4 days, or until pickled. Serve cold.

YIELD

1 dozen quail eggs

INGREDIENTS

per 1 dozen quail eggs

FOR THE QUAIL EGGS:

Water, as needed

1/4   cup (50 mL) white vinegar

1   tablespoon (15 g) kosher salt

FOR THE PICKLING LIQUID:

1   cup (250 mL) white wine vinegar

1   cup (250 mL) hot tap water

1½   ounces (40 g) kosher salt

2/3   ounce (20 g) granulated sugar

½   ounce (15 g) *pimentón picante*

½   ounce (15 g) *pimentón agridulce*

2   teaspoons (5 g) Four-Spice Blend (see recipe on p. 409)

1   fresh bay leaf

PIEL DE SANDIA

Early on in our ICEX Spanish culinary scholarship, my fellow cooks and I decided that it would be fun to make dinner for our teachers and the staff of the little school where we were enrolled in a language immersion course.

We each chose a course to cook, shopped for ingredients, and cooked our courses in turns. I drew the salad course and, given that it was the dead of a very hot summer in the Spanish countryside, I opted for a watermelon salad garnished with some lavender and other edible flowers that grew around the school.

When I served the salad—garnished with pickled watermelon rind—the Spaniards all started laughing and pointing at me and my salad. At first I thought I'd made some sort of social faux pas, until they explained to me that in Spain, pickled watermelon rind is an old-school aphrodisiac and sexual aid.

Ever since, I've called these pickles my very delicious recipe for Spanish Fly 2.0.

YIELD

1 Pickled rind of seedless watermelon

INGREDIENTS

per 1 seedless watermelon

	Water, as needed
	Apple cider vinegar, as needed
3½	ounces (100 g) granulated sugar
1	ounce (25 g) kosher salt
¼	ounce (8 g) dried or fresh lavender leaves
1	star anise
2	teaspoons (5 g) whole black peppercorns
1	teaspoon (2 g) whole allspice
10	sprigs fresh thyme
2	fresh bay leaves

1   Cut off the very top and bottom of the watermelon's rind, allowing it to stand flat and stable while peeling.

2   Using a very sharp knife, slice off the green outer skin of the watermelon in strips, exposing the white rind underneath. Discard the green skin. Then cut off the melon's white rind in long, wide strips, making sure to include as little of the pink flesh as possible.

3   Cut the strips of rind into large squares. Set aside. (And enjoy the delicious flesh of the watermelon.)

4   Place the rind squares in a jar or container that just holds them and cover with the water. Retain the rind squares in the jar and pour the water into a measuring cup. Note the amount. Discard half the water, and add an equal quantity of the vinegar. Set aside.

5   Place a large saucepan over medium–high heat. Add the sugar and 1 tablespoon of the water. Do not stir; instead, swirl the pan lightly and heat the mixture for 4 to 6 minutes, until a light amber caramel forms.

6   Add the reserved rind squares, the reserved vinegar and water mixture, the salt, the lavender, the star anise, the peppercorns, the allspice, the thyme, and the bay leaves to the saucepan. Bring the pickling liquid to a boil; then immediately remove from the heat. Taste and reseason as necessary, and then let the mixture steep in the water for 15 minutes. Set aside to cool to room temperature.

7   Transfer the mixture to a large foodsafe plastic container. Weigh down the rind squares as needed to keep them submerged in the liquid. Seal the container.

8   Store in the refrigerator for 1 to 2 days, or until pickled. Serve cold.

LA COCINA DE LOS CONVENTOS

THE MOST SECRETIVE kitchens in Spain aren't the laboratories of famous chefs like Ferran Adrià, Daní Garcia, or Juan Mari Arzak. Rather, the most clandestine Spanish cooking happens in kitchens behind stone walls and are run by women—Spanish *monjas,* or nuns—who practice closely guarded recipes dating back to the Middle Ages and Moorish invasions. The desserts made in these convent kitchens are as delicious as they are famous for the archaic hoops you have to go through to purchase them.

In many towns throughout Spain, nuns make a living by selling homemade sweets. We are talking about real-deal *comida casera* here: The eggs, which have yolks that are almost red in color, are typically from the nuns' own chickens; the almonds are hand-ground into almond flour; the fruit comes from their orchards; and the honey comes from their beehives.

But since the *monjas* have taken a vow of isolation, they sell their sweets through a quirky, archaic system called the *turno.* In most cases, you head to a room in the *convento* with a list of prices next to an ancient lazy Susan built into the wall (that's the *turno).* You make your choice and ring a bell, and the disembodied voice of a *monja* asks for your order. You tell her what you want,

and she sends the order to you via the *turno.* It's a process that keeps customers honest: If you don't want to go directly to Hell, you better slide your money back around the *turno* to complete the transaction. If only ATM banking could be so easy and charming.

These recipes come from several different *conventos,* as each order of *monjas* is known for different specialties. For example, the *monjas* of San Leandro in Seville are famous for their *yemas;* the Dominicans in Toledo are known for their marzipan; and the *convento* of Santa Paula in Seville is known for its jams and *mermeladas.*

So, no, these recipes aren't all that porky, but each is unique and delicious, and they'll make great accompaniments for your *charcuterie* projects.

MERMELADA DE TOMATE

Spain's love affair with the tomato extends into many different culinary dishes and expressions, including the *Mermelada de Tomate* made by the famous Convento Santa Paula in Seville.

This recipe is very similar to the version sold at the Convento. I learned it while cooking at Adolfo's in Toledo, but you can find variations of *Mermelada de Tomate* all over Spain. It's a favorite breakfast accompaniment.

The key here is to make this *mermelada* only when tomatoes are in season. Don't even bother with those nasty, mealy winter tomatoes. Mother Nature doesn't want you to eat tomatoes in winter anyway.

1   Using a very sharp paring knife or peeler, remove the lemon's zest in a single strip. Juice the lemon. Reserve the zest and juice and discard the lemon's pulp.

2   Cut a slit in the long side of the ½ vanilla bean. Slide the back of a paring knife down the length of the inside of the bean; the bean's seeds will accumulate on the knife's blade. Reserve the empty bean and the seeds separately.

3   Place a medium saucepan filled ⅔ of the way with the water over medium–high heat and bring to a boil. Fill a large bowl with ice and water and set aside.

4   Slice a small X in the end of each tomato opposite its stem. Place the tomatoes, 3 or 4 at a time, in the boiling water, and cook for 20 to 30 seconds, just long enough for the skin to begin to peel away from the fruit. Plunge the group of tomatoes into the ice bath to stop the cooking process, and then remove the tomatoes from the ice bath and set aside to cool to room temperature. Repeat until all of the tomatoes have been blanched.

5   Remove the skin from each tomato. Cut the tomatoes in half lengthwise and, using a spoon, remove and discard the seeds. Chop the tomatoes roughly and place them in a large saucepan with the lemon juice and zest, vanilla bean and seeds, and the brown sugar, Four-Spice Blend, and salt. Stir to combine.

6   Place the saucepan over medium–high heat and bring to a high simmer for 10 minutes. Reduce the heat to medium and cook, stirring frequently, for 30 to 40 minutes, until the mixture reduces down and takes on a jam-like consistency. Taste and reseason as necessary, and then set the *mermelada* aside to cool to room temperature.

7   Transfer the mixture to a large foodsafe plastic container and chill overnight in the refrigerator.

YIELD

2–3 cups (450–700 mL)

of *mermelada*

INGREDIENTS

per 2.2 pounds (1 kg) of fresh, in-season tomatoes

1	whole medium lemon
½	vanilla bean
	Water, as needed
1	cup (250 g) dark brown sugar, densely packed
2	teaspoons (5 g) Four-Spice Blend (see recipe on p. 409)
1	tablespoon (12 g) kosher salt

CABELLO DE ÁNGEL

The Monasterio de Santa María del Socorro in Sevilla is an interesting study in evolution and adaptation: The ladies there figured out quite some time ago that technology and the modern world can make excellent economic partners. Not only do these *monjas* now sell their *dulces* via the Web—they've even published a few cookbooks and have really done a great job of pulling back the veil (pun intended) on their *dulce*-making operations.

This quasi-*mermelada* is an exemplar of their good works. It's often used as a filling for *empanadas* that totally remind the fat kid deep inside of me of McDonald's apple pies. Fittingly, this jam is called *Cabello de Ángel* ("angel's hair"), since it's made from a special type of pumpkin that, when cooked, separates itself into filament-like strands. It's sometimes available in the United States at Asian or Latino markets; look for Siam pumpkin or vermicelli pumpkin, but you can candy any other kind of pumpkin as well (it just may not separate into strands like the more traditional varieties).

YIELD

2–3 cups (450–700 mL)
of *mermelada*

INGREDIENTS

1	whole medium lemon
1	vanilla bean
1	Siam pumpkin *(cucurbita ficifolia),* quartered and seeded
	Water, as needed
	Granulated sugar, as needed
1	cinnamon stick
1	tablespoon (12 g) kosher salt

1   Using a very sharp paring knife or peeler, remove the lemon's zest in a single strip. Juice the lemon. Reserve the zest and juice and discard the lemon's pulp.

2   Cut a slit in the long side of the vanilla bean. Slide the back of a paring knife down the length of the inside of the bean; the bean's seeds will accumulate on the knife's blade. Reserve the empty bean and the seeds separately.

3   In a large saucepan, place the pumpkin quarters, skin-side down. Add enough of the water to cover ⅓ of the quarters. Add the vanilla bean to the pan. Place the saucepan over medium heat and cook the pumpkin for 30 minutes, until it is very soft (cooking time will vary depending on the pumpkin's size). Drain and set aside to cool to room temperature. Save the vanilla bean.

4   Using a sharp paring knife, remove the pumpkin's skin, then, using your hands, separate the pumpkin's pulp into strands. Use a kitchen scale to weigh the amount of pumpkin pulp; note the weight.

5   Place the same saucepan over medium heat. Add the following to the saucepan: the pumpkin pulp, ¼ the pumpkin pulp's weight in the water, ½ the pumpkin pulp's weight in the sugar, the lemon juice, the lemon zest, the cinnamon stick, the reserved vanilla bean, the vanilla seeds, and the salt. Bring the mixture to a simmer and cook for 25 to 30 minutes, until the sugar dissolves completely, the mixture is syrupy, and the pumpkin is very soft. Remove from the heat. Set aside to cool to room temperature.

6   Transfer the mixture to a large foodsafe plastic container. The *mermelada* can be used immediately or canned in sterilized containers (see instructions on p. 130).

MERMELADA DE CEBOLLA

The *monjas* of Santa Paula do right by the humble onion. Their *Mermelada de Cebolla* is a rich, oniony mixture that gets an extra punch from wine and spices, but when I make it I like to use sherry since it really plays nicely with caramelized onions. In this recipe, I use very sweet Pedro Ximénez (PX) sherry, as well as PX sherry vinegar, for a sweet and sour effect, but using dry red wine and red wine vinegar is more traditional.

1   In a large sauté pan over medium–high heat, warm the oil for 4 to 5 minutes, until just rippling but not smoking. Add the onions and season with the salt. Sauté for 15 minutes, until the onions begin to wilt and take on a little color.

2   Add the brown sugar and bay leaves and reduce the heat to medium. Cook for 10 to 15 minutes, until the onions are uniformly light brown.

3   Raise the heat to high. Once the mixture begins to sizzle, add the sherry and vinegar to the sauté pan. Reduce the heat to medium and cook for 15 to 20 minutes, stirring frequently, until the liquid volume has reduced and the onions are very brown and wilted.

4   Taste the *mermelada* and season as necessary with the vinegar, salt, and black pepper. Set aside to cool to room temperature.

5   Transfer the mixture to a large foodsafe plastic container and chill overnight.

YIELD

Around 1 cup (250 mL)

of *mermelada*

INGREDIENTS

**per 2.2 pounds (1 kg) of
sliced onions**

¼   cup (60 mL) extra virgin olive oil

Kosher salt, as needed

4½ ounces (125 g) dark brown sugar, densely packed

2   fresh bay leaves

½   cup (100 mL) Pedro Ximénez (PX) sherry*

½   cup (100 mL) Pedro Ximénez (PX) sherry vinegar, plus more as needed

Freshly ground black pepper, as needed

* This type of sherry (and vinegar) is made from Pedro Ximénez white grapes. The varietal yields an intensely sweet dessert wine.

DULCE DE MEMBRILLO

One day while I was working in Toledo, we got in a gigantic load of quinces. Now, in the United States, we don't do much with this bizarre little fruit, but in the heart of Castilla–La Mancha, it was like discovering buried treasure! In that part of Spain, quince makes all of the locals dewy-eyed for one of their favorite childhood food memories: *Dulce de Membrillo* (quince paste).

All of the cooks gathered around, and we were each assigned tasks: sorting, rinsing, and then eventually cooking and puréeing the quinces into this delectable, sliceable jelly.

Serve *membrillo* as the Toledanos do, with a good Manchego cheese and a crusty loaf of bread, or do as the more progressive Spaniards do, and serve it with a nice slab of *Foie Salada en Torchon* (see recipe on p. 375).

YIELD

2–3 cups (450–700 mL)

of *membrillo*

INGREDIENTS

per 2.2 pounds (1 kg) of scrubbed quinces

1   whole medium lemon

1   vanilla bean

    Water, as needed

    Granulated sugar, as needed

1   tablespoon (10 g) kosher salt

    Unsalted butter, for greasing

1   Using a very sharp paring knife or peeler, remove the lemon's zest in a single strip. Juice the lemon. Reserve the zest and juice separately and discard the lemon's pulp.

2   Cut a slit in the long side of the vanilla bean. Slide the back of a paring knife down the length of the inside of the bean; the bean's seeds will accumulate on the knife's blade. Reserve the empty bean and the seeds separately.

3   Quarter and core the quinces, but leave their skins on (the skins will provide a lot of pectin for the jelly).

4   In a large stockpot, combine the quinces, lemon zest, and vanilla bean. Cover with the water and place over medium–high heat. Bring to a gentle boil, reduce to a simmer, and cook for 1 hour, until the quinces are very soft and easily pierced with a knife. Remove from the heat. Remove and discard the zest and vanilla bean, and drain the quinces.

5   Remove the peels from the quinces. Process the quinces through a food mill with a fine screen (if you don't have one, you can use a *chinois* or other fine strainer) into a large mixing bowl. If necessary, repeat until the purée is smooth.

6   Use a kitchen scale to weigh the amount of purée; note the weight. Measure out an equal weight of the sugar. Add the sugar, the reserved lemon juice and vanilla seeds, and the salt to the purée. Mix well.

7   Transfer the purée to a large saucepan. Place over medium heat and cook, stirring frequently, for 2 hours. Note that the thicker the jelly becomes, the pinker the color will be—and the easier it will be for the jelly to burn. The *membrillo* purée is ready when it becomes a deep orange-pink color and thickens considerably.

8   Preheat the oven to 225°F (110°C). Grease a sheet of parchment paper with the butter and place it on a baking sheet.

9   Pour the *membrillo* over the parchment paper and smooth it into an even ½-inch (13-mm) layer.

CONTINUED ON NEXT PAGE

DULCE DE MEMBRILLO CONTINUED FROM PREVIOUS PAGE

10 Place the baking sheet in the oven* and bake the *membrillo* for 4 to 5 hours, until its color deepens further and the texture becomes very tacky.

11 Remove the *membrillo* from the oven and set aside on the counter to cool to room temperature. Remove the *membrillo* from the baking sheet, cover it with another sheet of parchment paper, and place the *membrillo* in the refrigerator to cool overnight.

12 Slice the *membrillo* into pieces and serve cold, with a good Manchego cheese or some *foie gras*.

* I prefer to prop my oven door open slightly using a dry kitchen towel—this keeps the temperature low enough that the *membrillo* dehydrates evenly.

TRUCOS DE LA COCINA: MEMBRILLO VARIATIONS

MEMBRILLO: IT'S NOT just for quinces anymore. But coaxing different fruits into a semisolid, jelly-like state takes a little bit of skill and an understanding of our jellifying friend: pectin.

Pectin, a naturally occurring compound found in plant matter, is the substance that's responsible for making jams and jellies set up and become a quivering mass (and it's also at fault if your jam fails to set up and becomes a total jam-making failure; that's a typical speed bump on the path of almost any aspiring jam maven's journey).

The key to *membrillos* is figuring out how to manipulate pectin—and by that, I mean how much acid and sugar you need to add—in order to get the texture you're looking for. Or, for the science geeks, the pectin bonds with acid and sugar to form a gel, so if your *membrillo* doesn't have enough acid or sugar, you run the risk of it not gelling.

That's why quinces have been the *membrillo* fruit of choice for so long: Fresh quinces have high levels of naturally occurring pectin, so they're a no muss–no fuss option. And quinces are also the traditional secret weapon to help other flavors of *membrillo* gel up—when your fruit of choice for a *membrillo* is low in naturally occurring pectin, adding quinces and their skins might be just the thing to bring balance to the Pectin Force.

Or you can buy a type of sugar specifically used for jam making (not coincidentally called jam sugar) that contains added pectin. I'm not a fan of this option, though, since it makes a cloyingly sweet *membrillo,* but if you must: Add a 1:1 ratio of jam sugar to your fruit purée, and it should gel (bear in mind that this ratio depends on how much pectin the fruit you're using contains, any acids or enzymes present, etc).

The bottom line is this: Every fruit is a little bit different, and, depending on the time of year and freshness of the fruit, the amounts of sugar, acid, and pectin that you'll need to add will change. That's why I'm not going to include a specific recipe for each *membrillo* type that you could make: That would be a book all by itself.

Instead, check the chart on the next page to get a clearer picture of what types of fruit are high in pectin and low in pectin. Then, the suggested ratio can get you on your way to making your *membrillo* of choice.

Last, remember to always use *fresh* fruit with little to no bruising, since fresh fruit contains the highest levels of pectin. Always, always keep the fruit's peels on during the initial boiling stage, since pectin is mostly found in the skins.

HIGH PECTIN/ACID FRUITS						
ADDITIONS	CRAN-BERRIES	GREEN APPLES	GUAVA	PLUMS	PUMPKIN	BLACK-BERRIES
Quinces, per 2.2 pounds (1 kg) of fruit	n/a	n/a	n/a	n/a	n/a	n/a
Lemons, per 2.2 pounds (1 kg) of fruit	Zest only	1 each	2 each	2 each	2 each	2 each
Granulated sugar, per 2.2 pounds (1 kg)	2.2 pounds (1 kg)	2.2 pounds (1 kg)	2.2 pounds (1 kg)	2.2 pounds (1 kg)	2.2 pounds (1 kg)	2.2 pounds (1 kg)
Vanilla beans, per 2.2 pounds (1 kg)	½ each	½ each	½ each	½ each	½ each	½ each
Salt, per 2.2 pounds (1 kg)	⅓ ounce (10 g)	⅓ ounce (10 g)	⅓ ounce (10 g)	⅓ ounce (10 g)	⅓ ounce (10 g)	⅓ ounce (10 g)

LOW PECTIN/ACID FRUITS					
ADDITIONS	FIGS	PEACHES	CHERRIES	PEARS	STRAW-BERRIES
Quinces, per 2.2 pounds (1 kg) of fruit	2.2 pounds (1 kg)	2.2 pounds (1 kg)	2.2 pounds (1 kg)	2.2 pounds (1 kg)	2.2 pounds (1 kg)
Lemons, per 2.2 pounds (1 kg) of fruit	2 each	2 each	2 each	2 each	2 each
Granulated sugar, per 2.2 pounds (1 kg)	3⅓ pounds (1.5 kg)	3⅓ pounds (1.5 kg)	3⅓ pounds (1.5 kg)	3⅓ pounds (1.5 kg)	3⅓ pounds (1.5 kg)
Vanilla beans, per 2.2 pounds (1 kg)	½ each	½ each	½ each	½ each	½ each
Salt, per 2.2 pounds (1 kg)	⅓ ounce (10 g)	⅓ ounce (10 g)	⅓ ounce (10 g)	⅓ ounce (10 g)	⅓ ounce (10 g)

NOTE: For pumpkin *membrillo,* add a pinch (2 g) of pumpkin pie spice. Trust me: You'll dig it.

TOMATE FRITO

This standard Spanish tomato sauce plays an essential part in a lot of recipes so I always have some on hand. *Tomate Frito,* which basically translates as "fried tomato sauce," gets its unique flavor from frying ingredients in order to bring out their inherent sweetness and remove some of the acidity and canned flavor from the tomatoes.

Since this recipe uses just a few ingredients, it's imperative that everything is the best you can find. Look for ingredients like great olive oil and canned San Marzano tomatoes (yes, even the Spaniards bow to those great Italian tomatoes when their fresh crops aren't in season).

Otherwise, if you want to make this recipe with fresh tomatoes during tomato season, you can use an equal amount of high-quality fresh tomatoes like dry-farmed Early Girls (IMHO, these are just as delicious as San Marzanos!).

The resultant sauce—an umami-packed flavor bomb—was standard fare at our family meals in Toledo, usually served over pasta or as a component in other dishes. We always had a stash canned or in the freezer *(Tomate Frito* preserves really well), so consider making a lot at one time and then storing it for later.

1 Make a *sofrito:* In a large saucepan over medium heat, warm the oil for 4 minutes, until just rippling but not smoking. Add the onions and garlic and season with the salt. Cook for 20 to 25 minutes, stirring frequently, until the onions are very soft but have not taken on color.

2 Using a pair of kitchen shears, cut up the tomatoes into rough pieces. (If you are using fresh tomatoes, chop them roughly.)

3 Raise the heat to high. Add the tomatoes and season them to taste with the sugar, salt, and black pepper. Fry the tomatoes in the *sofrito* for 5 to 10 minutes.

4 Reduce the heat to medium and continue cooking, stirring frequently, for 30 to 40 minutes, until most of the water has cooked out of the tomatoes. Remove from the heat.

5 Process the mixture through a food mill with a fine screen (if you don't have one, you can use a *chinois* or other fine strainer) into a large mixing bowl. If necessary, repeat until the purée is smooth. Taste the sauce and reseason as necessary with the salt and black pepper.

6 If using the *Tomate Frito* immediately, transfer it to a large foodsafe plastic container and set aside to cool to room temperature. Cover and chill the sauce overnight. The *Tomate Frito* can also be canned in sterilized containers (see instructions on p. 130).

YIELD

Around 4¼ cups (1 L)
of *Tomate Frito*

INGREDIENTS

per 2.2 pounds (1 kg) of fresh tomatoes or canned San Marzano tomatoes

½   cup (125 mL) good Spanish extra virgin olive oil, such as piqual

1   medium yellow onion, sliced into thin julienne

10   cloves garlic, peeled, destemmed, and sliced thinly

Kosher salt, to taste

Granulated sugar, to taste

Freshly ground black pepper, to taste

PIQUILLO *CONFIT*

Piquillo peppers are a quintessentially Spanish ingredient, ranking right up there with *pimentón* and saffron as something that gives dishes a "Spanishy" edge. The fresh peppers are harvested in the fall and winter, charred over wood embers, and then canned, typically all in the area of Lodosa in Navarra.

When I cooked in José Andrés' kitchen a few years back, we made a delicious side dish of those piquillos by cooking them slowly in their own purée. Years later, I came across a similar dish while travelling around La Rioja. The dish, which was made with the piquillo's angrier, spicier cousin, the alegría Riojana (the name means "joyful," which is the opposite of how you'll feel if you don't like really spicy peppers), also featured peppers cooked in their own purée—but in that case, honey helped tame the peppers' inherent spiciness.

Though alegrías are very hard to come by in the States, you can find piquillos everywhere. With either pepper, this recipe is absolutely addictive.

1   In a blender or the bowl of a food processor fitted with the "S" blade, combine ¼ of the peppers, the oil, the honey, and the salt. Blend or process on high, scraping down the sides, until the *confit* mixture becomes liquid. Set aside.

2   Preheat the oven to 250°F (120°C).

3   Arrange the remaining peppers in a baking dish. Add the garlic, thyme, and bay leaf. Pour the *confit* liquid over the top. Cover with foil.

4   Bake the peppers for 2 hours. Remove from the oven and set aside to cool to room temperature. Serve warm.

NOTE: These are great with steaks and other meats, which is why nearly every Spanish steakhouse I know offers a roasted piquillo side dish. I also love these as a component of salads and other recipes, or as a great snack on their own.

YIELD

Around 2–3 cups (450–700 mL) of piquillos

INGREDIENTS

per 2.2 pounds (1 kg) of medium piquillo peppers

½   cup (125 mL) extra virgin olive oil

¼   cup (85 g) honey, such as rosemary, thyme, or orange blossom

⅔   ounce (20 g) kosher salt

1   ounce (25 g) granulated sugar

10   cloves garlic, peeled, destemmed, and minced

10   sprigs fresh thyme

1   fresh bay leaf

FOUR-SPICE BLEND

The Frenchies call this *quatre épices* (four spice), so I'm following suit even though almost every recipe that I have encountered has more than four ingredients despite the name.

Nevertheless, this is my go-to spice for *pâtés* and other *charcuterie*. Note that while the traditional recipe uses black pepper, I opt for the fruitier Balinese peppercorns. I love the flavor of this type of peppercorns, but more importantly, it doesn't add specks of color to my *foie gras torchon*.

1   Place all the spices, except the ginger, in a large sauté pan over medium heat. Toast for 3 to 4 minutes, until fragrant. Remove from the heat.

2   Using a mortar and pestle or spice grinder, grind the spices to a fine powder.

3   Add the ginger to the spices. Place in a sealed airtight container and shake well to combine.

4   Use immediately or store in a cool, dark place for up to 6 months.

NOTE: The Four-Spice Blend is pictured on pp. 378–379.

YIELD

Around ¼ cup (75 g)

INGREDIENTS

5	whole Balinese peppercorns
15	whole allspice berries
15	white peppercorns
2	whole cloves
½	cinnamon stick
1	tablespoon (5 g) ground ginger

VINAIGRETTA DE JEREZ

This is the Spanish dressing of choice. It's a 3:1 ratio of oil to vinegar, so if you cook on a regular basis, there's nothing in this recipe that you haven't seen before. I like to add a little honey, but if you're feeling frisky you can add a little Dijon mustard, chopped capers, shallots, or whatever you want to this very basic dressing—it's all about your whisking technique to hold a stable emulsion.

I'm also including some fun variations that I have used over the years. Give them a shot when you feel like pairing the dressing with different salads and meats. I'm very partial to using the Piquillo Vinaigrette with meats and seafood, and the *Membrillo* Vinaigrette goes great with grilled quail and other game birds.

1   In a small mixing bowl, whisk together the vinegar, garlic, honey, and salt.

2   Slowly drizzle in the olive oil, whisking constantly to form an emulsion. Taste and reseason with the salt as necessary.

VARIATIONS

For Pedro Ximénez (PX) Sherry vinaigrette: Substitute PX sherry vinegar for the sherry vinegar. Continue the recipe as written.

For *Membrillo* vinaigrette: Substitute ¼ cup (60 mL) *Dulce de Membrillo* (see recipe on p. 403) and 1 tablespoon Dijon mustard for the honey in Step 1. Whisk until the *Dulce de Membrillo* is dissolved. Continue the recipe as written.

For Moscatel vinaigrette: Substitute Moscatel vinegar for the sherry vinegar. Continue the recipe as written.

For Piquillo vinaigrette: In a blender, purée 4 Piquillos *Confit* (see recipe on p. 408). Add the purée to the vinegar and omit the honey. Continue the recipe as written.

For *Jamón* or *Chorizo* vinaigrette: In a saucepan over low heat, render 3½ ounces (100 g) chopped *Jamón* (see recipe on p. 143) or *chorizo* in a few tablespoons of olive oil for 6 to 8 minutes. Separate the rendered fat from the meat. Substitute ¼ cup (60 mL) of the oil with the rendered fat. Continue the recipe as written, stirring in the reserved meat at the end if desired.

YIELD

1 cup (250 mL)

INGREDIENTS

¼   cup (60 mL) sherry vinegar

1   clove garlic, peeled, destemmed, and grated using a Microplane

1   tablespoon (20 g) honey, such as acacia

1   tablespoon (20 g) kosher salt, or more to taste

¾   cup (175 mL) extra virgin olive oil

VINAIGRETTA DE JEREZ INVERTIDO

Take the basic vinaigrette and flip the proportions of oil and vinegar—that's an inverse vinaigrette. Instead of three parts oil to one part vinegar, you'll get a much more acidic dressing that's based on more vinegar than oil.

I love this dressing with really rich recipes that need the acid to cut through, but I also use it in *ceviche*-like applications like *Esqueixada* (see recipe on p. 173). The acid "cooks" the fish and veggies a little, and also imparts their flavor to the dressing. The basic vinaigrette's variations work well with this recipe, too.

1   In a small mixing bowl, whisk together the vinegar, garlic, honey, and salt.

2   Slowly drizzle in the olive oil, whisking constantly to form an emulsion. Taste and reseason with the salt as necessary.

YIELD

1 cup (250 mL)

INGREDIENTS

¾   cup (175 mL) sherry vinegar

1   clove garlic, peeled, destemmed, and grated using a Microplane

1   tablespoon (20 g) honey, such as acacia

1   tablespoon (20 g) kosher salt, or more to taste

¼   cup (60 mL) extra virgin olive oil

VARIATIONS

For Pedro Ximénez (PX) Sherry vinaigrette: Substitute PX sherry vinegar for the sherry vinegar. Continue the recipe as written.

For *Membrillo* vinaigrette: Substitute ¼ cup (60 mL) *Dulce de Membrillo* (see recipe on p. 403) and 1 tablespoon Dijon mustard for the honey in Step 1. Whisk until the *Dulce de Membrillo* is dissolved. Continue the recipe as written.

For Moscatel vinaigrette: Substitute Moscatel vinegar for the sherry vinegar. Continue the recipe as written.

For Piquillo vinaigrette: In a blender, purée 4 Piquillos *Confit* (see recipe on p. 408). Add the purée to the vinegar and omit the honey. Continue the recipe as written.

For *Jamón* or *Chorizo* vinaigrette: In a saucepan over low heat, render 3½ ounces (100 g) chopped *Jamón* (see recipe on p. 143) or *chorizo* in a few tablespoons of oil for 6 to 8 minutes. Separate the rendered fat from the meat. Substitute the oil with ¼ cup (60 mL) of the rendered fat. Continue the recipe as written, stirring in the reserved meat at the end if desired.

BASIC *ALIOLI*

Haters of garlic and tricky emulsification sauces: Beware! A true *alioli* is a fickle concoction of stank-breath beauty best made with a mortar and pestle and, despite what you may believe or have heard to the contrary, comprises exactly and only its etymological parts: garlic *(ali)* and oil *(oli). Y ya está,* so put down the egg, Chef—that's the gilded path to a mayonnaise.*

The trick to holding and forming the *alioli* emulsification lies in the technique of the cook, plus a little help from a splash of water, some salt, and—at least in modern times—a touch of science.

Thanks to Harold McGee, we know that pure, unrefined olive oil can be problematic for an *alioli* that you'll need to stay emulsified for longer than an hour—basically, the *alioli* splits after that time due to the nature of olive oil's emulsification properties. Older or improperly stored oils are also possible culprits, so I like to substitute in a little unflavored oil like grapeseed if I am making the *alioli* for longer periods of time—just as a precaution.

Otherwise—and I'm writing this so that my Spanish family and friends don't disown me—stick to 100 percent, properly stored, fruity Elixir-of-the-Gods Spanish olive oil. Just mix slowly and don't plan on using your arm for the rest of the day.

*Yes, I know that I've included some variations for this recipe that veer from my two-ingredient rule. Just don't go adding an egg, and we can still be friends.

1   Combine the oils in a measuring cup.

2   Using a mortar and pestle, crush together the garlic and salt to form an *ajosal.* Slowly drizzle in 2 tablespoons of the water, or just enough to help moisten the *ajosal* into a homogeneous paste. Using a spoon, begin stirring the paste.

3   While continuing to stir the moistened *ajosal,* add the oils in a slow drizzle, starting with just a few drops and continuing to a thin stream.

4   Midway through adding the oil, add a third tablespoon of water, which will help the emulsion hold. Continue adding the oil and stirring until the mixture has become a homogenous, silky white sauce.

5   Transfer to a bowl and serve.

YIELD

1 cup (250 mL)

INGREDIENTS

¾   cup (175 mL) extra virgin olive oil

¼   cup (60 mL) grapeseed or canola oil

6   cloves garlic, peeled, destemmed, and coarsely chopped

1   tablespoon (20 g) kosher salt

    Warm water, as needed

CONTINUED ON NEXT PAGE

BASIC *ALIOLI* CONTINUED FROM PREVIOUS PAGE

VARIATIONS

For Black Garlic *Alioli:* Substitute 4 black garlic cloves and 1 fresh garlic clove for the garlic in the base recipe. Continue the recipe as written.

For Green Garlic *Alioli:* Substitute ¼ cup (15 g) chopped green garlic and 1 fresh garlic clove for the garlic in the base recipe. Continue the recipe as written.

For Roasted Garlic *Alioli:* Substitute ¼ cup (150 g) roasted garlic and 1 fresh garlic clove for the garlic in the base recipe. Continue the recipe as written.

For Honey *Alioli:* Drizzle in 1¾ ounces (50 g) of honey once the emulsion starts to form—just before you add water the second time—and continue emulsifying. Continue the recipe as written.

For *Membrillo Alioli:* Pound 1¾ ounces (50 g) of *membrillo* into the *ajosal* in the first steps. Continue the recipe as written.

For Lemon *Alioli:* Juice and zest a lemon. Drizzle the juice of ½ of the lemon just as the emulsion starts to form, just before you add the water, and continue emulsifying. Continue the recipe as written, but stir in the lemon zest at the very end.

NOTE: To make the sauce in a food processor, follow the same procedure, but instead of mixing the *Alioli,* use the food processor's pulse setting. Be aware, however, that your *ajosal* won't be as finely textured in the food processor, so you may want to crush it using a mortar and pestle first.

ALIOLI WITH AN EGG
(AKA *SALSA MAHONESA*)

This sauce is a *salsa mahonesa,* aka Mayonnaise, aka the *Alioli* Walk of Shame. Sometimes the supercharged garlic flavor of a traditional *alioli* needs tempering, and sometimes, you just want a flavored mayonnaise.

Whatever. I won't judge—but it's not an *alioli.* Oh, and it's not French either...or at least that's what my friends from Mahón—a little town in Menorca that lays claim to the sauce—tell me.

1   Combine the oils in a measuring cup.

2   Using a mortar and pestle, crush together the garlic and salt to form an *ajosal.* Slowly drizzle in 2 tablespoons of the water, or just enough to help moisten the *ajosal* into a homogeneous paste. Using a spoon, begin stirring the paste.

3   In a separate mixing bowl, combine the egg yolks and sherry vinegar and whisk until thoroughly blended.

4   While continuing to stir the moistened *ajosal,* add the oils in a slow drizzle, starting with just a few drops and continuing to a thin stream.

5   Midway through adding the oil, add a third tablespoon of water, which will help the emulsion hold. Drizzle in the egg yolk–vinegar mixture and stir until thoroughly blended. Continue stirring until the mixture has become a homogenous, silky sauce. Taste and season as needed with the salt and white pepper.

6   Transfer to a bowl and serve.

YIELD

1¼ cups (300 mL)

INGREDIENTS

½   cup (120 mL) grapeseed or canola oil

½   cup (120 mL) extra virgin olive oil

6   cloves garlic, peeled, destemmed, and coarsely chopped

1   tablespoon (20 g) kosher salt, plus more as needed

    Warm water, as needed

3   large egg yolks

1   tablespoon (15 mL) sherry vinegar

    Freshly ground white pepper, as needed

CONTINUED ON NEXT PAGE

ALIOLI WITH AN EGG (AKA *SALSA MAHONESA*)

CONTINUED FROM PREVIOUS PAGE

> **VARIATIONS**

For Black Garlic *Alioli:* Substitute 4 black garlic cloves and 1 fresh garlic clove for the garlic in the base recipe. Continue the recipe as written.

For Green Garlic *Alioli:* Substitute ¼ cup (15 g) chopped green garlic and 1 fresh garlic clove for the garlic in the base recipe. Continue the recipe as written.

For Roasted Garlic *Alioli:* Substitute ¼ cup (150 g) roasted garlic and 1 fresh garlic clove for the garlic in the base recipe. Continue the recipe as written.

For Honey *Alioli:* Drizzle in 1¾ ounces (50 g) of honey once the emulsion starts to form—just before you add water the second time—and continue emulsifying. Continue the recipe as written.

For *Membrillo Alioli:* Pound 1¾ ounces (50 g) of *membrillo* into the *ajosal* in the first steps. Continue the recipe as written.

For Lemon *Alioli:* Juice and zest a lemon. Drizzle the juice of ½ of the lemon just as the emulsion starts to form, just before you add the water, and continue emulsifying. Continue the recipe as written, but stir in the lemon zest at the very end.

> **NOTE:** To make the sauce in a food processor, follow the same procedure, but instead of mixing the *Alioli,* use the food processor's pulse setting. Be aware, however, that your *ajosal* won't be as finely textured in the food processor, so you may want to crush it using a mortar and pestle first.

BASIC *SOFRITO*

Spanish recipes almost universally contain an opening commandment for imbuing food with the deep *sabor* of *la cocina Española,* and it almost always goes a little something like this: *"Hacer un sofrito"* ("Make a sofrito").

Sofrito, one of the keys to Spanish cuisine, is a word that pops up in every language spoken in the country. Whether it's called a *sofrito* in Castellano, a *sofregit* in Catalan, a *rustido* in Gallego, or *sueztitua* in Euskadi, it's going to be some combination of onions, garlic, and tomatoes cooked in fat to varying degrees of jam-like consistency. Stemming from the verb *sofreir,* which means "to fry lightly," the *sofrito* is an exercise in patience and finesse. It's all about listening, smelling, and slowly cooking the aromatics down in hot fat; about knowing when to add the various components; and about understanding the depth of flavor you want to achieve.

Or, as my ICEX compatriot Paras Shah likes to preach to his cooks, "It's the difference between romancing your food and simply fucking it."

Throughout this book, you will notice that many of the companion recipes start with a *sofrito.* So, to make life easier, I'm including a pretty standard base recipe here. From this base, you can add different ingredients or change the fat you are cooking with—depending on the cuisine of the region of Spain you're cooking—to impart characteristic flavors.

TO MAKE THE BASIC *SOFRITO:*

1   Cover the bottom of a medium saucepan with ¼ inch (6 mm) of the fat of your choice (basically, you want the entire bottom of the pan covered with a layer of fatty goodness). Place the saucepan over medium–high heat and warm the fat for 4 to 6 minutes, until rippling but not smoking and moving freely in the pan.

2   Add the onions and any of the optional ingredients from Group 1 to the saucepan. Season liberally with the salt.

3   Lower the heat to medium and cook the *sofrito,* stirring occasionally, for 15 to 45 minutes (depending on how far you wish to brown the onions: At 15 minutes, they're wilted, and at 45, they're wonderfully browned). Add small amounts of the water as needed to keep the browning consistent, or just adjust the heat accordingly.

YIELD

2 to 3 cups (450–700 mL) *sofrito*

INGREDIENTS

FOR THE BASE *SOFRITO:*

Extra virgin olive oil, unsalted butter, or melted *manteca* (pork lard), as needed

5   medium yellow onions, peeled, destemmed, and cut into small dice

Kosher salt, as needed

3   plum tomatoes, halved and grated on the medium holes of a grater, liquid reserved

Freshly ground black pepper, to taste

OPTIONAL INGREDIENTS—GROUP 1:

3   cloves garlic, grated on a Microplane

2   medium green bell peppers, seeded and cut into small dice

2   medium red bell peppers, seeded and cut into small dice

3   medium piquillo peppers, cut into small dice

1   medium chile pepper, such as Fresno or Anaheim, cut into small dice

2   medium leeks, cleaned and cut into small dice

3½ ounces (100 g) *Jamón* or *Panceta Curada* (see recipes on pp. 143 or 122)

OPTIONAL INGREDIENTS—GROUP 2:

1   fresh bay leaf

¼   cup (65 g) tomato paste

1   tablespoon (6 g) cumin seed, toasted and ground

1   cinnamon stick

1   tablespoon (3 g) dried oregano

1   tablespoon (6 g) fennel seed, toasted and ground

TRUCOS DE LA COCINA: COOKING FATS

DIFFERENT REGIONS IN Spain use different fats as a cooking medium, so *sofrito* recipes often change depending on where the recipe hails from. For example, recipes from the dairy-rich area of Asturias use butter as the fat of choice; recipes from the south use good olive oil; and in the land of the *Ibéricos,* Extremadura, *manteca* (pork lard) is king.

4   Once the onions have reached the desired color, add the tomatoes and their liquid and any of the optional ingredients from Group 2. Stir to incorporate.

5   Cook for 20 to 30 minutes, until all the liquid has evaporated and the mixture has a jam-like consistency. Taste the *sofrito* and season to taste with the salt and black pepper. Remove from the heat and set aside to cool to room temperature.

6   Chill in the refrigerator overnight. Once chilled, the *sofrito* can be refrigerated for a week or held frozen for up to 4 months.

NOTE: *Sofritos* are not just personal recipes. They're typically tailored to their end purpose, so feel free to add any of the optional ingredients, depending on what goes best with the recipe your *Sofrito* will play a part in.

CHAPTER

11

POSTRES Y LICORES

IN SPAIN, NO matter where you eat, and no matter how filling the meal is, you will always be offered some dessert, coffee, and liquor. It's an awesome tradition to signify the end of a meal.

That said, a cookbook on anything Spanish simply isn't complete without paying due homage to the desserts and liquors of Spain that incorporate both the techniques and ingredients of *charcutería*. So don't put away your pork fat just yet, and make sure you have some blood on hand: You're going to need both as we make our way through these traditional desserts found at *matanzas*.

And God knows there is always homemade hooch being poured during every part of the meal—so, yeah, I've got those recipes too.

PERUNILLAS

I'll admit it: I'm a bit of an elitist about cookies. Oatmeal raisin cookies are just granola bars with delusions of grandeur; peanut butter needs some chunks to be worthwhile; and the King of All Cookies will always be warm chocolate chip laced with sea salt. As for those dainty little fucking Frenchie *macarons* pronounced with a phlegm-y middle consonant, like you're about to pass a hairball...is their 15 minutes up yet? Please?

When the *sabias* at our *matanza* baked these delicate lemon and cinnamon cookies with *Ibérico* lard, I couldn't help but get excited. They're crumbly, rich, and make for life-altering results. Chocolate chip: Consider yourself on notice.

YIELD

Around 1 dozen cookies

INGREDIENTS

14	ounces (400 g) *manteca* (pork lard), preferably *Ibérico,* plus more for greasing, softened
10½	ounces (300 g) granulated sugar, plus more for sprinkling
4	whole large eggs, yolks and whites separated into bowls
¼	cup (50 mL) *aguardiente* or anise-flavored liqueur
	Juice and zest of 1 lemon
28	ounces (800 g) unbleached all-purpose flour
1	tablespoon (10 g) kosher salt
1	tablespoon (15 g) baking powder
1	tablespoon (7 g) ground cinnamon

CONTINUED ON NEXT PAGE

TRUCOS DE LA COCINA: THE GLORY OF IBÉRICO FAT

MANTECA IS ONE of the secrets to why these cookies are so amazing. The cookies' crumbly texture is a product of the natural properties of pork lard so don't try to substitute butter or shortening for it. The cookie you'll get will be completely different.

While any *manteca* from a pig treated with respect will do here, if you really want to be authentic you're going to have to source some *Ibérico* fat. And I've got good news: This glorious product is getting easier to come by in the United States now that a few specialty purveyors are importing it (the purveyors are listed in the back of this book).

Just don't go using that supermarket-shelf lard that's pumped with nasties like BHT, BHA, and hydrogenated jibber-jabber. I mean, seriously—who wants to eat that shit?

PERUNILLAS CONTINUED FROM PREVIOUS PAGE

1   In a large mixing bowl, cream together the *manteca* and sugar until the mixture resembles a fluffy little pork cloud.

2   One by one, whisk the egg yolks into the *manteca*–sugar mixture. (Place the bowl containing the egg whites in the refrigerator for later use.) Whisk in the *aguardiente,* lemon juice, and lemon zest until thoroughly combined. Set aside.

3   In another mixing bowl, sift together the flour, salt, baking powder, and cinnamon.

4   While whisking constantly, add the dry-ingredient mixture, bit by bit, to the wet-ingredient mixture until a smooth dough forms.

5   On a clean work surface, roll the dough into a ball.

6   Preheat the oven to 400°F (200°C). Grease a baking sheet with some *manteca.*

7   Break off pieces of the dough weighing about 3½ ounces (100 g) each. Roll each piece into a ball and press lightly, forming them into 2-inch to 3-inch (5-cm to 8-cm) wide discs.

8   Place each disc on the prepared baking sheet. Brush with the egg whites and sprinkle a little sugar over the top of each disc.

9   Bake for 15 to 20 minutes, rotating the baking sheet halfway through baking, until the cookies are golden brown. Remove from the oven and set aside on a rack to cool.

10   Serve warm or store for up to 4 days in a sealed container.

FILLOAS DE SANGRE

I promised you a blood dessert, and here it is.

In Galicia, *filloas* (crepes filled with fresh fruit, pastry cream, or powdered sugar) are a popular dessert. But, come *matanza* season, those crepes get an extra ingredient added to them—pork blood.

The blood turns the crepes jet black after cooking, and the flavor profile is pretty much in line with the *Morcilla Dulce* or *Butifarra Dulce* (see recipes on pp. 265 and 233) *embutido* recipes—Medici spices with pine nuts, raisins, some sweet sherry, and sometimes a little citrus.

I'm a huge fan of filling these crepes with a pastry cream made with an added winter-appropriate flavor. Chestnut works really well, but even simple whipped cream and berries are a great option.

1   In a small mixing bowl, soak the raisins in the PX sherry for 30 minutes, until plumped. Set them aside (and try not to snack on all of them).

2   In a blender, pulse the milk, flour, *sangre*, eggs, granulated sugar, anis dulce, oil, lemon zest, salt, Four-Spice Blend, and nutmeg until it is thoroughly combined. (It should resemble a thin pancake batter.)

3   Transfer the batter to a large mixing bowl. Stir in the reserved raisins and sherry mixture and the pine nuts. Refrigerate for 30 minutes.

4   Warm a medium nonstick skillet over medium heat. Set up a work station by the cooktop with a plate and plastic wrap or parchment paper for the finished crepes.

5   Once the pan is hot, add the *manteca*. After it has melted, wipe down the entire pan using a paper towel (this will provide a light, yet complete, coating of fat).

6   Stir the batter well, helping the raisins and pine nuts rise to the top.

7   Using a ¼ cup (60 mL) measure, pour some of the batter in the skillet. Tip the skillet in a circular motion, coating the entire bottom of the skillet with a thin layer of batter. After 30 seconds, small bubbles should begin to form. When you see them, gently lift the edges of the crepe with a large

YIELD

25–30 crepes

INGREDIENTS

1½   ounces (40 g) golden raisins

¼   cup (50 mL) Pedro Ximénez (PX) sherry

2   cups (500 mL) whole milk

9   ounces (250 g) unbleached all-purpose flour

7   fluid ounces (200 mL) *sangre* (pork blood), whisked to prevent coagulation

4   whole large eggs

1¾   ounces (50 g) granulated sugar, plus more to taste

¼   cup (50 mL) *anis dulce* or brandy

2   tablespoons (30 mL) extra virgin olive oil

    Zest of ½ lemon

1   tablespoon (20 g) kosher salt, plus more to taste

1   teaspoon (2 g) Four-Spice Blend (see recipe on p. 409)

½   teaspoon (1 g) freshly grated nutmeg

1¾   ounces (50 g) toasted pine nuts

1   ounce (25 g) *manteca* (pork lard)

    Confectioners' sugar, for sprinkling

    Whipped or pastry cream, for filling (optional)

spatula to loosen it from the pan. Flip the crepe and cook for 30 seconds. (Don't worry if the first one is more crap than crepe; this always happens.) Remove from the pan and place on the prepared plate.

8   Sprinkle the confectioners' sugar on top of the crepe and then place a sheet of plastic wrap or parchment paper over it. Taste the first crepe and reseason the batter as necessary with more sugar or salt.

9   Repeat with the rest of the batter, stacking the crepes one on top of another, until all of the batter has been used.

10   Fill the crepes with the pastry cream, if using, and sprinkle with more of the confectioners' sugar. Serve.

"SALCHICHOC" CHOCOLATE SAUSAGE

Don't freak out: This isn't a recipe for pork sausage with chocolate—it's just a typically Spanish play on words.

I first saw this "sausage" in a pastry shop in Madrid and loved the ingenuity of it. It's a super-simple-to-make dessert and winds up looking exactly like an *embutido*. You can, of course, add or substitute the nuts or dried fruits as you wish. I have a thing for Raisinettes, so for me it's all about the flavor of chocolate and raisins together.

1 Place a double boiler containing 1 inch (2.5 cm) of water (in the bottom pan) over medium heat. Bring the water to a boil. Break up the chocolate into 1-inch (2.5-cm) pieces and place the pieces, along with the butter, in the upper pan of the double boiler. (If you don't have a double boiler, you can just microwave the chocolate and butter on low for 5 to 8 minutes, stirring the mix every 2 minutes, until melted.)

2 Allow the chocolate and butter to melt together and then transfer the mixture to a room-temperature large mixing bowl. Stir well and add the PX sherry.

3 Add the cookies, nuts, raisins, and vanilla extract to the chocolate mixture and stir well, making a dough.

4 Cover the bowl and refrigerate the dough for 1 hour to overnight. (The cookies will absorb all the liquid.)

5 Place a large piece of parchment paper on your counter. Place ½ of the dough on the paper and reserve the other ½ in the refrigerator.

6 Spread the dough out across the paper. Using the paper to help you roll, shape the dough into a compact "sausage" shape.

7 Repeat the process with the other ½ of the dough and set both rolls of dough in the refrigerator to chill for at least 1 hour to harden.

8 Place a fresh sheet of parchment paper on the counter. Pour the confectioners' sugar on it. Roll the "sausages" in the confectioners' sugar until completely coated.

9 Slice the "sausages" on the bias into 1-inch to 2-inch (3-cm to 5-cm) rounds. Transfer to a platter and serve chilled.

NOTE: These are especially good with a little flaky sea salt on top.

YIELD

2 *Salchichocs* (sausages)

INGREDIENTS

9 ounces (250 g) dark chocolate (60% cacao or higher)

3½ ounces (100 g) unsalted butter

¼ cup (50 mL) Pedro Ximénez (PX) sherry

10½ ounces (300 g) shortbread or vanilla wafer cookies, crushed to a powder

5¼ ounces (150 g) finely chopped and skinless almonds, walnuts, or cashews

2½ ounces (75 g) golden raisins

2 teaspoons (10 g) pure vanilla extract

1 ounce (25 g) confectioners' sugar, for dusting

TOCINO DE CIELO

Tocino de Cielo loosely translates to "fatback from heaven"—which is just about the best dessert name ever, even though no pork is involved. The name supposedly comes from a time when the sherry-producing town of Jerez started using egg whites to clarify the wine produced there. Since that meant a surplus of egg yolks, *Tocino de Cielo* was born to put those extra yolks to good use. It quickly became popular throughout southern Spain.

This dessert is loosely based on the same technique as for making a *flan*. A *flan* is made with whole eggs and typically includes milk, but *Tocino de Cielo* is made with only yolks, sugar, and water. That's because it needed to be cheap and easy to make, and, most importantly, was a means to an end for using up surplus egg yolks.

The trick here is in getting the right temperature on the caramel—you are looking for around 220°F to 225°F (104°C to 106°C), which is just at the very beginning of the thread stage of cooking sugar.

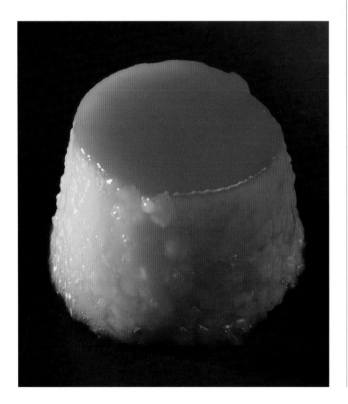

YIELD

4 servings

INGREDIENTS

1	vanilla bean
7	ounces (200 g) granulated sugar, divided
¾	cup (175 mL) water, divided
¼	cup (50 mL) oloroso sherry
	Zest of 1 lemon, removed in a single strip and free of pith
	Zest of 1 orange, removed in a single strip and free of pith
6	large egg yolks
	Sherry, for serving

1   Cut a slit in the long side of the vanilla bean. Slide the back of a paring knife down the length of the inside of the bean; the bean's seeds will accumulate on the knife's blade. Reserve the empty bean and the seeds separately.

2   In a medium saucepan over medium–high heat, place ½ the sugar and 1 tablespoon of the water. Do not stir; instead, swirl the pan lightly and heat the mixture for 4 to 6 minutes, until a light amber caramel forms. Remove from the heat and divide the caramel into 4 6-fluid-ounce (175-mL) ramekins. Set aside.

3   Preheat the oven to 350°F (176°C).

4   In a saucepan over medium–high heat, combine the remaining vanilla bean and seeds, the sugar and water, the sherry, and the lemon and orange zests. Bring to a boil and cook at a steady boil for 10 to 15 minutes, until the mixture's temperature reaches 220°F to 225°F (104°C to 106°C).

5   Meanwhile, in a large mixing bowl, whisk together the egg yolks. Set aside.

6   Remove and discard the vanilla bean and lemon and orange zests from the saucepan. While whisking the yolks, slowly and carefully drizzle in the sugar mixture. Whisk until well combined and the mixture has lightened in color.

7   Pour the contents of the mixing bowl into the prepared ramekins (they will not fill completely, as the custards will rise in the oven).

8   Place the ramekins in a large, high-sided baking dish and place the dish on the middle rack of the preheated oven. Pour hot tap water into the dish until it reaches halfway up the side of the ramekins. Cover the baking dish with aluminum foil and bake for 45 minutes, until the custards are firm.

9   Remove from the oven and set aside on a rack to cool to room temperature. Place the *tocinos* in the refrigerator overnight to chill and set.

10 Run a hot, wet knife around the inside of each ramekin to loosen the custard. Unmold each custard onto a serving plate, allowing the caramel from the cups to drizzle on and around the *tocino*. Serve alongside a glass of sherry.

EL RESACÓN

"SEE YOU TOMORROW." Amongst our group of ICEX cooks, these three little words marked the beginning of the end and the end of the beginning to our evenings in Spain. It was a toast that we cooks gave to each other over potent glasses of homemade Spanish *licores* at our local bar in Madrid, a little place called La Soberbia that steeps its own in house and isn't shy about a heavy pour.

Depending on what region in Spain you visit, you'll likely have a few different homemade liquors thrust into your hands at the end of your meal as *digestivos*. These beverages, for which I'm providing a few recipes in this section, are steeped in tradition and history in addition to fruits and spices, so they're definitely worth the time to seek out and/or make.

The first is for *Patxaran*. She is a sneaky—and easily drinkable—temptress that's high in Vitamin C...the perfect excuse to imbibe! Made from *endrinas* (sloe berries) harvested each September in Navarra where this drink hails from, *Patxaran* is made using a process that's not entirely dissimilar to Italy's famous *limoncello* but with a flavor that is closer to Fernet Branca. Though sloe berries are hard to come by in the United States, I have some friends who use East Coast beach plums to make a very similar and delicious drink.

The second recipe is for *Licor de Guindas,* a liquor made from sour cherries that is very lightly sweetened. It's mostly served in the northern regions of Asturias and Galicia where a large proportion of cherry trees grow.

The third recipe is for *Licor de Hierbas,* a bright yellow liquor that is truly an acquired taste. You can find it all over Spain, but this particular recipe comes from Galicia, which is known for all sorts of flavored *aguardientes* using herbs, spices, and coffee.

PATXARAN

1   Combine the liquor and sloe berries, plus any of the
    optional ingredients, in a large, well-sealed jar or bottle.
    (See Note.)

2   Set aside the jar in a dark, cool place for 3 months, until the
    liquid's color has taken on a permanent shade of red. It's
    hard to oversteep the alcohol, so err on the side of more,
    rather than less, time to make sure you get the full flavor
    extraction.

3   Transfer the jar to the refrigerator or freezer for 1 month,
    until mature.

4   Serve over ice or in any of these traditional drinks:

    ➡ *San Fermín: Patxaran* topped off with cava

    ➡ *Butano:* A 50/50 mix of *Patxaran* and orange juice

    ➡ *Pachalón: Patxaran* topped off with lemon soda

    ➡ *Patxaran del Bosque:* A 50/50 mix of *Patxaran* and the
       juices of mixed berries such as pomegranate or a fruit
       cocktail mix

NOTE: Anise liqueur comes in sweet and dry forms. If you choose
sweet anise, don't add any sugar, since the anise will be sweet enough.
However, if using dry anise or *orujo,* add 1 ounce (25 g) of granulated
sugar per 1 quart (1 L) of alcohol.

YIELD

4¼ cups (1 L)

INGREDIENTS

4¼ cups (1 L) *orujo* or anise-flavored
      liqueur (see Note)

9   ounces (250 g) sloe berries or
    beach plums

OPTIONAL

1   ounce (25 g) granulated sugar
    (see Note)

1   cinnamon stick

10 coffee beans

    Zest of 1 lemon, removed in a
    single strip and free of pith

    Zest of 1 orange, removed in a
    single strip and free of pith

4   fresh manzanilla flowers

1   vanilla bean, split open

2   whole cloves

1   whole star anise

LICOR DE GUINDAS

1   Combine the liquor, sugar, and cherries, plus any of the optional ingredients, in a large, well-sealed jar or bottle. (See Notes.)

2   Set aside the jar in a dark, cool place for 3 months, until the liquid's color has taken on a permanent shade of red. It's hard to oversteep the alcohol, so err on the side of more, rather than less, time to make sure you get the full flavor extraction.

3   Transfer the jar to the refrigerator or freezer for 1 month, until mature.

4   Serve over ice or top with club soda.

NOTES: *Aguardiente* is a type of clear brandy found in Spain. The highest quality types, from Galicia, are similar to Italian grappa. It's a little hard to come by in the United States, so feel free to substitute pisco, from South America, or even vodka.

This drink is traditionally made with sour cherries found in the north of Spain, but you can use a sweeter variety if that's all you can find.

If using sour cherries, add 9 ounces (250 g) of sugar to the *aguardiente* in the first step and continue the recipe as written. If using sweet cherries, add 3½ ounces (100 g) of sugar instead.

YIELD

4¼ cups (1 L)

INGREDIENTS

4¼ cups (1 L) *aguardiente* liquor

Granulated sugar, as needed (see Notes)

18   ounces (500 g) sour cherries (see Notes)

OPTIONAL

1   sprig fresh lemon verbena

Zest of 1 lemon, removed in a single strip and free of pith

Zest of 1 orange, removed in a single strip and free of pith

4   fresh manzanilla flowers

½   vanilla bean, split open

2   whole cloves

1   whole star anise

LICOR DE HIERBAS

1   Combine the liquor, plus any of the optional ingredients, in a well-sealed 2-quart (2-L) or larger jar or bottle. (See Notes.)

2   Set aside the jar in a dark, cool place for 2 weeks.

3   In a large saucepan over medium heat, combine the water and sugar. Cook the mixture for 8 to 10 minutes, until the sugar dissolves and a *jarabe* (simple syrup) forms. Remove from the heat and set aside to cool to room temperature.

4   Transfer the *jarabe* into the jar containing the liquor. Return the jar to the dark, cool place for 2 weeks.

5   Transfer the jar to the refrigerator or freezer for 1 month, until mature.

6   Serve over ice or top with club soda.

NOTES: *Aguardiente* is a type of clear brandy found in Spain. The highest quality types, from Galicia, are similar to Italian grappa. It's a little hard to come by in the United States, so feel free to substitute pisco, from South America, or even vodka.

This drink was traditionally made with whatever wild herbs were around, so add as many or as few of the optional spices and herbs in the recipe as you would like.

If lemon tree leaves are hard to come by, substitute the zest of 2 lemons per quart (liter) of liquor. Just make sure to remove any bitter white pith.

YIELD

2 quarts (2 L)

INGREDIENTS

4¼ cups (1 L) *aguardiente* liquor

2   cups (500 mL) water

18 ounces (500 g) granulated sugar

OPTIONAL

1   sprig fresh lemon verbena

10 fennel fronds

10 mint leaves

5   fresh bay leaves

5   fresh lemon tree leaves

5   fresh sage leaves

5   manzanilla flowers

3   fresh basil leaves

3   sprigs fresh rosemary

3   juniper berries

3   whole cloves

    Pinch saffron

1   cinnamon stick

NOTES

1   I use the term *charcuterie* interchangeably with the Castellano word *charcutería* throughout this book. They both refer to the same spectrum of techniques for salting, smoking, curing, and preserving meat and animal products.

2   Ana Castañer and Teresa Fuertes, *El Libro del Jamón y la Matanza* (Madrid: Alianza Editorial, 1988), 31.

3   Eric Solsten and Sandra W. Meditz, eds., *Spain: A Country Study* (Washington, DC: GPO for the Library of Congress, 1988).

4   Marcella Trujillo Melendez, "The Ancient Cultural History of the Matanza," New Mexican Hispanic Culture Preservation League, accessed September 19, 2013, http://www.nmhcpl.org/uploads/the_matanza.pdf.

5   Solsten and Meditz, *Spain*, "Iberia."

6   Antonio Blanco Freijeiro, "Cultura y Simbolismo del Cerdo," *Historia* 16 (January 1983): 105.

7   Antonio Gázquez Ortiz, *Porcus, Puerco, Cerdo* (Madrid: Alianza Editorial, 2000), 27–28.

8   Antonio Gázquez Ortiz, "El Jamón en la Gastronomía Española," *A Fuego Lento*, January 20, 2002.

9   Manuel Torres, et al., *España Visigoda* (Madrid: Espasa-Calpe, 1940), 162.

10 Torres, et al., *España Visigoda*, 162–65.

11 Darío Vidal, *Siete Ensayos Aragoneses y un Apócrifo* (Alcañiz, Spain: Ayuntamiento, 1986), 61–62.

12 Mick Vann, "A History of Pigs in America," *Austin Chronicle*, April 10, 2009.

13 Teresa Barrenechea, *The Cuisines of Spain: Exploring Regional Home Cooking* (New York: Ten Speed Press, 2005).

14 Barrenechea, *Cuisines of Spain*, 2–3.

15 Barrenechea, *Cuisines of Spain*, 3.

16 Daniel Arzamendi, "El Chorizo Español," *Diario de Tarragona*, July 28, 2009.

17 Chris Meesey, "On the Range: Chorizo," *Dallas Morning Observer*, November 18, 2009.

18 Amanda Hesser, "Truly Spanish Chorizo in America, At Last," *New York Times*, March 6, 2002.

19 Hesser, "Truly Spanish Chorizo," 2.

20 Glenn Collins, "A Ban on Some Italian Cured Meat Is Ending," *New York Times,* April 29, 2013.

21 The USDA has only recently dropped its pork cooking guideline to 145°F (63°C), which is definitely a step in the right direction and a victory for chefs everywhere (who have been pretty much cooking their pork that way for years anyway).

22 Paul Richardson, *A Late Dinner: Discovering the Food of Spain* (New York: Scribner, 2007).

23 "Swine and Pork," USDA, Foreign Agriculture Service Report.

24 F. Jiménez-Colmenero, J. Ventanas , and F. Toldrá, "Nutritional Composition of Dry-Cured Ham and Its Role in a Healthy Diet," *Meat Science* 84 (April 2010).

25 Gayle Hartley, "Andalucían Matanza: A History," Orce Serrano Hams, 2006, http://www.orceserranohams.com/andalucian-matanza.

26 As an American cook raised in a culture of stringent health codes, I took temperature readings in hanging rooms like this one many times during my trips. The rooms never registered above 40°F (4°C) in the morning and never rose higher than 50°F (10°C) during the day—right in line with modern US health code standards.

27 Truth be told, this is a tradition left over from the days when eating pork of dubious origin commonly caused illness. Nowadays, and especially given the care and attention to feeding and fattening Ibéricos, illness is nearly nonexistent. Case in point: I never saw a problem during all of my time with the matanzas, and El Maestro can't even remember the last time he saw a bad liver.

28 Sorry, ladies: Butchery was exclusively an occupation for men in those days.

29 Jose Manuel Iglesias, in discussion with the author, January 21, 2013.

30 If you're really into calculating things like parts per million (ppm), nitrates/nitrites, and other more technical aspects of *charcuterie,* you'll find a list of great literature on the topics on p. 450. These are books I greatly respect and have learned a lot from, so I heartily encourage you to check them out. The more you know, the more educated decisions you can make in your own meat-curing endeavors.

31 American Medical Association. *Report 9 of the Council on Scientific Affairs (A-04): Labeling of Nitrite Content of Processed Foods,* accessed September 30, 2013, http://www.ama-assn.org/resources/doc/csaph/a04csa9-fulltext.pdf.

PURVEYORS AND OTHER COOL PEOPLE

UNITED STATES

Spanish products, charcuterie, and fresh Ibérico meat

La Tienda
1325 Jamestown Road
Williamsburg, VA 23185
(800) 710-4304
www.tienda.com

Despaña Foods
408 Broome Street
New York, NY 10013
(212) 219-5050
www.despanabrandfoods.com

La Española Meats
25020 Doble Avenue
Harbor City, CA 90710
(310) 539-0455
www.laespanolameats.com

D'Artagnan
280 Wilson Avenue
Newark, NJ 07105
(800) 327-8246
www.dartagnan.com

Wagshal's
4845 Massachusetts Avenue NW
Washington, DC 20016
(202) 363-0777
www.wagshals.com

Jamón Ibérico and Ibérico embutidos

Fermín USA
1001 Avenue of the Americas,
Suite 1205
New York, NY 10018
(212) 997-3161
www.fermínibérico.com

5 Jotas
641 Lexington Avenue,
Suite 1425
New York, NY 10022
(855) 368-6603
www.cincojotas.com

Charcuterie equipment and products

Butcher and Packer Supply Company
1780 E. 14 Mile Road
Madison Heights, MI 48071
(248) 583-1250
www.butcher-packer.com

Chr. Hansen, Inc.
9015 West Maple Street
Milwaukee, WI 53214
(888) 289-2218
www.chr-hansen.com

Anova Inc.
PO Box 66
Stafford, TX 77477
(281) 277-2202
info@waterbaths.com

Organizations, sites, and craftsmen

ICEX USA
405 Lexington Avenue,
44th Floor
New York, NY 10174
(212) 661-4959

Windrift Hall
256 Smith Road
West Coxsackie, NY 12192

Jean Louis Frenk Ceramics
jeanlouis@frenk.info
(917) 496-7008

Awesome restaurants, bars, and markets

Mockingbird Hill
1843 7th Street, NW
Washington, DC 20001
(202) 316-9396
www.drinkmoresherry.com

Jaleo
480 7th Street NW
Washington, DC 20004
(202) 628-7949
www.jaleo.com

Manzanilla
345 Park Avenue S
New York, NY 10010
(212) 255-4086
www.manzanillanyc.com

SPAIN AND EUROPE

Spanish culinary tours and matanzas

Finca Montefrio
Armando and Lola Lopez-Sanchez
Carretera del Corte
21230 Cortegana,
Huelva, Spain
(+34) 959 50 32 51
www.fincamontefrio.com

A Taste of Spain
Alonso Cano 8
Cádiz, Spain 11010
(+34) 856 079 626
www.atasteofspain.com

Tenedor Tours
San Sebastián, Spain
(+34) 609 46 73 81
www.tenedortours.com

Where Is Asturias?
info@whereisasturias.com
www.whereisasturias.com

Awesome restaurants, bars, and markets

La Soberbia
Calle de Espoz y Mina, 1 28012
Madrid, Spain
(+34) 913 19 82 15
www.lasoberbia.es

Restaurante Adolfo
Calle del Hombre de Palo, 7
45001 Toledo, Spain
(+34) 925 22 73 21
www.grupoadolfo.com

Restaurante Calima
Calle de José Melia, 4
29602 Marbella, Spain
(+34) 952 76 42 52
www.restaurantecalima.es

Restaurante Arzak
Avda. Alcalde Jose Elosegui, 273
20015 Donostia, San Sebastián
(+34) 943 278 465
www.arzak.info

Portal de Echaurren
Calle del Padre José García, 19
26280 Ezcaray, La Rioja, Spain
(+34) 941 35 40 47
www.echaurren.com

Bar Pinotxo
Rambla de Sant Josep, 91
08002 Barcelona, Spain
(+34) 933 17 17 31
www.pinotxobar.com

Asador Etxebarri
Plaza San Juan, 1
48291 Atxondo- Bizkaia
(+34) 946 58 30 42
www.asadoretxebarri.com

Quimet i Quimet
Carrer del Poeta Cabanyes, 25
08004 Barcelona, Spain
(+34) 934 42 31 42

Taverna Ca L'Espinaler
Av. del Progrès nº 47
08340 Vilassar de Mar
(+34) 937 50 25 21
www.espinaler.es

Awesome markets, purveyors, and specialty shops

Cerato Fratelli (C3)
Via Peveragno, 113 - 12012
Boves (Cuneo), Italy
(+39) 0171 38 87 93
www.c3-boves.it

La Botifarreria
Calle Santa Maria, 4
08003 Barcelona, Spain
(+34) 933 19 91 23
www.labotifarreria.com

La Patería de Sousa
Calle Real, 31
06240 Fuente de Cantos,
Badajoz, Spain
(+34) 626 82 51 90
www.lapateria.eu

Cecinas Nieto
Av Madrid-Coruña
24714 Pradorrey, León, Spain
(+34) 987 61 86 05
www.cecinasnieto.com

Hortelanos de la Vera S.L.
Calle Badajoz 8
10460 Losar de la Vera
Cáceres

Order de Sabadiego
Calle de Libertad, 4
Aptdº de Correos nº 45
33180- Noreña

Mercado San Miguel
Plaza de San Miguel, s/n
28005 Madrid, Spain
(+34) 915 42 49 36
www.mercadodesanmiguel.es

Mercat de Sant Josep
La Boqueria Rambla, 91
08002 Barcelona, Spain
(+34) 933 18 20 17
www.boqueria.info

Mercat Sta. Caterina
Avinguda de Francesc Cambó, 16
08003 Barcelona, Spain
(+34) 933 19 57 40
www.mercatsantacaterina.net

KITCHEN LINGO

A Glossary of *la Cocina Española*

Adobo: A thick, paste-like marinade.

Alioli: A temperamental emulsion sauce made with garlic and olive oil. These days, its definition has expanded to include a garlic-flavored mayonnaise made with egg yolks (bringing stability to the traditional recipe).

Aliño: An oil-based marinade that typically has lots of garlic and other seasonings.

Amigo: Buddy, pal, friend.

Asadura: Typically refers to livers, kidneys, and gizzards. It is typically chopped up and fried, as in *Cachuela* (see recipe on p. 367).

Baño maria: Known in French as a *bain marie,* this technique involves placing the pan containing the item you are cooking inside a larger pan filled halfway with water. The method is typically used for melting chocolate or gently heating something you don't want to overcook.

Bocadillo: A sandwich typically made with a demi-baguette and some meats like Pamplona-Style *Chorizo* (see recipe on p. 306).

Bomba rice: A specific kind of short-grain rice typically grown around the area of Calasparra in Murcia. The rice is unique because it absorbs a high volume of liquid compared to the weight of each individual grain—somewhere around three times its own weight.

Cachuela (see recipe on p. 367): A mixture of *asaduras,* blood, and spices that's roughly chopped to form a coarse *pâté.* It is standard breakfast fare at many *matanzas* in the area around Badajoz in Extremadura.

Caldo: Broth or stock.

Caña: A short beer.

Casco viejo: The old quarter of a city. Most cities in Spain have old sections that are historic districts and city centers. These areas are distinctly different from the newer areas, where expansion has pushed out into the city's suburbs.

Cecina (see recipe on p. 115): A cured beef leg, made in a similar fashion to *jamón.* Traditionally, the muscle that is cured comes from one of four major Spanish beef butchery cuts of the leg (the *tapa, babilla, contra,* or *cadera).*

Charcuterie/charcutería: The lineage of culinary products made from smoking, salting, curing, and preserving meat. Though traditionally this word referred only to cured products made from pork, today the term has a much broader use.

Chinchón: A sweet, potent anise liqueur consumed in shots or poured into coffee at the start and throughout the day of a *matanza*.

Choricero pepper: A long, slightly spicy Spanish pepper. They are available either dried or in an *adobo*. The choricero is one of the peppers used to make *pimentón*.

Chorizo (see recipe on p. XXX): The Spanish catch-all word for "sausage," it's typically a *pimentón*- and garlic-spiked *embutido* that can be fresh, semicured, or dry cured. Some of the more popular *chorizos* are *Chorizo Fresco* (see recipe on p. 226), Cantimpalos-Style *Chorizo* (see recipe on p. 299), Riojano-Style *Chorizo* (see recipe on p. 302), *Chorizo Asturiano* (see recipe on p. 308), Pamplona-Style *Chorizo* (see recipe on p. 306), and *Patatera* (aka *Morcilla Patatera*) (see recipe on p. 317).

Chuleton: A porterhouse steak; one of the best in the world can be found at Asador Etxebarri in Axpe, Spain.

Ciego de cerdo: A special type of large sausage casing taken from the bung (rectum) of a pig.

Cocción: The cooking liquid in which something was poached. For example, the *cocción* for blood sausage is called *cachundo;* drinking it is thought to be a powerful aphrodisiac.

Cocido Madrileño (see recipe on p. 332): A famous boiled stew from Madrid that is typically served on Sundays; it's closely related to the *pot au feu* of France and the *bollito misto* of Italy and is considered a great hangover cure.

Comarca: A local government area or region.

Comida casera: Literally translates as "home cooking"; a restaurant that offers *comida casera* typically serves rustic, homestyle dishes typical to a region.

"Con carino sin miedo": Literally translates as "With love and absent fear"; the only way to make proper *croquetas*.

Confit: A traditional technique of preservation in which a food is cooked slowly in its own fat and then stored under that fat for an indeterminate period of time. In modern times, the term's definition has expanded to include any food cooked under any type of fat.

Cortador: Someone who cuts ham. A *maestro cortadero* is a master ham slicer, and a *maestro jamonero* determines if a *jamón* is ready to be cut and is in charge of ham production.

Dehesas: Areas in central and southern Spain that are the domain of the *Ibérico* pigs. They include oak and cork tree groves and meadows and are typically private or community property that farmers either own or lease as grazing land for their pigs.

Endrinas: Sloe berries picked in early autumn around the area of Navarra for making *Patxaran* (see recipe on p. 433), one of the greatest alcoholic beverages to ever grace the hands of man.

Euskal Txerria: A heritage breed of pig native to País Vasco known for its short legs, small size, and multicolored pelt.

Family meal: In restaurant parlance, this is the meal that cooks make for the staff; in Spain, "family meal" is typically a humble dish made with rice, beans, or other comfort foods.

Gochu Asturcelta: A heritage breed of pig native to Asturias typified by its gigantic hooves, floppy ears, and high proportion of water in the muscle, which makes it great for smoking.

Grelos: Turnip greens popular all over Galicia as a herald of winter stew season.

Guindas: Baby sour cherries steeped in liquor to make a popular beverage in the cherry-growing regions of the north.

Ibérico **pig:** The king of the *dehesa.* These black pigs are native to the Iberian peninsula and are a crossbreed with the local wild boar. The most famous product made from these pigs, *jamón Ibérico,* is a highly cultivated form of *charcuterie.*

Indicación Geográfica Protegida (IGP) / Protected Geographic Identification (PGI): An official recognition designated by the European Union that protects and identifies a product as being from a place, region, or country and having a specific quality, reputation, or other characteristic attributable to its geographical origin.

Jamón (see recipe on p. 143): A cured ham.

Jarabe: A simple syrup, typically a 1:1 ratio of sugar to water.

Jefe de la cocina: Literally translated as "chief of the kitchen," it's the Spanish term for *chef de cuisine.*

Lacón (see recipe on p. 151): A ham from the shoulder muscle from white pigs; it can be cooked, smoked, or dry cured.

Maestro cortadero: A master ham slicer. A *cortador* is someone who cuts ham, and a *maestro jamonero* determines if a *jamón* is ready to be cut and is in charge of ham production.

Maestro jamonero: A person who determines if a *jamón* is ready to be cut and is in charge of ham production. A *cortador* is someone who cuts ham, and a *maestro cortadero* is a master ham slicer.

Manchego: A term to describe either someone from La Mancha or a famous cheese from the region.

Manteca: Pork back fat that has been melted down into lard.

Matanza: The traditional ritual of killing a pig and harvesting its meat, fat, blood, bones, and organs for various preparations. In Spain, *matanza* season typically occurs in the winter months, from December until around March, depending on the temperature.

Mercados: The local markets of Spain.

Mermeladas: Marmalades and jams.

Monjas: Nuns; they famously run secluded kitchens in their convents that sell sweets, marmalades, jams, and sauces from ancient stores and *turnos.*

Montanera: The acorn-feeding months prior to slaughter when *Ibérico* pigs are allowed to graze freely on the *dehesa.* During the *montanera,* pigs must gain a specific amount of weight in order to be ready for slaughter.

Morcilla: The catchall name for blood sausages in Spain. Some popular ones are: *Morcilla de Burgos* (see recipe on p. 258), *Morcilla de Cebolla* (see recipe on p. 260), *Morcilla Asturiana* (see recipe on p. 263), *Morcilla Achorizada* (aka *Morcilla de Jaen)* (see recipe on p. 322), *Morcilla de Patatera* (aka *Patatera)* (see recipe on p. 317), *Morcilla Dulce* (see recipe on p. 265), and *Morcilla Blanca* (see recipe on p. 273).

Morcillo: Another name for a shank of beef, it is typically used for preparations like *Cocido Madrileño* (see recipe on p. 332).

¡Oido!: A Spanish kitchen affirmation analogous to the French *"Oui, chef!"*

País Vasco: Basque country.

Paletilla: A cured shoulder ham of a pig, typically an *Ibérico.*

Pastel: A type of cake.

Patxaran (see recipe on p. 433): An alcoholic beverage made by steeping sloe berries and herbs in grain alcohol, vodka, or grappa (aka liquid deliciousness).

Pellicle: A tacky layer that forms when meat is left uncovered in the refrigerator or in front of a fan for 30 to 45 minutes. Pellicles are necessary for hot smoking, since the pellicle helps attract and retain the smoke particulates.

Picada: A nut and bread-based thickening agent used in sauces typical of the Catalan regions.

Pimentón: Spanish paprika that comes from two major regions (Murcia and La Vera) and is available either smoked or unsmoked.

Pluma: A cut including the *rhomboideus* muscle of a pig, located at the top of the loin.

Potajes: Porridges or stews commonly found in the poorer kitchens of Spain, they comprise the backbone of true *comida casera.*

Presa: This pork cut, known in the United States as the Boston Butt, is part of the pig's *serratus ventralis* muscle.

Puntillitas: Literally, this means "fringe," but it colloquially refers to the crispy, browned area on the bottom and edges of a properly fried egg.

Puta máquina: The highest compliment you can receive in a Spanish kitchen; if you hear it, you are a "cooking machine."

Ración: A large portion of food typically shared by several people. It's basically the opposite of a *tapa*.

Resacón: A very bad hangover. For cooks visiting Spain, this is an unfortunate rite of passage.

Romesco: A Catalan sauce based on tomatoes, peppers, garlic, and a nut-based *picada*.

Sabor: Literally, this translates as "flavor." If something is *sabroso,* it is utterly delicious.

Sal: Salt.

Secreto: A cut of pork located between the *latissimus dorsi* and *pectoralis profundi* muscles of the pig. The muscle is located at the animal's armpit, but is sandwiched between layers of fat. It is sometimes referred to in the United States as a pork skirt steak.

Tapa: A small portion of food, typically served alongside an alcoholic beverage at bars in Spanish cities like Madrid. The cost of the *tapa* is typically included in the price of the drink.

Tapeo: The act of going out with friends to drink and nibble on *tapas* at different bars over the course of an evening. It's typical—especially in Madrid—to go on a *tapeo,* hopping from bar to bar, nibbling on different *tapas,* and drinking *cañas.*

Tocino: Pork back fat.

Tocino de Cielo (see recipe on p. 430): A custard-like dessert.

Tomate Frito (see recipe on p. 407): The tomato sauce of Spain. It's not entirely dissimilar from a traditional tomato sauce, except *Tomate Frito* has much more onions and garlic.

Turno: A lazy Susan used to transfer money and food between the *monjas* (nuns) who made the food and their customers. The *monjas* were not permitted to see the buyers, so this antiquated form of transaction was used to sell sweets and marmalades. Today, it is rarely used.

White pig: The most common pigs found in Spain and most of Europe. Most are various breeds of Duroc, Pietrain, or Landrace hog.

Zurrapa: Essentially the same recipe as *Cachuela* (see recipe on p. 367), except that with *Zurrapa,* you purée the mixture until it is very smooth.

KNOWLEDGE IS POWER

Where to Learn More about *Charcuterie* and Butchery

Bruce Aidells with Lisa Weiss, *Bruce Aidells's Complete Book of Pork*, 2004

Jessica and Joshua Applestone, *The Butcher's Guide to Well-Raised Meats*, 2011

Paul Bertolli, *Cooking by Hand*, 2003

Taylor Boetticher, *In the Charcuterie*, 2013

Ryan Farr, *Whole Beast Butchery*, 2011

Stephen and Adam Marianski, *The Art of Making Fermented Sausages*, 2009

Michael Ruhlman and Brian Polcyn, *Charcuterie: The Craft of Salting, Smoking, and Curing*, 2005

Michael Ruhlman and Brian Polcyn, *Salumi: The Craft of Italian Dry Curing*, 2012

USDA Food Safety and Inspection Service: www.fsis.usda.gov

ACKNOWLEDGMENTS, PROPS, AND RESPECT

This book would still be some scattered ideas and inspirations floating around my laptop without a whole lot of *puta máquinas*.

In the United States:

Mom & Dad: Thank God you guys can't cook and have always supported my attempts at doing so.

Julie & Olga: *Mis hermanas Españolas.*

Lori, Pam, Bill, & Jonathan: Thanks for your support from afar.

Gary, Marge, & Esther: Thanks for your encouragement and a place for this book to start.

Chef Steve Chan: I will always be your student.

Chef José Andrés: Thanks for putting me on the path to Spain and supporting my journey.

Secret Agent Sally Ekus: Thanks for many more sausage jokes to come.

Doug & the team at Agate: Thanks for helping guide this vision and believing in it.

Perrin: Thanks for taking a bunch of jumbled thoughts and turning them into a book.

Photographer Nathan Rawlinson: The only things more amazing than your photos are your patience with me and your capacity for PBR.

Artistic divas Mariana Cotlear and Erin Merhar: You made this book come to life.

ICEX & Foods from Spain: Thanks for the opportunity to learn the beauty of Spanish cuisine and culture.

My ICEX 2009 crew: See you tomorrow, *amigos*.

Papí Chulo Paras: You are one of the most talented cooks that I know. Your time is now.

Fermín USA and 5 Jotas: Thanks for your porky support throughout this process.

Mission College & the Cornell SHA faculty and staff: Thanks for giving me the education to reach my dreams.

Dr. David Meisinger & Iowa State University: Thanks for your help with pork butchery specifics.

Davey Griffin & Texas A&M University: Thanks for your help with beef butchery specifics.

Harold McGee: Thanks for making cooks smarter cooks.

Susan & Jean Louis Frenk: Thanks for a wonderful weekend photo shoot at your beautiful home.

Christopher Ager: Thanks for keeping me sane, organized, and surrounded by good food and drink.

Paul Weipert, Jason Molinari, and other meathead *amigos* from afar: Keep fighting the good fight!

In Spain:

ICEX Spain: *Gracias por la oportunidad de aprender de la cocina y cultura de España.*

Gavin, Dolo, & the Where Is Asturias crew: Thanks for new friends and helping show me the beauty of Asturias.

Miguel Ullibarri: Thanks for sharing your passion and knowledge of *jamón Ibérico.*

Saul Aparicio-Hill: Thanks for showing our crew your passion for Spain, drinking us under the table, and then picking us up to do it again.

Janet Mendel: Thanks for your inspiring words and support.

Elaine Hill: Your *cocido* has no equal.

Adolfo Muñoz *y familia:* *Estáis mi familia Manchega. Gracias por compartiendo el alma Manchega conmigo.*

Carlos Tristancho *y familia:* *Porque no pude entender la matanza aparte de la vida Extremeña.*

Daní Garcia: *Gracias por la oportunidad de cocinar contigo y tu equipo.*

Angel Correas: *Gracias por una matanza inolvidable.*

Margarita Torres-Vidal *y familia:* *Gracias por nos mostrando la cocina Gallega.*

INDEX

ABOUT THE AUTHOR

JEFFREY WEISS is currently the chef at jeninni kitchen + wine bar in Pacific Grove, California. Previously, he was an elite figure skater who represented the United States as a national and international medalist before retiring to pursue a life in the world of hospitality. It was a decision, he knew full well, that would mean getting his balls busted by every line cook from here to Madrid and back again for his past life of spandex and sequins...but it's one he is thankful for every day.

Fortunately, Jeffrey's early culinary years were shaped under the tutelage of Chef Steve Chan, a mentor who intoned daily the importance of being a well-rounded cook with a strong understanding of both the culinary arts *and* the restaurant business. Steve's lessons helped Jeffrey learn and appreciate all aspects of the hospitality industry, as it was under Steve's watchful eye that Jeffrey built the foundation of skills he would later need for his schooling at Mission College and Cornell University's School of Hotel Administration, as well as his hands-on education in the kitchens of chefs like José Andrés, April Bloomfield, and others during his time cooking in Spain and the US.

Today, many of the most popular items on the menu at jeninni kitchen + wine bar include the *charcuterie* recipes you'll find in this book, including chicken liver *pâté,* numerous fresh and semi-cured sausages, and other preparations that have been a part of Jeffrey's culinary journey. Jeffrey lives in Monterey, California, but his heart resides variously with his *amigos y familia* in Madrid, La Mancha, Extremadura, Asturias, and Andalucía.

ABOUT THE PHOTOGRAPHER

NATHAN RAWLINSON is a James Beard–nominated photographer based in New York City. He graduated from the Culinary Institute of America and has worked at numerous restaurants including Michelin-starred Eleven Madison Park.